This book presents a detailed analysis of changes in world energy use over the past twenty years. It considers the future prospects of energy demand, and discusses ways of restraining growth in consumption in order to meet environmental and economic development goals. Based on a decade of research by the authors and their colleagues at Lawrence Berkeley Laboratory, it presents a wealth of information on energy use and the forces shaping it in the industrial, developing, and formerly planned economies. The book presents an overview of the potential for improving energy efficiency, and discusses the policies that could help realize the potential. While calling for strong action by governments and the private sector, the authors stress the importance of considering the full range of factors that will shape realization of the energy efficiency potential around the world.

ENERGY EFFICIENCY AND HUMAN ACTIVITY: PAST TRENDS, FUTURE PROSPECTS

CAMBRIDGE STUDIES IN
ENERGY AND THE ENVIRONMENT

Editors

Chris Hope, *Judge Institute of Management, University of Cambridge*
Jim Skea, *Science Policy Research Unit, University of Sussex*

We live at a time when people are more able than ever to affect their environment, and when the pace of technological change and scientific discovery continues to increase. Vital questions must continually be asked about the allocation of resources under these conditions. This series is intended to provide readers interested in public policies on energy and the environment with the latest scholarship in the field. The books will address the scientific, economic and political issues which are central to our understanding of energy use and its environmental impact.

ENERGY EFFICIENCY AND HUMAN ACTIVITY: PAST TRENDS, FUTURE PROSPECTS

Sponsored by the Stockholm Environment Institute
Stockholm, Sweden

Lee Schipper and Stephen Meyers
with
Richard B. Howarth and Ruth Steiner
International Energy Studies Group, Lawrence Berkeley Laboratory,
University of California at Berkeley

Prologue by
John Holdren
University of California at Berkeley

CAMBRIDGE
UNIVERSITY PRESS

Published by the Press Syndicate of the University of Cambridge
The Pitt Building, Trumpington Street, Cambridge CB2 1RP
40 West 20th Street, New York, NY 10011-4211, USA
10 Stamford Road, Oakleigh, Victoria 3166, Australia

First published 1992

Printed in Great Britain at the University Press, Cambridge

A catalogue record for this book is available from the British Library

Library of Congress cataloguing in publication data
Schipper, Lee, 1947–
Energy efficiency and human activity: past trends, future
prospects/Lee Schipper and Stephen Meyers with Richard B. Howarth
and Ruth Steiner: prologue by John Holdren.
p. cm. – (Cambridge energy studies)
"Sponsored by the Stockholm Environment Institute, Stockholm, Sweden."
Includes index.
ISBN 0 521 43297 9
1. Energy conservation. 2. Energy consumption. I. Meyers,
Stephen, 1955– . II. Stockholm Environment Institute.
III. Title. IV. Series.
TJ163.3.S35 1992
333.79′16 – dc20 92–14112 CIP

ISBN 0 521 43297 9 hardback

Contents

Acknowledgements

This book is the fruit of years of work to which many people have contributed. Colleagues in the International Energy Studies Group who have made important contributions include Andrea Ketoff, Jayant Sathaye, Dianne Hawk, Sarita Bartlett, Adam Kahane, Deborah Wilson, and R. Caron Cooper. In addition, we are indebted to colleagues around the world who have provided information and assistance over the years.

In the production of this book, Anne Sprunt and Claudia Sheinbaum provided research assistance for the chapters on the service and residential sectors, respectively. Charles Campbell and Feng An provided assistance with data and graphics for all chapters. Karen Olson managed the overall production of the manuscript.

Gerald Leach (SEI) provided us a tremendous amount of support in Berkeley, Stockholm, and in England. We acknowledge the timely advice and assistance of Prof. Steinar Strøm (University of Oslo) and Prof. Marc Ross (University of Michigan), who worked with us here in Berkeley. Thanks also to Prof. Thomas Sterner (University of Gothenberg, Sweden), Gordon McGranahan (SEI), and Ms Genevieve McKinnes (International Energy Agency, Paris), who provided helpful comments on the first draft.

This book would have never been possible without the support of the Stockholm Environment Institute. Its first Director, Dr Gordon Goodman, now retired, proposed the book to us, while Lars Kristofferson, who acted as the Deputy Director after Dr Goodman's retirement, continued to support us during the long creative process. The new director, Dr Michael Chadwick, has also been supportive of the final birth pangs of this book. To these three colleagues we offer our deep gratitude.

Finally, we mention the loss of a good friend and colleague, Prof. David Wood of the Massachusetts Institute of Technology, who passed away shortly before this book was finished. For 15 years, Dave supported the efforts of one of us (LS) to weave the economists' and physical scientists' perspectives on energy use into a view that both disciplines could accept.

Units of measurement

Energy units

The energy content of various fuels, heat, and electricity is expressed in different units around the world. In the course of our work with numerous sources from many countries, we have converted all units into Joules, the basic unit of the SI system. The units that we commonly use, and their equivalence to other units often found, are as follows:

EJ (exajoules) = 10^{18} Joules = 0.948 quads (10^{15} Btu) = 240×10^6 toe = 239 10^{12} kcal

GJ (gigajoules) = 10^9 Joules = 0.948 million Btu = 0.024 toe = 239 $\times 10^3$ kcal

MJ (megajoules) = 10^6 Joules = 0.948 thousand Btu = 0.024×10^{-6} toe = 239 kcal

When referring specifically to electricity, we present data in watt–hours (Wh) or watts (W):

kWh (kilowatt–hours) = 10^3 watt–hours = 3.6 MJ

TWh (terawatt–hours) = 10^{12} watt–hours = 3.6 PJ

kW (kilowatts) = 10^3 watts

MW (megawatts) = 10^6 watts

TW (terawatts) = 10^{12} watts

"Commercial energy" refers to all forms of energy other than biomass fuels (fuelwood, agricultural residues, and dung), traditional uses of wind and solar energy (e.g., water pumps and solar drying), and animal and human power. The term "commercial energy" is actually misleading, since much biomass fuel is traded in commercial markets or is used by industries, in many cases substituting for fossil fuels. It is still the most commonly used term for "modern" fuels, however, so we use it despite its flaw.

"Primary energy" includes losses and own-use in the production of fuels, district heat, and electricity and in the delivery of district heat and electricity. "Final energy" refers to actual consumption by end users. (In some international data compilations, what is reported as final energy often includes losses in district heat delivery.)

Monetary units

When comparing monetary units among countries, we usually make use of Purchasing Power Parities (PPP) rather than currency exchange rates to convert local currencies to a common unit. Use of purchasing power parity is designed to equalize purchasing powers of currencies in the respective countries. It is defined as the number of units of a country's currency required to buy the same amounts of goods and services in the domestic market as one dollar would buy in the United States. Thus, the unit in which GDP or energy prices is expressed is not a dollar per se, but rather a dollar-equivalent.

Other units

Measures of weight are given in metric tons (tonnes); one tonne = 1000 kilograms (kg). Measures of volume are given in liters (l) or US gallons; one gallon = 3.785 liters. One US gallon = 0.833 Imperial gallons.

Measures of distance are usually given in kilometers (km); one kilometer = 0.62 miles.

Prologue

The transition to costlier energy
JOHN P. HOLDREN

Civilization is not running out of energy resources in any absolute sense, nor running out of technological options for transforming energy resources into the forms our patterns of energy use require. What is running out, rather, is the capacity to expand energy supply at low cost – a capacity which was fundamental to the growth of material wealth in today's industrial nations and which had been the basis of expectations that today's less developed countries would be able to follow a similar path to prosperity.

In this connection as in others, "cost" must be understood to include not only monetary but also environmental and sociopolitical components. Thus the declining availability of low-cost energy is not simply a matter of the most accessible and cheaply producible oil and natural gas having been depleted, or of the most cost-effective hydroelectric sites already having been used – even though the first situation does indeed prevail in most parts of the world and the second situation prevails in many. It is also a matter of the substantial economic risks and potential political costs suffered by countries that depend too heavily on imports of oil or gas or hydropower from countries where these energy sources still can be cheaply harvested. And it is a matter of paying the higher environmental costs of harvesting and using more widely distributed fossil fuels (coal and offshore oil, for example) or paying increased monetary costs for advanced fossil-fuel technologies that reduce these environmental impacts – or for nuclear and renewable energy options that avoid the environmental and political liabilities of fossil fuels but create others.

If the world is really embarked on a transition to costlier energy, as opposed to experiencing only a temporary aberration in the costs of energy supply, then the implications will be profound for the technological and economic strategies of industrial and developing countries alike – and

1

above all for the role to be played by increasing the efficiency of energy end use, which is the central theme of this book. The remainder of this prologue explores in some detail the basis for believing that a fundamental transition is indeed underway, discusses the implications of the changing circumstances, and surveys some of the ingredients of a strategy for dealing with the situation. Thus is the groundwork laid for the analysis, in subsequent chapters, of the role we must hope that increased energy efficiency will be able to play.

Energy and well-being in historical perspective

The heart of the energy problem – the reason not just energy experts but also politicians and citizens need to be concerned about it – is the two-sided character of the connections between energy and human well-being: the balance between energy benefits and energy costs. The energy benefits that contribute positively to human well-being include such consumer services as heating, lighting, and cooking, and energy's role as an input to economic production. The energy costs that subtract from well-being include the capital, labor, and raw materials devoted to obtaining and exploiting energy – and hence not available for other purposes – and the environmental and sociopolitical impacts of energy supply and use. (See Box A.)

For most of human history, the dominant concerns about energy have centered on the benefit side of the energy/well-being relation. Inadequacy of energy resources – or, more often, of technology and organization for harvesting, conversion, and distribution of energy – has resulted in insufficient energy benefits. Such problems have ranged from gasoline lines to electrical brownouts to chronically unmet demands for basic energy services in Third World villages. And their consequences have ranged from inconvenience to material deprivation and severe constraints on industrialization and economic growth. Problems of this sort remain the principal preoccupation today in the rural areas of the less developed countries, where energy for food production and other basic human needs is the key issue. They are also of concern in the modern regions of LDCs and in intermediate-income countries, where the issue is energy for production and growth.

In addition to the benefit-side problem of having too little energy, people can also suffer on the cost side from paying too much for it. The price may be paid through excessive diversion of capital, labor, and income from

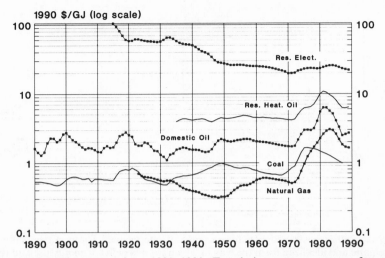

Fig. 1. US Real energy prices, 1890–1990. Trends in monetary costs of energy during the last 100 years are illustrated by these data on real prices of domestic petroleum, natural gas, coal, heating fuel, and electricity in the United States. Fuel oil and electricity prices are those paid by residential consumers. Natural gas and oil prices are for domestic production, at the wellhead. Coal prices are at the mine mouth. The effect of the oil price shocks of 1973–74 and 1979 are striking in the cases of the fuels, less so for electricity (because of the limited role of oil in electricity generation and the modest contribution of fuel costs, in general, to the total costs of electricity supply). The figures are three-year running averages, expressed in all cases in constant (1990) dollars per gigajoule. Conversion from current dollars was based on the Consumer Price Index (Bureau of the Census, 1990). The raw data are from Bureau of the Census (1972, 1990) and Energy Information Administration (1990).

non-energy needs. Or it may be paid in the forms of excessive environmental and sociopolitical impacts of energy supply not reflected in energy prices. For most of the past century, however, problems arising from high monetary costs of energy were, on the whole, in retreat: between 1890 and 1970, the real (inflation-corrected) monetary costs of supplying energy and the prices paid by consumers stayed more or less constant or declined. (Fig. 1 illustrates this point for the case of the United States.) And during this same period, the environmental and social costs of energy were regarded more as local nuisances or temporary inconveniences than as pervasive and persistent liabilities which seriously subtracted from energy's benefits.

All this changed in the 1970s. The oil-price shocks of 1973–74 and 1979 doubled and then quadrupled the real price of oil on the world market. In 1973, oil constituted nearly half the world's annual use of "industrial"

Prologue

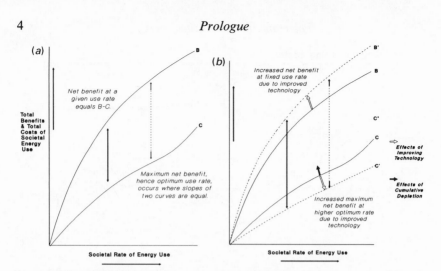

Fig. 2. Energy costs and benefits vs. rate of energy use. These curves show in schematic and idealized form the way in which total benefits of energy use and total costs of energy supply in a society can be expected to grow with the society's rate of energy use. Drawing such curves on a single scale implies that all the benefits and all the costs are commensurable (expressible in a single unit, such as dollars). Although this is not really the case, pondering what such curves might look like if it were is instructive. For any fixed set of energy-supply options, energy-use technologies, and end-use pattern, one would expect the curves of total cost and total benefit versus energy use rate to have the general shapes shown at the left (*a*). The slope of the benefit curve represents the "marginal" benefit (in economists' terms) of an increment of energy use; it declines with increasing energy use because the most productive applications of energy are exploited first. The slope of the cost curve, representing the marginal cost of each new increment of energy, declines at first because of economies of scale in energy supply, then at some higher use rate begins to increase as economies of scale are exhausted, more expensive sources must be added to the mix, and environment costs increase because of saturation and threshold effects. Energy's net benefit to society is represented by the difference between the two curves. The net benefit has a maximum where the slopes of the curves are equal (marginal cost equals marginal benefit). This is the optimum use rate; at higher rates, the costs are rising more steeply than the benefits so the net benefit to society is shrinking. (Note that these curves omit the phenomenon in which high rates of growth of the energy use rate tend to aggravate both economic and environmental costs by forcing hasty adoption of higher-cost options than those than could be brought to bear if growth were slower.) Resource depletion, technological change, and altered end-use patterns can shift these curves as indicated in the figure on the right (*b*), thus altering the net benefit at any given energy use rate and the maximum net benefit available at the optimum use rate. Technological improvements in end-use possibilities shift the benefit curve upward, as in B to B'. Cumulative depletion of energy resources and of environmental absorptive capacities tend to shift the cost curve upward, as in C to C", while technological improvements in supply options work to shift it downward, as in C to C'. One would not expect such changes to alter the overall shape of the curves, which means there will always be a finite optimum use rate and a possibility of

energy forms (oil, natural gas, coal, nuclear energy, and hydropower, as opposed to "traditional" energy forms like fuelwood, crop wastes, and dung). Not surprisingly, then, the rise in oil prices pulled up the prices of other industrial energy forms too. The results vividly illustrated the perils of excessive monetary costs for energy: worldwide recession, spiralling debt, a punishing blow to the development prospects of oil-poor LDCs and the imposition of disproportionate economic burdens on the poor in the industrial nations.

The late 1960s and early 1970s also marked a transition in the prospects for and perceptions of increased environmental and sociopolitical costs. Problems of air and water pollution, many of them associated with energy supply and use, were being recognized not as local nuisances, but as regional threats to human health, economic well-being, and environmental stability. The 1969 MIT-sponsored Study of Critical Environmental Problems, which convened prominent environmental scientists from around the world, called attention to more than a dozen environmental problems of global scope – including particulate pollution, greenhouse gases, acid precipitation, oil spills, and radioactive wastes – nearly all of them caused wholly or partly by energy supply (SCEP 1970).

In the United States, the passage of the National Environmental Policy Act, the Coal Mine Health and Safety Act in 1969, and the Clean Air Amendments in 1970 signalled national recognition that environmental costs of energy supply that had been tolerated previously would now have to be reduced, even if energy had to become more costly in monetary terms as a result. The 1972 United Nations Conference on the Human Environment in Stockholm underlined the relevance of energy-related environmental problems for both poor and rich countries by stressing the impacts on soils, forests, fisheries, and water supplies. Consciousness of the sociopolitical costs of energy supply also grew in this period, as increasing dependence on Middle East oil brought constraints on foreign policy while increasing the possibility of war, and as India's nuclear explosion in 1974 brought wider awareness that spreading competence in nuclear energy can provide countries with weapons as well as electricity.

The 1970s, then, seem to have been a turning point in the total costs of energy supply. After decades of constancy or decline in energy's monetary

Caption for Fig. 2 (*cont.*)

overshooting it. The proper tasks of energy-technology development and energy-policy formation are to increase the net benefit available at any given use rate and to approach the use rate that provides the maximum net benefit without overshooting it.

costs – and of relegation of environmental and sociopolitical costs to secondary status – energy was seen to be getting costlier in all respects. It became plausible for the first time that the problem of excessive energy costs could pose threats on a par with those of insufficient supply. And it became possible to think that some forms of expansion of energy supply could create marginal costs exceeding the marginal benefits, and thus should be foregone.

As indicated in Figs. 2a and b and the arguments in the caption, it is hardly surprising that civilization would arrive eventually at the end of the era of low-cost energy – more precisely, the end of the era in which the incremental benefits of further increases in energy supply clearly exceeded the incremental monetary and nonmonetary costs. But if this transition from low-cost to costly energy is really occurring *now*, that changes everything.

Low-cost energy meant that the relation between energy and human well-being was simple and positive: the value to society of expanding the supply of energy was obvious and not in dispute; the principal "energy problem" was the loss of potential benefits arising from failure to expand energy supply as rapidly as society needed it; and the proper aim of energy strategy was to facilitate the expansion of supply. With costly energy, by contrast, the relation between energy and human well-being is more complex. Societies may suffer from "paying too much" as well as from "having too little," and disputes about whether and how to expand energy supply are inevitable. It is then the task of energy assessment to determine when expanding energy supply will generate greater marginal costs than marginal benefits. And the task of energy strategy, in those circumstances, is to avoid expanding supply while seeking through technology and policy to increase the benefits and reduce the costs.

The key question at this point in history is whether the trend toward increasing energy costs will be temporary or permanent. Are there any reasons, in other words, for thinking that the events of the past twenty years really represent a fundamental transition, rather than a temporary, less threatening aberration? Is the era of low-cost energy really over, or will some combination of new resources, new technologies, and new geopolitics bring it back? And why, after all, should a permanent transition to costlier energy have begun in the 1970s, and not sooner or later?

To shed some light on these questions, we need to look in more detail at the evolution of energy demand and supply over the past century or so, and at future energy-supply options.

Energy demand and patterns of supply

Evidence of a fundamental shift in the energy picture can be found in the enormous increase in the scale of energy demand since the middle of the last century. That increase is the result of a combination of unprecedented population growth and equally remarkable growth in per-capita use of industrial energy forms. World population grew 3.2-fold between 1850 and 1970; per-capita use of industrial energy forms increased about 20-fold; and total world use of industrial and traditional energy forms combined increased more than 12-fold (Table 1). Growth slowed but did not stop in the 1970s. Between 1970 and 1990, population increased another 47%, industrial energy use per capita rose some 7%, and total use of industrial and traditional energy increased 58%.

To supply energy at rates in the range of 10 terawatts[1] – first achieved in the late 1960s – is an enormous enterprise. In 1970, it required the harvesting, processing, and combustion of some 3 billion metric tonnes of coal and lignite, some 17 billion barrels of oil, more than a trillion cubic meters of natural gas, and perhaps 2 billion cubic meters of fuelwood (see Table 2). It required a hydropower effort equivalent to running a third of the stable runoff from the world's continents through a drop of 100 meters. It entailed the use of dirty coal as well as clean, undersea oil as well as terrestrial, deep gas as well as shallow, mediocre hydroelectric sites as well as good ones, and deforestation as well as sustainable fuelwood harvesting. The capital investment to maintain and expand this massive energy-supply system amounted in 1970 to perhaps $250 billion (adjusted to 1990 dollars). The corresponding figure for 1990, based on a study of global energy-system investments for the period 1980–2000, approximates $700 billion (Schneider & Schulz, 1987).

The industrial growth of the past 100 years was powered above all by fossil fuels – initially coal and subsequently oil and natural gas. In 1880, world coal use at 0.3 TWyr/yr just equalled that of fuelwood; by 1910, the use of coal had tripled to 1 TWyr/yr while use of fuelwood declined slightly. Between 1910 and 1945, coal use fluctuated in the range 1 to 1.5 TWyr/yr, but the use of oil and natural gas grew dramatically, from a combined contribution of 0.08 TWyr/yr in 1910 to about 0.6 TWyr/yr in 1945. The convenience and versatility of oil and natural gas – above all the attractiveness of the former for transport fuel and space heating and of the latter for space heating, water heating, and industrial processes – continued to drive their rapid growth after World War II. The contribution

[1] One terawatt (TW) $= 10^{12}$ W $= 10^{12}$ J/s $= 31.5$ EJ/yr.

Table 1. *World population and energy use, 1850–1990*

	World population (billions)	Energy use per person (kW)		World energy use (TW)			Cumulative use of industrial energy forms since 1850 (TWyr)
		Industrial forms	Traditional forms	Industrial forms	Traditional forms	Total	
1850	1.13	0.10	0.50	0.11	0.57	0.68	0.0
1870	1.30	0.16	0.45	0.21	0.59	0.79	3.2
1890	1.49	0.32	0.35	0.48	0.52	1.00	10.1
1910	1.70	0.64	0.30	1.09	0.51	1.60	25.7
1930	2.02	0.85	0.28	1.71	0.56	2.28	53.7
1950	2.51	1.03	0.27	2.58	0.68	3.26	96.6
1970	3.62	2.04	0.27	7.38	0.98	8.36	196.3
1990	5.32	2.19	0.29	11.66	1.54	13.20	386.7

Industrial energy forms are mainly coal, petroleum, and natural gas, with modest contributions from hydropower and, after 1970, from nuclear energy. (Geothermal energy would be included in this category, but its use is negligible on a global scale.) Traditional energy forms are fuelwood and charcoal, crop wastes, and dung. An energy use rate of a kilowatt is equivalent to consumption of a metric ton of high-grade coal per year; a terawatt (10^{12} watts) is equivalent to a billion tonnes of coal, or 700 million tonnes of oil, per year. Nutritional calories and the energy contributions of work animals are not included; the nutritional energy requirement of an average human being is just over 0.1 kilowatt. Population figures are from Bogue (1969) and Population Reference Bureau (1990). Figures for industrial energy use are derived from Darmstadter (1968), Hubbert (1969), Cook (1976), Haefele (1981), British Petroleum (1990), and Energy Information Administration (US EIA, 1989); the 1990 figures are author extrapolations from 1989 data in British Petroleum (1990) and partial 1990 data in US EIA (1990). Figures for traditional energy use are the author's estimates based on a variety of sources; see, e.g., Hughart (1979); World Bank (1983); Hall et al. (1982); Goldemberg et al. (1987).

Table 2. *Pattern of world energy supply in 1970 and 1990*

	1970		1990	
	Terawatts	*Share*	*Terawatts*	*Share*
"Industrial" energy forms	7.3	88.0%	11.7	88.6%
Petroleum	3.3	39.7%	4.5	34.1%
Coal	2.2	26.8%	3.2	24.2%
Natural gas	1.4	16.4%	2.5	18.9%
Hydropower	0.4	4.7%	0.8	6.1%
Nuclear fission	0.03	0.4%	0.7	5.3%
"Traditional" energy forms	1.0	12.0%	1.5	11.4%
Fuelwood	0.6	7.2%	0.9	6.8%
Crop wastes and dung	0.4	4.8%	0.6	4.5%
TOTAL	8.3	100%	13.2	100%

Sources for the data are the same as indicated in Table 1.

of oil and gas combined had passed that of coal by 1960, and by 1990 oil and gas together were delivering some 6 TWyr/yr compared to 3 TWyr/yr for coal.

This tremendous growth in the use of oil and natural gas was, naturally, supplied from the most accessible and inexpensively exploitable deposits of these – the most versatile, concentrated, and transportable chemical fuels on the planet. The century's cumulative consumption of some 200 terawatt–years of oil and gas represented 20% to 30% of the ultimately recoverable portion of the Earth's initial endowment of these fuels in their conventional forms. In the context of growing demand, this is a large proportion, not a small one. If the cumulative consumption of oil and gas continued to double every 15 to 20 years, as it did for the 100 years preceding 1973, the initial endowment would be depleted by 80% in another 30 years or so. More probably, the rate of consumption would peak and then begin to decline when cumulative consumption reached about half the initial endowment. We can expect this to happen between the years 2000 and 2010 (Hubbert 1969).

Except for the huge pool of oil underlying the Middle East, the cheapest oil and gas are already gone. The trends that had kept costs down until the beginning of the 1970s in spite of cumulative depletion – new discoveries and economies of scale in processing and transport – have played themselves out. Discoveries of oil and gas per kilometer of exploratory drilling had fallen substantially by 1970 and have stayed down. It is possible that no new giant fields remain to be discovered in convenient locations, and even several more giants would make little difference against consumption at today's scale. A new Prudhoe Bay would provide about 6 months' oil supply for the world at the 1990 use rate; a new North Sea, 3 years' supply.

For most countries, future oil and gas will come increasingly from smaller and more dispersed fields, from offshore and Arctic environments, from deeper in the Earth, from secondary and tertiary recovery, and from unconventional deposits (e.g., heavy oils, methane clathrates) – all of which are more difficult and more costly than oil and gas production used to be – or they will come from imports whose reliability and affordability (in political as well as monetary terms) cannot be guaranteed. The timing of the oil shocks of the 1970s was not a mere happenstance, the result only of the whims of the OPEC (Organization of Petroleum Exporting Countries) cartel. OPEC could do what it did only because, in much of the world, the cheapest oil and gas were gone. The peak oil prices of 1980 and 1981 were a cartel artifact, and they came down because, at those levels, coal, conservation, and non-OPEC oil supplies were cheaper. Still, real prices for crude oil and natural gas did not fall quite to their early 1970s levels, and very possibly they never will.

Other energy resources, of course, are more physically abundant than oil and gas. Coal, oil shale, solar energy in its diverse manifestations, geothermal energy, and the fission and fusion fuels are the most important ones. But all of these require elaborate and expensive transformations into electricity or fluid fuels in order to meet society's end-use needs. None has very good prospects for delivering large quantities of fuel at costs comparable to those of pre-1973 oil and gas, or large quantities of electricity at costs comparable to those of the cheap coal-fired and hydropower plants of the 1960s. In the case of electricity, which has always been expensive energy compared to other forms, it may or may not be possible for the long-term options to match the costs per kilowatt–hour that prevailed in the 1970s and 1980s; but it is extremely unlikely that they will be significantly cheaper. (It is the perverse nature of all of the most abundant energy sources that the technologies needed to make electricity from them tend to be technologically complex, or materials intensive, or labor intensive, or a combination of these, all of which translates into high costs even if the raw energy is free.)

This overview of the prospects for energy supply suggests – although of course it cannot prove conclusively – that monetary costs of energy substantially higher than those prevailing before 1973 will turn out to be a long-term phenomenon, not a temporary aberration. This would be so even without considering the ways in which future attempts to reduce environmental costs of energy supply are likely to boost the monetary costs. Let us now turn to this environmental dimension.

Environmental and sociopolitical costs of supplying energy

The capacity of the environment to absorb the effluents and other adverse impacts of energy technologies without significant damage is itself a finite resource. This finitude manifests itself in an array of environmental costs, which may be "internal" or "external" and which may or may not be expressible in monetary terms.

> "External" environmental costs are those imposed on members of society other than the producers and users of the energy that causes the costs; these costs may be monetary (as in lost farmers' income as a result of crop damage from air pollution, or loss of tourism revenues because of an oil spill) or nonmonetary (such as reduction of life expectancy from exposure to toxic substances, or loss of bio-diversity in impacted ecosystems). Some of the nonmonetary external costs can be expressed, with some effort and controversy, in monetary terms (e.g., the idea that a year's reduction in life expectancy can be valued approximately at the individual's annual income).
> "Internal" environmental costs are those borne by the producers and users of the energy. They may take the form of environmental damages – monetary or magnetizable or not – that the producers and users happen to suffer in connection with their energy production and use; or they may take the form of increases in the monetary costs of energy imposed by measures aimed at reducing environmental damages.

Both "external" and "internal" environmental costs have been rising, not only in the aggregate but even, for some technologies, per unit of energy supplied.

There are some fundamental reasons why such increases in environmental costs with increased energy use rates (and increased cumulative use) should be expected for any given array of energy-supply technologies (Holdren, 1987). These reasons apply with particular force to the fossil fuels, on which we remain so heavily dependent, although they also influence renewable and other sources.

First, all else being equal, the quality of the fuel deposits and energy-conversion sites a society exploits will decline as its energy use rate and cumulative use increase. This means more material moved or processed at the site, bigger facilities for a given output, longer transport distances, and so on (Hall et al., 1986). These factors tend to increase external or internal environmental costs or both.

Second, environmental systems able to absorb moderate amounts of effluents or other insults without disruption can become saturated as energy use grows and sites and resources decline in quality. In effect, the environment's absorptive capacity gets consumed. At that point, additional inputs to the environment begin to generate disproportionate damages (Holdren & Ehrlich, 1974; Myers, 1984; Ehrlich, 1986). The rapid acidification of lakes and streams following cumulative consumption of their chemical buffering capacity by years of acid precipitation is a prime example (Harte, 1988; Stigliani & Shaw, 1990). The result of such threshold effects is often a sharp increase in some combination of external and internal costs.

Third, for a given state of control technology, the monetary costs of pollution control per gram of pollutant removed from effluents tend to increase with the removal percentage that is required (Ehrlich et al., 1977). When higher energy use rates, lower resource quality, and the enhanced sensitivity of an already stressed environment combine to require a higher degree of pollution control just to hold external damages constant, the rising marginal costs of control tend to force the internalized environmental costs sharply upward. This phenomenon is reflected in the acrimonious debate between regulators and electric utilities over what degree of control over emissions of oxides of sulfur and nitrogen is to be required.

In practice, at least one further phenomenon has acted to increase the component of energy's costs attributable to environment. Growing public and political concern with environmental impacts – and associated debates about who is to bear the external and internalized environmental costs – have lengthened the time required for siting, construction, and licensing of energy facilities (adding to interest on investment during construction as well as other cost factors), have increased administrative complexity (to deal with permits and regulations), and have led to greater frequency of expensive mid-project changes in design and specifications.

It is difficult to quantify the contribution of environmental costs to the monetary costs of energy supply, in part because factors not related to the environment are often entwined with environmental ones. The longer construction times for energy facilities, for example, have been caused not only by environmental concerns, but also by management and quality-control problems, as well as by engineering problems due to the increasing scale of facilities. In the case of the United States since 1970, however, it seems likely that costs related to actual or attempted internalization of environmental impacts have increased the monetary costs of supplying petroleum products by at least 25 or 30% over what they would otherwise be, and the costs of electricity supply from coal and nuclear power by 40% or more (Holdren, 1987). Similar figures probably apply in Western Europe and Japan.

Despite these expenditures, the environmental costs that have not been internalized have been substantial. In many cases they are growing (Brown et al., 1990; World Resources Institute, 1990). There are many varieties of such environmental costs, as suggested by the categorization in Box A. Attention is focused in what follows on three classes of these problems: deaths, injuries, and illnesses suffered by workers in energy-supply industries; risks of death and disease to the public as a result of emissions from and accidents at energy facilities; and impacts of energy supply on ecosystem function and international relations. As shown below, the occupational hazards generally are not a large contributor to the total costs and risks of energy supply and, therefore, not usually an appropriate basis for altering preferences for one energy option over another. The other two classes of problems can, as will be seen, generate far larger costs and can be expected to exert a correspondingly bigger influence on energy choices.

Occupational hazards

Occupational injuries and illnesses in energy-supply industries in industrial nations account, for the most part, for 5 to 50 deaths per exajoule of energy delivered to the final consumer. (Coal-fired electricity generation in countries with poor underground-mine safety practices reaches as high as 100 deaths per exajoule.) (See Table 3.) Nonfatal injuries and illnesses account for 10,000 to 150,000 additional lost workdays per exajoule. Overall, the figures for different energy technologies are remarkably similar.

How should one assess the significance of such numbers? One appropriate yardstick is the level of occupational hazard in the workforce in general. With the exception of the most dangerous underground coal

Table 3. *Occupational hazards of energy
supply (worker deaths per exajoule
of delivered energy)*

Energy source	Deaths
Heat and fuel	
Refined petroleum products	1–2
Synfuels from coal	15–30
Biomass wastes	1–10
Fuelwood from forestry	20–40
Flat-plate solar collectors	20–30
Electricity	
Coal-fired	30–100
Nuclear (light-water reactor)	10–40
Hydropower	5–10
Windpower	3–10
Photovoltaic cells	10–30

mines and some forestry occupations, the hazards experienced by energy workers, measured per worker per year, are not much different than those in industry as a whole (Holdren et al., 1983). Another measure, favored by environmental economists, is the economic value of the lost life and health, which can be compared with the economic value of the energy. If one suppresses one's natural uneasiness about monetizing human life for long enough to multiply by $0.5–$1 million per death and $100–$200 per lost workday (loosely defensible, in the industrial-nation context, in terms of lost economic output), one obtains monetary values for the occupational hazards in the range of 0.1% to 2% of the energy's monetary cost. (Part of this, moreover, has already been internalized through wage premiums for hazardous occupations, company-paid insurance and sick-leave, and so on.)

Even if these figures underestimate the costs of lost life and health by a factor of two or three, due to the omission of unquantified occupational illnesses, the occupational hazards cannot be considered a major external cost of energy supply. Nor can the difference in costs be the basis for a society's choosing between two energy sources. Of course, there is a good case for cleaning up the workplace in general, and for carefully monitoring practices in the least safe industries in particular (coal mining in some countries has been made as safe as the average manufacturing job); but that is not just a problem of the energy industry per se, and solving it could hardly add as much as 10% to the monetary costs of energy.

Public health and safety

The adverse impacts of energy-supply technologies on public health and safety are much more problematical in several respects. The health impacts involve, generally, more subtle effects in much larger populations than is the case in the energy workforce, so the problems are much harder to identify and quantify. In the case of accidents, the most worrisome problems are usually matters of a small chance of a big disaster, for which neither accumulated experience nor analysis is adequate to pin down the probabilities convincingly. As a result, the uncertainties about the magnitudes of expected damages to public health and expected risk to public safety from energy systems can be enormous – factors of hundreds or even thousands between the lowest plausible and highest plausible estimates. Such uncertainties, for example, plague the estimates of the impacts on the public of both fossil-fuel-derived air pollution and radiological hazards from nuclear power. Figure 3 shows some of the relevant ranges of estimates, with comparisons to other hazards experienced by the public.

The dominant threat to public health from fossil-fuel-derived air pollution has been thought to be particulate sulfates formed from sulfur dioxide emissions. Here, the huge range of estimates results from variations in fuel composition, air-pollution control technology, power-plant siting (in relation to population distribution), meteorological conditions affecting sulfate formation, and, above all, uncertainties in the dose–response relation for harm caused by exposure to sulfates. In particular, some reputable analysts think that sulfates from fossil-fuel combustion (largely from coal) contribute to the causation of respiratory and cardiovascular diseases (of the order of tens of thousands of deaths per year in the United States, for example, with loss of life expectancy averaging perhaps 10 years per death). Other, equally reputable analysts believe that available data does not support such a conclusion, and that the only certain conclusion is that air pollution at high concentrations increases the death rates among people who were already very ill from other causes (about ten times fewer deaths per year, and an average loss of life expectancy per death of a month or less) (National Research Council, 1975; Morgan et al., 1978; Wilson et al., 1980).

In the case of nuclear fission, a huge range in estimates of the expected health damage results from variations among sites and in the routine

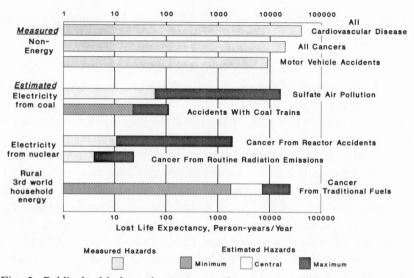

Fig. 3. Public health hazards: energy and nonenergy. The bars show on a logarithmic scale, the relative magnitudes of some energy and nonenergy risks to public health. The units are years of lost life expectancy (annual deaths times lost life expectancy per death) in a population of one million people. The top three bars are based on actual US death rates in the late 1980s and provide a basis for comparison with the calculated risks from electricity generation (in the United States) and household use of traditional fuels (in the Third World). The bars for coal-fired and nuclear electricity generation in the United States were calculated by assuming, in each case, that all of the electricity for a population of one million people at 1988 US per capita use rate—11,000 kilowatt–hours per year per person or 1.25 gigawatt–years per year per million persons—is supplied by the indicated source, and that all of the public health impacts are experienced in the population of one million for whom the electricity is provided. The "maximum" estimates for the effects of sulfates from coal burning and emissions of radioactivity from nuclear operations are based on unfavorable plant sites and pessimistic dose–response relations, while the "central estimate" figures are based on best-estimate dose–response relations (in the author's judgment) and US-average sites (see Holdren, 1987). In the nuclear case, a range of estimates of accident probabilities has also been taken into account (Nuclear Regulatory Commission, 1975; National Academy of Sciences, 1979; Holdren, 1987). "Minimum" figures for both coal and nuclear energy, based on optimistic dose-response relations and favorable sites are not visible here because they are so close to zero. The figures for cancer from rural Third World indoor air pollution from combustion of traditional fuels are based on the estimates of Smith (1987, 1988) for total exposure and a range of dose-response relations (Smith, 1987; Holdren, 1987). Deaths from cancer and cardiovascular disease are assumed to cause the loss of 10 years of life expectancy per death and deaths from accidents 45 years per death. Effects experienced by future generations are not shown here, although these might be significant for all of the energy sources. Note that the amount of energy per person associated with the Third World traditional fuels is much smaller than that associated with electricity use in the United States, so that the traditional fuels are even more hazardous per exajoule than the graph seems to indicate.

emissions of radioactivity from different reactors; from uncertainties in the routine emissions from fuel-cycle steps that are not yet fully operational (especially fuel-reprocessing plants and the management of uranium-mill tailings); from uncertainty about the dose-response relation for exposure to low-dose radiation; and, above all, from uncertainties relating to the probabilities and average consequences of large accidents at reactors and reprocessing plants, and in the transport and storage of wastes. The question of reactor-accident probabilities, by itself the focus of a huge literature (see, e.g., US Nuclear Regulatory Commission 1975; National Academy of Sciences 1979; US OTA 1984) is obscured not only by the difficulty of predicting internally generated accident sequences in extremely complex systems, but also by the problem of accounting for human frailties – including those of reactor designers, builders, operators, safety analysts, and possibly saboteurs (Holdren, 1976).

As indicated in Fig. 3, the ranges of expected hazards to public health from both coal-fired and nuclear electricity generation extend from negligible to substantial in comparison with other risks to the population. If expressed in monetary terms, as described above for workers, these ranges extend from a few tenths of a percent to 10% or more of the monetary value of the electricity. One cannot conclude with any confidence, then, either that the external public-health damages from coal-fired and nuclear-electricity generation are very small, or very large. Nor can one conclude that there is much basis, from this evidence, for preferring one over the other. For both energy sources, then, the very size of the uncertainty about these external costs is itself a significant liability.

Considerably more certain, although unmentioned by most energy analysts, is the tremendous public-health menace represented by indoor air pollution created by the traditional fuels widely used for cooking and water heating in the rural Third World. About 80% of global exposure to particulate air pollution (measured as number of persons times their average annual exposure to respirable suspended particulate matter) occurs indoors in developing countries; and the smoke inhaled is heavily laden with carcinogenic benzpyrene and other dangerous hydrocarbons (Smith et al., 1983; Smith, 1987, 1988). A disproportionate share of this enormous burden is borne, moreover, by women (who are doing the cooking) and small children (who are indoors with their mothers). Although the public health statistics for developing countries are not adequate to sort out fully the consequences of this exposure (Pandey et al., 1989), a figure in the range of 100,000 premature deaths per exajoule of traditional fuel is not implausible. By any measure – whether contribution

Table 4. *Energy's contributions to global environmental impacts*

Indicator of impact	Natural baseline	Human disruption index	Share of human disruption caused by:			
			Industrial energy supply	Energy supply	Agriculture	Manufact., other
Lead emissions to atmosphere	25,000 tonnes/yr	15	63% (fossil-fuel burning, incl. additives)	Small	Small	37% (metal processing, manufact., refuse burning)
Oil added to oceans	500,000 tonnes/yr	10	60% (oil harvesting, processing, transport)	Negligible	Negligible	40% (disposal of oil wastes)
Cadmium emissions to atmosphere	1,000 tonnes/yr	8	13% (fossil-fuel burning)	5% (burning traditional fuels)	12% (agricultural burning)	70% (metals processing, manufact., refuse burning)
Sulfur dioxide emissions to atmosphere	50 million tonnes/yr (S content)	1.4	85% (fossil fuel burning)	0.5% (burning traditional fuels)	1% (agricultural burning)	13% (smelting, refuse burning)
Methane stock in atmosphere	800 parts per billion	1.1	20% (fossil fuel harvesting & processing)	3% (burning traditional fuels)	62% (rice paddies, domestic animals, land clearing)	15% landfills
Mercury emissions to atmosphere	25,000 tonnes/yr	0.7	20% (fossil-fuel burning)	1% (burning traditional fuels)	2% (agricultural burning)	77% (metals processing, manufact., refuse burning)
Land use or conversion	135 million km² ice-free land	0.5	0.2% (occupied by energy facilities)	6% (to supply fuelwood use sustainably)	88% (grazing, cultivation, cumulative desertification)	6% (lumbering, towns, transport systems)

	Natural baseline	Index				
Nitrogen fixation (as NO_x, NH_4)	200 million tonnes/yr	0.5	30% (fossil fuel burning)	2% (burning traditional fuels)	67% (fertilizer, agricultural burning)	1% (refuse burning)
Nitrous oxide flows to atmosphere	7 million tonnes/yr (N content)	0.4	12% (fossil fuel burning)	4% (burning traditional fuels)	84% (fertilizer, land clearing aquifer disruption)	small
Carbon dioxide stock in atmosphere	280 parts per million	0.25	75% (fossil fuel burning)	3% (net deforestation for fuelwood)	15% (net deforestation for land clearing)	7% (net deforestation for lumber, cement mfg)
Particulate emissions to atmosphere	500 million tonnes/yr	0.25	35% (fossil fuel burning)	10% (burning traditional fuels)	40% (agricultural burning, wheat handling)	15% (smelting, non-agric. land clearing, refuse)
Ionizing radiation dose to humans	800 million person-rem per year (300 from whole-body equivalent of radon lung dose)	0.20	1% (half from nuclear energy, half from radon in coal)	unquantified extra radon release from soil disturbance	unquantified extra radon from soil disturbance	99% (medical X-rays, fallout, air travel)
Nonmethane hydrocarbon emissions to atmosphere	800 million tonnes/yr	0.13	35% (fossil fuel processing & burning)	5% (burning traditional fuels)	35% (agricultural burning)	20% (non-agric. land clearing, refuse burning)

Some impacts are most appropriately characterized as alterations of natural inventories, or stocks, others as alterations of natural flows. The Human Disruption Index is the ratio of the size of the human alteration to the size of the undisturbed stock or flow, denoted the "natural baseline". The figures for the shares of human disruption accounted for by different classes of activities are based on current conditions. Estimates are the author's based on a variety of sources and are very approximate; see, e.g., Holdren (1987), Lashof & Tirpak (1989), Graedel & Crutzen (1989), Intergovernmental Panel on Climate Change (1990), and World Resources Institute (1990).

to total societal risk or magnitude in comparison to the value of the energy
– this is a very large cost.

Climate and ecosystems

The threats posed by energy supply to climate and ecosystems are even
more difficult to analyze and quantify than the more direct threats to
human health and safety from effluents and accidents. Nonetheless, enough
is known to suggest that they embody the potential for even larger damages
to human well-being.

This potential arises from the combination of two circumstances. First,
civilization continues to be heavily dependent on environmental conditions
and processes for the provision of a wide array of services indispensable to
human well-being: building and fertilizing the soil; regulating and
distributing runoff; controlling the populations and distribution of most
crop pests and disease vectors; and limiting the deviations of climate from
patterns on which the productivity of agriculture, forestry, and fisheries
depend (Holdren & Ehrlich, 1974; Myers, 1984; Ehrlich, 1986). These
environmental services have not been supplanted by technology – and
cannot be on any time scale of practical interest.

Second, global human activities in general, and energy supply in
particular, have grown to a size and character demonstrably capable of
disrupting at global scale the environmental conditions and processes that
provide these services (SCEP, 1970; Ehrlich et al., 1977; Seiler & Crutzen,
1980; Graedel & Crutzen, 1989). Humans use 35% of the planet's land
area for agriculture and grazing, and have transformed another 10% from
forest and grassland into desert. Human activities have increased global
erosion about 5-fold over prehistoric rates. And human activities are
dispersing a wide variety of biologically and climatologically active
elements and compounds into atmosphere, surface waters, and soil
worldwide, at rates far above the natural flows of these substances. The
results of these alterations include a 10-fold increase in the acidity of rain
and snow over areas of millions of square kilometers, and significant
changes in the global composition of both the stratosphere and the
troposphere.

As indicated in Table 4, activities connected with supplying both
industrial and traditional energy forms are responsible for a striking share
of these environmental impacts of human activity at a global scale. It
has been in the course of the last century that most of these phenomena
have grown from mere local perturbations to global disruptions. The en-

vironmental transition of the last 100 years – driven above all by 20-fold growth in the use of fossil fuels and augmented by a tripling in the use of the traditional energy forms – has amounted to no less than the emergence of civilization as a global ecological and geochemical force.

Of all of the global environmental problems, climate change is the most pervasively threatening to human well-being and in many respects the most intractable. It is pervasively threatening because climate affects – and climate change can disrupt – most of the other environmental conditions and processes on which the well-being of the world's population critically depends: those governing the fertility of soils and the availability of water, those determining the productivity of forests and ocean fisheries, those regulating sea level and the frequency and intensity of storms, and those affecting the distribution and populations of crop pests and disease organisms (Schneider & Londer, 1986; Schneider, 1989). The problem is intractable because the "greenhouse" gases mainly responsible for the danger of rapid climate change over the next several decades come largely from human activities too massive, widespread, and central to the functioning of our societies to be easily altered: carbon dioxide from deforestation and above all the combustion of fossil fuels; methane from rice paddies and cattle guts, and the harvesting and transport of oil and natural gas; and nitrous oxides from fuel combustion and fertilizer use. (See Table 5.)

Political hazards

The only other "external" costs of energy that might match its climato-logical and ecological risks are those associated with causation or aggravation of large-scale military conflict. The most obvious such threat is the potential for conflict over access to petroleum resources (Deese & Nye, 1981, Yergin, 1988). This danger had been thought to be declining throughout much of the 1980s, as oil-importing countries diversified their sources of imports and reduced their overall dependence on oil through increased reliance on energy efficiency, coal, natural gas, nuclear energy, and renewable energy sources. By the end of the 1980s, however, the dependence of the United States and some other major industrial powers on oil from the politically volatile Middle East was again growing, increasing the danger of linkage between energy interests and conflict. The invasion of Kuwait by Iraq at the beginning of August 1990, and the military response led by the United States, moved such linkage out of the realm of the theoretical and into world headlines.

This is not to say that oil was the only reason for the war in the Persian

Table 5. *Sources of principal greenhouse gases, late 1980s*

Gas	Share of total warming potential of all late 1980s emissions	Sources of emissions
Carbon dioxide	66%	Coal-burning 32%, oil-burning 31%, net deforestation 22% (of which about $\frac{1}{6}$ for fuelwood), gas-burning 13%, cement manufacturing 2%. About $\frac{3}{4}$ of fossil fuel and $\frac{1}{4}$ of fuelwood are burned in industrialized nations by 23% of world's population.
Methane	17%	Rice cultivation 25%, domestic animals 22%, fossil fuels 20%, biomass burning 18% ($\frac{1}{6}$ for fuelwood), landfills 15%.
Chlorofluorocarbons	12%	Refrigeration and air-conditioning, plastic foams, solvents, aerosol cans
Nitrous oxide	5%	Land transformations and fertilizer use 64%, biomass burning 24% ($\frac{1}{6}$ for fuelwood), fossil-fuel burning 12%.

Figures shown are for anthropogenic emissions only; for relation of anthropogenic to natural emissions or stocks, see Table 4. Share of warming potential depends on time horizon: figures shown here are based on warming potential over the next 100 years. It has been assumed that half of land-clearing and fuelwood use represents net deforestation. Combining figures in the table reveals that fossil fuels are contributing 53% of the warming potential and fuelwood another 3%. Data are from Houghton et al. (1990), Lashof and Tirpak (1989), British Petroleum (1990), and the author's calculations based on these.

Gulf, or that fighting over oil makes economic sense. The importance of Middle Eastern oil to the world economy is entangled with many other rationales for major-power reaction to the Iraqi invasion – principle, precedent, desire to disarm an increasingly dangerous and aggressive dictator, resistance to changes in the regional power balance that could threaten the existence of Israel, and the idea that a superpower must act like one (using its military power to remind others of its prerogatives) in order to remain one. There is no evidence that the United States concluded, by any rational calculus, that the energy/economic benefits of preventing Iraqi control of Kuwaiti and Saudi oil were by themselves sufficient to justify the costs and risks of military action; indeed, a serious attempt at such a calculation would probably find the opposite. (This is consistent with past experience, in which rhetoric about resources as a rationale for major-power military and foreign policies has greatly exceeded any basis in analysis or logic for resources to play a determining role. See, e.g., Lipschutz, 1989.) Nonetheless, it is beyond dispute that an important rationale for Iraq's invasion of Kuwait was to gain control over the vast oil resources underlying the latter country; and the perception that the oil reserves of the entire Middle East were at risk unquestionably contributed to persuading publics and policy makers around the world that a forceful response was appropriate.

A second energy/conflict connection that is plausibly as dangerous as fighting over oil is the link between nuclear energy and nuclear weapons. The essence of this problem is that the spread of nuclear energy technology spreads access to nuclear-explosive materials and related capabilities in ways that make it easier for additional countries to acquire nuclear weapons (Holdren, 1983; Sweet, 1984; Holdren, 1989). In the future, as subnational criminal groups become more sophisticated, the related threat that these, too, might acquire nuclear bombs or radiological weapons by misusing nuclear-energy technologies may grow in importance (Willrich & Taylor, 1974).

It is generally conceded that the knowledge needed to design and fabricate fission bombs is available by now to almost every nation. Although the diffusion of knowledge that produced this situation was partly due to the international promotion of nuclear fission as an energy source, the diffusion is already so complete that it is no longer given much weight as a liability of nuclear energy. Lack of access to nuclear explosive materials, not lack of knowledge, has been the main technical barrier to the spread of nuclear-weapons capability for the past two decades and more. The continuing major threat of nuclear weapons proliferation posed by fission power is that it tends to provide this missing ingredient, either in the form of uranium enrichment capability or in the form of plutonium extractable from spent reactor fuel by means of chemical reprocessing.

Three main arguments have been used over the years by people contending that the link between fission power and fission weapons is not so serious (see, e.g., Spinrad, 1983). The first argument is that plutonium of the sort produced by reactors optimized for electricity generation   is unsuitable for the fabrication of high-quality nuclear weapons. This view has been discredited publicly and repeatedly by professional weapons designers and by officials of the US defense establishment: there is some performance penalty associated with using "reactor-grade" rather than "weapons-grade" plutonium to construct a weapon, but the penalty can be made small by means of clever design (Gilinsky, 1977; US OTA, 1977).

The second argument is that there are more direct and cheaper ways for a country to acquire nuclear bombs than via its own commercial nuclear energy facilities. Centrifuges for uranium enrichment and special reactors dedicated to plutonium production are usually mentioned. This argument is technically correct but seriously misleading. Countries can get nuclear explosive materials by these other means, but doing so is made much easier if the requisite technical skills and infrastructure are already in place, courtesy of a nuclear power program. The existence or prospect of commercial nuclear power in a country, moreover, provides a legitimating cover for nuclear activities that, without electricity generation as their manifest purpose, would be considered unambiguously weapons oriented and thus subject both to internal dissent and external sanctions and countermeasures. Even countries that initially have no intention of acquiring nuclear weapons might later find the built-in weapons capability that comes with nuclear power too tempting to resist, particularly if their internal or external political circumstances change.

The third argument is that weapons proliferation is basically a political problem that must be and is being handled by such political measures as the Non-Proliferation Treaty (NPT). That the problem is partly political is true, but the NPT offers only limited reassurance. Many key countries have not ratified it; any of those that have can abrogate it legally with a few months' notice; the safeguards that have been implemented to detect violations are not adequate; and the treaty is in danger of collapse because of the failure of the existing nuclear-weapons states (above all the United States and the Soviet Union) to live up to their own NPT obligations to make real progress toward nuclear disarmament. (The persistent and deliberate failure to agree on a comprehensive ban on nuclear explosive testing is a particular problem in this last regard. See, e.g., Spector & Smith, 1990.)

It is true that additional countries inevitably will acquire nuclear weapons, whether or not commercial nuclear power is available to facilitate the process; but the key issue is the rate of spread of nuclear weapons capability. If the proliferation problem is viewed as a race between the spread of nuclear weapons, on the one hand, and the growing effectiveness of political and moral barriers against the use of the weapons, on the other, then nuclear power's contribution to the rate of spread of nuclear weaponry must be considered a serious cost.

Prospects for reducing environmental and sociopolitical costs

What are the prospects for reducing the environmental and sociopolitical costs summarized here? Clearly, the two basic approaches for doing so are to "fix" the present energy sources or to replace them with others having lower external costs.

Fixing fossil fuel technologies

Many of the environmental impacts of fossil fuels (including the accident hazards of underground coal mining and the emissions of oxides of sulfur, oxides of nitrogen, and particulate matter) could be substantially abated without increasing the monetary costs of fossil fuels or electricity derived from them by more than 30% or so above current figures. Doing so would involve, among other measures, the use of coal-gasification/combined-cycle and fluidized-bed technologies for new power plants, improved scrubbers for existing industrial and electric power plants, and advanced automotive emission controls (Rubin, 1989; Fulkerson et al., 1990). Implementing these possibilities, however, will entail a massive investment in retrofitting or replacing existing facilities and equipment. This will represent a major barrier even in the United States, Western Europe, and Japan, and a larger barrier still in parts of the world where capital is scarce and existing equipment is far below current Western standards.

The carbon dioxide problem is harder. Some progress is possible for a time by substituting natural gas for coal, since the former emits only 60% as much CO_2 per gigajoule as the latter; but this is only a medium-term (as well as only partial) solution since world resources of gas are less than a tenth the size of those of coal. Some progress can also be made through improvements in coal-to-electricity conversion efficiencies. For example, increasing the conversion efficiency from 36%, typical for modern coal-burning plants in operation today, up to 45%, attainable in integrated coal–gasification/combined-cycle plants that could be in operation by the year 2000, would decrease emissions of CO_2 per kilowatt-hour by 20%. Reaching a conversion efficiency of 60%, which may be possible by 2020 in commercial coal-fueled power plants based on combining coal gasification with molten-carbonate fuel cells (Douglas, 1990), would drop the CO_2 emissions per kilowatt-hour by 40% from today's figure. The economic costs of the higher-efficiency coal-fueled electricity technologies are not yet established.

A further and more drastic approach to controlling CO_2 emissions would be to modify the largest CO_2-emitting facilities, such as coal-burning electric power plants, to capture the carbon dioxide from the stack gases and then sequester it in depleted natural gas wells or in deep ocean waters (Okken et al., 1989). This will be difficult and expensive because the volume and mass of CO_2 involved are so large – about 3 tonnes of CO_2 for every tonne of coal burned, nearly 10 million tonnes of CO_2 from a 1-million-kilowatt coal-burning power plant in a year.

Still more difficult will be reducing CO_2 emissions from the dispersed uses of fossil fuels in vehicles, homes, commercial buildings, and industry. Some of these uses can be replaced by electricity, which can be generated by non-fossil means or by burning fossil fuels in centralized facilities which sequester the CO_2. (In 1990, about 60% of the world's electricity generation was based on fossil fuels.) The rest of the dispersed uses of fossil fuel could be replaced, in principle, by converting the relevant devices to burn hydrogen and alcohol fuels instead of petroleum products, coal, and natural gas. As long as the hydrogen and alcohol were made from non-fossil sources – or, in the case of hydrogen, if it were made from fossil fuels in a way that permitted capturing and sequestering the associated CO_2 – there would be no net CO_2 addition to the atmosphere from their use (Ogden & Williams, 1989; Williams, 1990).

Fixing nuclear-fission technologies

Nuclear fission is incomparably less disruptive of global climate and geochemistry than fossil fuels are. But its expanded use is unlikely to be accepted unless a new generation of reactors with demonstrably improved safety characteristics is developed, unless radioactive wastes are shown to be manageable in the real world and not just on paper, and unless the proliferation issue is decisively resolved.

It seems likely that alternative fission reactor types designed from the outset to minimize safety hazards can reduce reactor accident risks considerably from those of the conventional pressurized-water and boiling-water reactors dominating commercial nuclear power today (Spiewack & Weinberg, 1985; Forsberg & Weinberg, 1990). Such reactors would not be "inherently" safe as some of their promoters have argued (MHB

Associates, 1990), but they would rely more heavily than today's reactors do on "passive" mechanisms of heat removal and radioactivity containment.[2] As a result, they should not only be safer but easier to prove to be adequately safe.

Still, to develop fully these alternative reactors and to demonstrate to the technical community and the licensing authorities that they are an improvement over today's reactors will be a major task, not accomplishable overnight. It is difficult to believe that any new reactor type could be in commercial operation much before 2005. As for the monetary costs of safer reactors, the lower power densities and smaller outputs per nuclear core associated with greater reliance on passive safety mechanisms probably will increase monetary costs per installed kilowatt and, thus, per kilowatt–hour of generation. Conceivably, some of these increases will be offset by lower costs for active safety systems, containment buildings, and licensing and construction delays. My own guess is that, on balance, new and safer converter (nonbreeder) reactors will increase the costs of nuclear-generated electricity by 10 to 30% above those associated with the most successful of today's light-water reactors (which does, however, mean lower costs than those of the least successful reactors today).

If one accepts, as the most informative measure of hazard, the product of probability times consequences summed over all possible adverse events – which is the formal definition of "risk" – then there are reasons to think that the hazards of any reasonable scheme for long-term storage of radioactive wastes are not as great as the accident risks at nuclear reactors. With respect to probability, the chance of a large release is diminished in the case of the wastes by the absence of the concentrated sources of stored energy that, in a reactor, provide the means for opening a path by which large quantities of radioactivity can reach the environment. (This statement assumes that the wastes are not stored in the form of the chemically explosive nitrates that result from some fuel-reprocessing schemes.) With respect to consequences of possible releases, the wastes are less dangerous once in the ground than they were in the reactor, both because the radioactivity is in less mobile forms and because a substantial part of the radioactivity has had time to decay away. The most likely pathway for release from a waste repository is into groundwater, moreover, in which the solubility of the wastes would be limited and dispersion rather slow.

---

[2] "Passive" mechanisms are those based on readily verified materials properties and heat transfer by natural convection and thermal radiation, rather than on "active" systems that include sensors, valves, pumps, and the need for correct and timely actions by reactor operators.

These advantages in probability and consequences of accidents at waste repositories, compared to the case of reactor accidents, are partly offset by the long time period over which the wastes remain dangerous – albeit less and less so as time goes on – compared to the time they spend in the reactor. On balance, however, I believe that more analysis and more experience both are likely to show that radioactivity from fission reactors is more dangerous *before* it is in the ground – that is, while it is in the reactor, in spent-fuel storage pools, in reprocessing plants, and in transit – than after it is emplaced in geologic storage.

This is not to say there is no risk in geologic storage, or that a satisfactory scheme for long-term management of waste is already in place. Quite the contrary, there is no system for long-term management of high-level radioactive wastes in commercial operation anywhere in the world. Moreover, the combination of a growing public aversion to the sating of toxic-waste facilities of any kind combined with a history of bungling and misrepresentation surrounding radioactive wastes in particular makes it possible that lack of a publicly accepted solution to the radioactive-waste problem will block the expansion of nuclear power even if the technical community agrees that satisfactory solutions are available (Carter, 1987; Colglazier & Langum, 1988).

Concerning possible reductions in the proliferation risks of nuclear fission, some improvements clearly are possible but there are no panaceas. On the technical side, much effort has been devoted over the years to the search for "proliferation resistant" nuclear fuel cycles (see, e.g., Feiveson et al., 1979; U.S. Department of Energy [US DOE], 1980; Holdren, 1989); but no combination of reactor type and fuel cycle has been found that avoids both the need for enriched uranium and the production of plutonium in the reactor core. (Some schemes avoid one or the other; no scheme avoids both.) Proliferation hazards can be reduced by avoiding fuel cycles that use highly enriched uranium and those that actually recycle the plutonium that is produced – the combination of contemporary light-water reactors using low-enriched uranium and a "once-through" fuel cycle that leaves the plutonium mixed with fission products in unrepro-cessed spent fuel is a good example of this – but the mere existence of enrichment facilities and chemically separable weapons-usable plutonium represent some proliferation risk. Increasing the authority and the resources available to the International Atomic Energy Agency (IAEA) for monitoring enrichment plants and spent fuel is the only readily available way to reduce this residual risk further, and that approach will work only for countries voluntarily subjecting themselves to the IAEA's safeguards.

What is worse, the recycling of plutonium – with attendant increased risks of diversion of this material by national and even subnational groups – cannot be avoided in the long run if fission is to play a significant role in a sustainable global energy future. Such a role will require the use of breeder reactors; they can extract about 100 times as much energy from a kilogram of uranium as do the nonbreeders on which the world mainly relies today, but they are able to do so only by recycling plutonium bred from uranium-238 (or, equivalently from the proliferation standpoint, by recycling uranium-233 bred from thorium-232). One can get a rough sense of the time scale over which breeding could be postponed by noting that a fission energy enterprise of 5 TW (instead of today's 0.7 TW) would consume the world's estimated ultimately recoverable resources of high-grade uranium (25 million tonnes) in only 50 years if the energy were supplied by today's light-water reactors operating on once-through fuel cycles. (For a more detailed discussion, see Albright & Feiveson, 1988).

In the long run, it may be that the only adequate approach to reducing the proliferation/diversion risks of nuclear fission is to cluster all the sensitive facilities – enrichment plants, reactors, reprocessing plants, fuel fabrication plants – in heavily guarded "nuclear parks" under international control. There is no doubt that this is technically feasible and that it would reduce proliferation and diversion dangers substantially. What is uncertain, however, is the extent to which it is politically realistic to expect all the world's countries to place a major component of their electricity supply under international control and to agree on the administrative arrangements for doing so.

Reducing the impacts of current biomass fuels

Current uses of biomass for energy exploit a wide variety of types of biomass (e.g., trees, shrubs, grains, sugar crops, crop residues, wastes from pulp and paper manufacturing, municipal trash, animal dung) and an even wider array of techniques and technologies for harvesting, processing, and burning the biomass material. All use of biomass energy, given only that the biomass fuel is not burned faster than it grows, avoids fossil fuel's problem of net carbon dioxide production; and biomass use creates no political problems comparable to those associated with nuclear energy and imported oil. But the local and regional environmental impacts of some approaches to biomass-energy supply – e.g., deforestation and erosion from overharvesting of fuelwood, loss of soil fertility when crop-waste nutrients go up in smoke rather than being returned to the land, indoor and outdoor air pollution from burning biomass fuels without adequate emission controls – can be both devastating and difficult to remedy.

Both the deforestation and indoor-air-pollution impacts of current biomass energy use in LDCs could be somewhat abated, in principle, by introduction of more efficient wood- and waste-burning stoves with adequate exhaust systems (Ramakrishna et al., 1989; Ahuja, 1990). But the staggering number and vast dispersal of households requiring this change, and the inability of most of these households to pay for it, are formidable obstacles. The longer-term solution for indoor air pollution in rural areas must be a transition to cleaner-burning fuels for household use (Smith, 1988). In order of decreasing in-house emissions and increasing cost, the main possibilities are kerosene, gas (including gas derived from biomass), and electricity. It is not clear, however, whether the transition up this "ladder" of higher-technology household energy options can proceed much faster than the transition of economic development itself, and it seems likely to involve very substantial increases in the monetary costs of household energy use in the Third World even as the huge environmental cost of indoor air pollution is being reduced. Among the options less polluting of indoor air, moreover, use of kerosene, natural gas, or fossil-fuel-generated electricity would contribute to the global-warming problem.

Expanding lower-impact energy supplies

The principal energy-supply options that might be used to reduce reliance on fossil fuels, nuclear fission, and the high-impact/low-efficiency uses of biomass energy that prevail today are readily listed: direct harnessing of sunlight; "indirect" harnessing of sunlight as hydropower, wind, ocean heat, and (with improved technology) biomass; geothermal energy; and nuclear fusion. The magnitudes of some of these sources are summarized and compared to those of the fossil and fission fuels in Table 6.

All of these options have promise, but all of them have liabilities as well (see, e.g., National Research Council, 1980; Lashof & Tirpak, 1989; Solar Energy Research Institute, 1989; Brower, 1990; Flavin & Lennsen, 1990).

• Direct harnessing of sunlight for heating buildings is widely practised and cost-effective today in the form of "passive solar" design. Solar water heating, although also fairly widely practised in hot climates, remains more expensive than water heating with natural gas and electricity in most parts of the world; and solar heat for most industrial applications is much more expensive than use of fossil fuels at current

Table 6. *Estimates of world energy resources*

Stock resources ("nonrenewables")

	Probable remaining recoverable resources (TWyr)	
	United States	*World*
Petroleum	40	600
Natural gas (conventional)	40	400
Coal	1,000	5,000
Heavy oils, tar sands, unconventional gas	200?	1,000??
Oil shale	5,000	30,000
Uranium (in conventional reactors)	200	2,500
Uranium (in breeder reactors)	200,000	3,000,000
Lithium (for 1st generation fusion)	140,000,000 (oceans)	
Deuterium (for 2nd generation fusion)	250,000,000,000,000 (oceans)	

Flow resources ("renewables")

	Total flow	*Plausibly harnessable flow*
Sunlight	88,000 TW at Earth's surface, 26,000 TW on land	Converting insolation on 1% of land area at 20% efficiency yields 52 TW (electric)
Biomass	100 TW global net primary productivity, 65 TW on land	Biomass fuels from 10% of land area at 1% efficiency yields 26 TW (chemical)
Ocean heat	22,000 TW absorption of sunlight in oceans	Converting 1% of absorption at 2% efficiency yields 4 TW (electric)
Hydropower	13 TW potential energy in run-off	Using all feasible sites yields 2–3 TW peak, 1–1.5 TW average (electric)
Wind	1000–2000 TW driving winds worldwide	Using all cost-effective terrestrial sites may yield 1–2 TW (electric)

Stock resources are measured in terawatt–years (TWyr). $1\,\text{TWyr} = 31.5 \times 10^{18}$ joules. Flow resources are measured in terawatts (or terawatt–years per year). Compare 1990 world energy use of 13.2 TW. Estimates are the author's based on a wide variety of sources; see, e.g., Hubbert (1969), Brobst and Pratt (1973), Hughart (1979), Haefele (1981), World Energy Conference (1983), British Petroleum (1990).

prices for the same purposes. The environmental impacts of solar-heating technologies are modest, consisting largely of the damages associated with mining and processing the materials they require.

- Using sunlight to make electricity with photovoltaic cells remains 3 to 5 times more expensive than fossil-fueled electricity generation, despite very substantial reductions in the costs of photovoltaics over the last two decades. Solar electricity generation with focusing collectors and heat engines is 1.5 to 2 times as expensive as fossil-fueled electricity generation, in favorable climates. The environmental impacts include land use in the case of centralized power plants and, for the case of photovoltaic cells, toxic substances; such substances, which may be embodied in the cells or merely used in their manufacture, can pose hazards to workers, contaminate water in manufacturing areas, or be released to the atmosphere from overheated or burning cells.

- Hydropower could be expanded to a few times its current global contribution of 0.25 terawatts of electricity (0.8 TW thermal equivalent). Most of the cost-effective unexploited potential is in the less developed countries; the cost-effective hydropower sites in the industrial nations are mostly already being exploited, so expansion generally will entail increased costs. Environmentally, hydro reservoirs destroy riverine ecosystems, flood fertile land, block the migrations of anadromous fish, and alter downstream ecological conditions; also, large dams can pose a hazard of catastrophic collapse from miscalculation, earthquake, or sabotage (Holdren et al., 1983).

- Windpower at the best sites has monetary costs not dissimilar to those of new hydro development – in the range of 1 to 1.5 times the costs of electricity generation from fossil fuels – but much smaller ecological impacts. Some modest environmental hazards are associated with obtaining and processing the materials for windmill construction, but the land occupied by windmills can be used simultaneously for grazing and some other purposes. Wind's global contribution ultimately will be limited by the number of suitable sites to the same 1–2 TW (electric) range as the hydro potential.
Given the severe environmental problems associated with today's biomass energy use, an expanded contribution from this source is only plausible if based on more sophisticated approaches to growing, harvesting, processing, and burning the biomass fuels. Various techniques for production of biogas and alcohols from biomass have promise as sources of fuels that can be burned relatively cleanly (Miller

et al., 1986) – perhaps at 1 to 1.5 times today's fossil-fuel costs – but growing and harvesting the raw materials without excessive deforestation, soil depletion, or competition with other biomass needs could become a problem at use rates much bigger than today's.

- The forms of geothermal energy being exploited today, which altogether account for only a fraction of a percent of world energy supply, are isolated deposits of steam or hot water that become depleted by use within a few decades; nonetheless, the total energy resource represented by these deposits may be in the range of 4000 terawatt–years. Tapping the hot rock that is ubiquitous in the Earth's crust offers even larger potential supplies – perhaps in the millions of terawatt–years – but the technology for this remains to be demonstrated and the cost is correspondingly uncertain. Environmentally, today's geothermal technologies entail some water pollution by dissolved salts and toxic metals in geothermal water, as well as worker and public exposure to airborne hydrogen sulfide and radon gases. The environmental impacts of hot-rock geothermal technologies remain to be explored.

- Nuclear fusion offers a possibility for reducing the safety, waste-management, and proliferation hazards of nuclear fission by very substantial margins (Holdren et al., 1988), but it will require at least another 20 to 30 years of research and development effort to be certain that this possibility can actually be realized. Electricity from fusion is unlikely in any case to be significantly less expensive than that from fission breeder reactors and might be more expensive.

The use of hydrogen as a chemical fuel is often mentioned in the context of long-term energy options (Ogden & Williams, 1989). Hydrogen is not, however, a primary energy source; it must be produced either from fossil fuels or by splitting water molecules (using another energy source – e.g., solar or nuclear energy – to pry the hydrogen away from the oxygen). Hydrogen is relatively clean burning: the only direct product of its combustion is water, although nitrogen oxides are formed if it is burned at high temperature in air. Using the hydrogen in fuel cells can greatly increase the efficiency with which the fuel is converted to electricity or mechanical work, while minimizing emissions. In no event, however, will hydrogen made by splitting water using solar- or nuclear-generated electricity be cheap: under the optimistic assumption that long-term solar and nuclear sources will be able to make electricity for 5 cents per kilowatt–hour, and assuming an electrolysis efficiency of 75%, the contribution of the electricity cost to the cost of the hydrogen would be

$18.50 per gigajoule, equivalent to $2.30 per gallon of gasoline. (This neglects capital and operating costs of hydrogen production other than the electricity cost; and it should be compared to gasoline production costs, not retail prices.)

Some of the liabilities of these long-term energy options will prove to be reducible with time and effort; others will prove resistant. We will not know which is which without a substantial investment in research. It is safe to say, however, that there is no panacea among these energy options. None offers good prospects of making energy abundant, cheap, and completely free of significant environmental impacts. And those with the greatest promise of abundance and low impact seem quite likely to be the most expensive.

The global picture in summary

There is much reason to think that the energy circumstances of civilization have been changing recently in fundamental rather than superficial ways. For most parts of the world, the era of low-cost energy has ended – a consequence mainly of rising environmental and political costs of the energy sources on which civilization relies most heavily today, coupled with the prospect of high monetary costs for alternative sources with lower environmental and political impacts.

Quite probably, many industrial nations are already near or even beyond the point where further energy growth based on existing energy-supply systems, end-use technologies, and end-use patterns begins to create greater marginal costs than marginal benefits. In that situation, "full speed ahead" using the existing approaches cannot be a solution. We are likely to need transitions in energy-supply systems and patterns of end use, in fact, just to maintain current levels of well-being, because cumulative consumption of high-grade resources and of environmental capacity to absorb energy's impacts will tend to produce rising total costs even at constant rates of use. Providing for economic growth without environmental costs that undermine the gains will require even faster transitions to low-impact energy-supply technologies and higher end-use efficiency.

Although this situation poses formidable challenges, the advanced industrial nations are probably rich enough and technologically capable enough to master most of the problems. The richest countries could, if they chose, tolerate low or even negative energy growth by extracting increases in economic well-being from efficiency increases, and they could afford to pay considerably higher energy prices to finance a transition to energy-supply technologies that are less disruptive environmentally. So far, however, there is little sign of this actually happening. And whether it could be managed in the Soviet Union and Eastern Europe, even in principle, without massive assistance from the West is doubtful.

Still more difficult is the situation in the less-developed countries. These countries would like to industrialize the way the rich did, on cheap energy, but they see the prospects of doing so undermined by high energy costs – whether imposed by the world oil market or by a transition to cleaner energy options. An acute shortage of capital accentuates their tendency to choose the options that are cheapest in terms of monetary costs, and they see the local environmental impacts of cheap and dirty energy as a necessary trade for meeting basic human needs (with traditional energy forms) and generating economic growth (with industrial ones).

While the LDC share of world energy use is modest today, their population size, population growth rates, and economic aspirations represent a huge potential for future energy growth. If this growth materializes and comes mainly from fossil fuels, as most of these countries now plan, it will generate huge additions to the atmospheric burdens of carbon dioxide and other pollutants of regional and global concern (Sathaye & Ketoff, 1991). And while they resent and resist the go-slow approach to energy growth that global environmental concerns have fostered in many industrialized nations, the LDCs are, ironically, the more vulnerable to global environmental change: they have smaller food reserves, more marginal diets, poorer health, and more limited resources of capital and infrastructure with which to adapt to altered conditions. Global climate change brought on by carbon dioxide and other greenhouse gases could have devastating consequences for them: more dry-season heat and drought; more wet-season floods and storm victims; more famine and disease; perhaps hundreds of millions of environmental refugees.

The industrial countries should not expect to be able to isolate themselves from these phenomena. Even assuming that these countries will suffer less from the direct effects of climate change because of their greater capacities to adapt and adjust, the world is too interconnected by trade, investment, finance, resource interests, politics, porous borders, and possibilities for venting frustrations militarily for the worst consequences of climate change to remain confined to the LDCs. All too plausibly, the future political stresses arising from climate change could grow into an energy-conflict connection comparable in importance to the oil and nuclear proliferation issues discussed above (see, e.g., Gleick, 1989, 1990).

This global energy-environment-development predicament would be frightening enough even if the population of the world could be frozen at its 1990 level of 5.3 billion people. But the population cannot be frozen. Indeed, short of catastrophe, it can hardly be levelled off below 9 billion.[3] Indeed, without a global effort at population limitation far exceeding anything that has materialized so far, the population of the planet could soar to 14 billion or more by the year 2100.

The implications of such population growth for energy demand and associated environmental impacts are appalling. Supplying 5.3 billion people in 1990 with an average energy use rate of 2.5 kilowatts per person – a total of 13.2 terawatts – was placing severe strains on the planet's technological, managerial, and environmental resources; and crucial human needs were going unmet. Yet for every billion people added to the world's population at that same level of energy use per person, new energy supplies capable of sustaining an additional continuous drain of 2.5 terawatts must be mobilized, paid for, and their environmental impacts somehow absorbed.

That figure understates the problem, of course, because in all likelihood the world-average use of energy per person will rise even if there are Herculean accomplishments in increasing end-use efficiency. Indeed, to avoid a large increase in world-average energy use per person as we meet the aspirations of the developing countries, energy use per person in today's industrialized nations will have to fall substantially. The problem is illustrated by the scenario presented in Box B, which I regard as the most optimistic that is currently defensible in terms of improvements in energy efficiency and progress in redirecting economic growth toward narrowing the rich–poor gap. The scenario is based on the premise that a standard of living somewhat higher than that of the United States today – presumed high enough to satisfy expectations for a long-term global average – can be delivered by the middle of the next century with a rate of energy use averaging 3 kilowatts per person, just over a quarter of the current US rate. A second premise is a global compact to reduce the rich–poor gap as rapidly as practical. A third is a trajectory of world population growth that corresponds to achieving replacement fertility worldwide by the year 2020.

What is most striking about the figures in Box B is that even the most optimistic assumptions about "early" population stabilization, increased energy efficiency, and narrowing the rich–poor gap still lead to world energy use in 2050 more than double that of 1990. Yet, as suggested by the foregoing survey of the nature of current energy-supply problems and the

---

[3] To stabilize the population at 9 billion, it would be necessary for the 1990 global average total fertility rate of 3.5 (live births per woman) to fall to the "replacement level" of 2.1 by the year 2000 (Ehrlich & Ehrlich, 1990; Population Reference Bureau, 1990).

prospects for alternative sources, to provide twice 1990's energy use rate at tolerable costs will be a formidable challenge, likely to strain even the increased ingenuity presumably available a half-century from now. The higher the ultimate rate of energy use, moreover, the less flexibility society will have in its choices about how the energy is to be supplied, and the greater will be the cost of energy supply – not only in the aggregate but also per gigajoule and hence per capita. The more numerous we are, therefore, the lower will be our level of well-being.

Managing the energy transition

How should society respond to the changing and increasingly alarming interaction between energy and human well-being? How, and at what level, can the trend of rising internal and external costs of energy supply be restrained or brought to a halt? How can the transition towards more expensive energy, on which civilization has embarked largely unaware, be transformed into a transition toward supportive and sustainable relations between energy, the economy, and the environment?

Understanding the predicament

An essential step is to develop an improved and shared understanding of where we are, where we are headed, and where we would like to go instead. There needs to be an extended public and, indeed, international debate on the connections between energy and well-being, supported by a greatly expanded research effort to clarify the evolving pattern of energy benefits and costs. Of special importance will be the sharpening and concretizing of our conceptions of the meaning of sustainability, in all of its resource-supply, economic, environmental, and sociopolitical dimensions. (A brief overview of some elements of sustainability is provided in Box C. For some further thoughts on sustainability in a broader context, see, e.g., Daly, 1974, 1988; Hueting, 1980; Page, 1980)

Of course, the study and debate that are called for will take time. Large uncertainties attend many of the important issues. Some of these will require decades to resolve. Perhaps, with more information, the situation will seem less threatening and difficult than suggested here; it could also turn out to be even more threatening and difficult. In any case, we face the usual dilemma of action versus delay in an uncertain world. That is, if we wait, our knowledge will improve but the effectiveness of our actions may shrink. Damage may become irreversible, dangerous trends more entrenched, and our technologies and institutions even harder to steer and reshape.

A reasonable response to this dilemma is a two-pronged strategy consisting of "no regrets" elements and "insurance policy" elements. No-regrets actions are those that have high leverage against the dangers we fear, but will turn out to be useful even if those dangers do not fully materialize. Insurance-policy actions offer high potential leverage against uncertain dangers in exchange for only modest investments now, although they might end up not being needed.

Internalizing and reducing environmental costs

One essential no-regrets policy is a program to internalize and reduce the environmental and sociopolitical costs of existing energy sources. When these costs are significant, as we have seen some are, they generate both inefficiency and inequity in society's energy choices: energy suppliers and users who are not paying the full costs of these activities will supply and use more than is socially efficient; and the imposition of significant costs on people other than the beneficiaries of the activities generating those costs is inequitable. Even when environmental costs of energy are borne by its beneficiaries – as with consumers who live near the power plants that supply them electricity or Third World villagers who breathe the smoke from their own cookfires – those damages should be reduced wherever they are larger than the costs of abating them.

Priority should be given to abating the emissions of oxides of sulfur, oxides of nitrogen, hydrocarbons, and particulate matter from fossil-fuel combustion, and the emissions of hydrocarbons and particulates from the combustion of traditional fuels. Technologies for substantially controlling , these emissions exist and will more than repay their costs through reduced damages to health, property, and ecosystems.

This program should also include a carbon tax, as a way of partly internalizing the still unquantifiable risks posed by the emissions of CO_2. The revenues from such a tax could then be used to develop and finance technologies for reducing CO_2 emissions from fossil-fuel burning – and for reducing dependence on fossil fuels altogether – around the world. Every dollar per tonne in carbon tax would raise about 6 billion dollars per year at current consumption rates of fossil fuels.

More effort is also warranted on increasing the safety and decreasing the weapons-proliferation potential of the contemporary (nonbreeder) nuclear power system, including the development of more forgiving reactor designs and internationalizing proliferation-vulnerable fuel-cycle steps. Neither the nuclear industry nor policy makers have yet demonstrated the commitment that these tasks will require. But if they are not undertaken seriously and successfully, there is little chance of nuclear fission's gaining the public acceptance needed for it to make a significantly expanded contribution to world energy supply.

Accelerating development of sustainable energy options

Another crucial element of a sensible energy strategy is to accelerate research and development on the long-term energy sources that must eventually replace fossil fuels: sunlight, wind, ocean heat, and biomass; the "hot rock" form of geothermal energy that is ubiquitous in the Earth's crust at great depth; fission breeder reactors; fusion. This research should stress not only the attainment of economic ways to harness these resources, but also the prospects for minimizing their environmental costs.

Making the R&D investments needed to accelerate the development of these options qualifies as an "insurance" approach, because we do not yet know which of the options will be needed, or how soon. Indeed, some of the money will be "wasted", in that some options will never be used. But the amounts of money required to make these options available are modest compared to the potential costs of not having them when needed, or choosing foolishly.

Unfortunately, decision makers in many countries seem insufficiently aware of the merits of this "insurance" argument for energy R&D. The US Federal government, for example, was spending less than half as much in Fiscal Year 1990 on energy research, development, and demonstration as it had been spending 10 years earlier (see Fig. 4), pleading fiscal constraints as its excuse. Restoring federal support for US energy R&D to its FY1980 level could be funded by increasing the federal gasoline tax by a mere 3 cents per gallon. Meanwhile, the US energy community is so busy with divisive arguments about which energy options should have the largest share of the budget that it is failing to make the fundamental point on which all should agree – that the energy R&D budget as a whole is much too small.

Building international cooperation

Building East–West and North–South cooperation on energy and environmental problems is a "no regrets" strategy that has many benefits no matter how the energy future unfolds. It could begin with increased

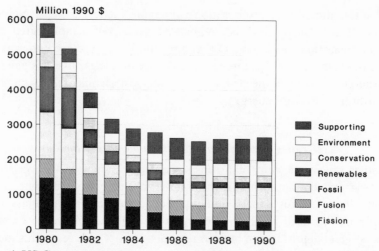

Fig. 4. US Government energy research, 1980–1990. The graph shows US federal budget authority for energy research, development, and technology demonstrations (RD&D) from Fiscal Year 1980 through Fiscal Year 1990, adjusted to constant FY1990 dollars using the GNP deflator. Budget authority shown covers nuclear fission, nuclear fusion, fossil fuels, renewables (including geothermal and energy transmission and storage), efficiency ("conservation"), environment, and basic energy sciences ("supporting"). Excluding the last two items (top two sections of the stacked bars) to focus only on spending for RD&D on specific supply and conservation options shows a drop of more than 3-fold, from $5.1 billion to $1.6 billion, over this decade. US DOE "general science" research and its state and local grants for low income weatherization are not included in any of these figures, which are derived from the annual budget documents of the Office of Management and Budget (US OMB).

cooperation on energy research and development. This would relieve some of the constraints imposed by inadequate energy R&D budgets worldwide by eliminating needless duplication, sharing diverse specialized strengths, and dividing the costs of the large projects that some avenues of energy R&D require. (Until now, nuclear fusion has been the only area of energy research that has enjoyed really major international cooperation.)

It is especially important that international cooperation in energy research be extended to include North-South collaborations on energy technologies designed for application in developing-country contexts. It is shameful that the industrialized countries – the only ones that can afford to do much energy research – have confined their efforts until now almost entirely to technologies tailored to industrial-nation economic and cultural contexts.

International cooperation on understanding and controlling the environmental impacts of energy supply is also extremely important, since many of the most threatening environmental problems respect no boundaries. Air and water pollution from Poland, Czechoslovakia, Hungary, East Germany, and the Soviet Union reach across Europe and into the Arctic; and the environmental impacts of energy supply in China and India, locally debilitating at today's levels of energy use, could become globally devastating at tomorrow's. Pleas from the rich countries that global environmental problems require global energy restraint will fall on deaf ears in the less-developed and economically intermediate countries unless the former find ways to help the latter achieve increased economic well-being and environmental protection at the same time. Cooperative and comparative assessments of energy/environmental problems and requirements are clearly needed, as are: the transfer and financing of existing technologies for pollution control, cleaner energy supply, and increased energy efficiency; cooperative development of improved technologies in these categories; and international agreements on standards and targets (the last being sensible only if the other ingredients are in place to make meeting the standards and targets feasible).

Full internalization of environmental costs will also require international cooperation, not only because of the transboundary nature of many environmental impacts but also because of the potentially adverse effects of country-to-country variations in approaches to internalization. At present there exists in different countries a wide array of taxes on fuels to be used in households and vehicles, but far less in the way of taxes or fees applied to fuels for manufacturing or air travel. This is because countries are reluctant to burden their own firms with environmental taxes that might cause the firms to lose international competitiveness or move elsewhere. International agreement on such taxes may be the only effective remedy for this difficulty.

The most difficult agreement to reach and implement will be one to stabilize and begin to reduce global emissions of carbon dioxide. "No regrets" approaches to CO_2 reductions – such as cost-effective increases in energy end-use efficiency, fuel-switching, and conversion-efficiency improvements that have economic or environmental benefits besides their effects on CO_2, and afforestation and reforestation programs – will make substantial contributions to slowing the accumulation of CO_2 in the atmosphere. But few analysts, and even fewer policy makers, are confident that these "no regrets" approaches will be sufficient to bring about the stabilization or even reduction of CO_2 emissions that many environmental scientists believe is needed. The policy makers' concern is that actually stabilizing or, worse, reducing CO_2 emissions could be very costly economically. They are reluctant to pay these costs without unambiguous evidence that the costs of failing to stabilize or reduce CO_2 emissions would be even bigger.

Whether a global compact to reduce CO_2 emissions can be drafted in the near future may depend largely on the severity, over the next year or two, of weather anomalies plausibly attributable to global climate change. But it would be a wise investment, in any case, to improve the analytical foundation for the debate about a global climate agreement by beginning now to develop a set of detailed plans – carefully researched, cooperatively developed, and continuously updated – by showing how the world could most cost-effectively constrain CO_2 emissions to varying degrees (e.g., stabilization by 2010, reduction by 10% per decade, reduction by 20% per decade, and so on). The development of such plans would not be very expensive compared to the benefits of having them ready for discussion and adoption if rapid growth in the evidence for and consequences of global warming overcomes the current reluctance of policy makers to act.

Limiting population growth

In the long run, the world will not be able to have an effective energy strategy without also having an effective population strategy. Quite probably the best that can now be expected is that population growth might be halted at around 10 billion – an accomplishment that would require reducing the global-average total fertility rate to the replacement level by the year 2025. Achieving that much would be a tremendous challenge, requiring, in all likelihood, massive development assistance and other forms of international cooperation. But as difficult as this will be, it will nonetheless be easier than coping with the consequences of a world population soaring to 12 billion, 15 billion, or more. That is true not only in terms of energy problems, but also in terms of food, forests, water, biodiversity, and much else (Ehrlich et al., 1977; Ehrlich & Ehrlich, 1990).

Conclusion: the need to restrain energy use

Everything that has been said in this Prologue leads to the conclusion that considerable restraint in energy use will be essential if the interacting problems of energy, environment, and economic well-being are to prove soluble. The centerpiece of such restraint must be very large increases in the efficiency of energy end use. Using energy more efficiently in response to rising energy costs needs to be pursued in parallel with attempts to reduce those costs through improvements in energy-supply technology. In the longer term, however, reducing the amount of energy needed to provide each person with a high material standard of living is clearly crucial. Population growth cannot be quickly halted, the economic aspirations of the majority of the world's population who are now poor – and of their descendants – cannot safely or morally be frustrated, and no combination of energy technologies in hand or in prospect appears able to meet, at tolerable costs, the levels of energy demand that the prospective levels of population and income per person imply if current ratios of energy use to GNP continue to prevail.

The goal of providing a high standard of living with lower energy use will require accelerated penetration of energy-efficient technologies throughout the world, as well as progress in developing new techniques. But improving efficiency alone may well be insufficient. Moving toward less energy-intensive activities, and even reducing certain types of activities, may be needed as well.

What are the prospects for achieving the restraint in energy use that may be required? What are the historic and current trends in energy use? What are the potentials, and what are the obstacles, across the diverse array of nations East and West, North and South? And what are the ways in which international comparisons and international cooperation can facilitate the process of realizing the potentials that exist? It is to these crucial questions about energy use that the remainder of this book is addressed.

Box A: Types of costs from energy supply and use

MONETARY COSTS: money paid by suppliers of energy for, e.g., land used and rights of way, construction of energy-supply facilities and equipment, insurance and taxes on these, labor and materials for operation and maintenance of the facilities and equipment, interest on the debt incurred for these purposes, rate of return on investors' equity, taxes on income derived from energy sales; money paid for energy by users. Monetary costs to users and suppliers would be equal in a completely market-based system without subsidies or other distortions.

OPPORTUNITY COSTS: the value of alternative uses to which resources devoted to energy supply and use (including capital, land, labor, water, materials and so on) could have been put had they not been used for energy. Opportunity costs would be entirely captured in monetary costs in a completely marketized system without distortions, hence would not have to be counted separately, but this idealization is never realized in practice. Some facets of energy supply are not marketized – consider the labor of a Third World villager gathering fuelwood for her own cooking and water heating, or the work of a homeowner cleaning his rooftop solar collector. In many other cases, imperfections in markets cause opportunity costs (e.g., the use of capital or land or water) to be incompletely reflected in monetary costs.

ENVIRONMENTAL COSTS: (1) deaths, injuries, and illnesses suffered by workers in energy-supply industries; (2) deaths, injuries, and illnesses suffered by members of the public as a direct result of effluents and accidents associated with energy supply: (3) damage to economic goods and services (including, e.g., buildings, agricultural productivity, fisheries, tourism) directly attributable to such effluents and accidents; (4) damage to human health, safety, and other aspects of well-being as a result of energy-associated disruption of biogeophysical conditions and processes (e.g., climate, hydrology, nutrient cycles, biotic diversity, stratospheric ozone); and (5) nuisance from energy-related noise, dirt, smells, congestion, visual blight. Parts of these costs may be incorporated into monetary costs of energy ("internalized") through, e.g., higher wages for energy-industry workers in dangerous occupations, employer-paid worker insurance programs, effluent fees, and nuclear-reactor-accident insurance pools; or, in another form of internalization, they may be avoided through pollution-control and risk-abatement measures whose monetary costs are paid by suppliers and users. Environmental costs that remain external are paid (or experienced) by various publics, including citizens of other countries and members of future generations.

SOCIOPOLITICAL COSTS: undesired impacts of energy supply and use on social and political conditions and processes, including settlement patterns, distribution of wealth and income, organization of societal decision making, interregional and international relations (including the chance of war), and individual, family, and community values. Costs in this category may sometimes be brought into monetary balance sheets through payments by energy-supply enterprises to affected individuals or communities; but most such costs remain external to monetary energy accounts.

Box B: Population and energy use in the long run

The scenario is constructed, for simplicity, using just two subpopulations, consisting in 1990 of 1.2 billion "rich" and 4.1 billion "poor." (The dividing line is an average GNP per person of 4000 US dollars per year.) I assume that energy use per person among the population of the rich countries can be reduced by 2% per year between 1990 and 2025, with gains in economic well-being to come from increases in energy efficiency exceeding 2% per year. (For example, energy efficiency gains of 3% per year would permit per capita real economic growth of 1% per year combined with a 2% per year decline in energy use per person.) For the poor countries, I assume that the rate of energy use per person increases at 2% per year, which, together with efficiency improvements, would yield a much higher rate of increase in economic well-being. The result is a halving of energy use per person in the rich countries between 1990 and 2025 and a doubling in energy use per person in the poor countries. After another 25 years, during which rich-country energy use per person falls at around 1% per year and poor–country energy use per person grows at just over 1% per year, the rich–poor distinction has disappeared. Because of the momentum built into the age structure of the world population, the population does not actually stabilize, at around 10 billion people, until after the year 2100. I assume that energy use per person holds constant at 3 kilowatts per person after 2050, with continuing gains in economic well-being coming from innovations that further increase energy efficiency.

If replacement fertility is not achieved until 2060, world population stabilizes at 12.5 billion rather than at 10.0 billion; just the difference in energy use between these two population figures, at the hypothesized 3 kilowatts per person, is equal to the world's 1970 use rate of all industrial energy forms. If a satisfactory standard of living in the long run turns out to require closer to half the 1990 US rate of energy per person, say 5 kilowatts, then a population of 10 billion would use energy at nearly four times the world's 1990 rate.

		Population (billions)	Energy/person (kilowatt/person)	Total energy* (terawatts)
1990	Rich	1.2	7.5	9.0
	Poor	4.1	1.0	4.1
		5.3		13.1
2025	Rich	1.4	3.8	5.3
	Poor	6.8	2.0	13.6
		8.2		18.9
2050	(converged)	9.1	3.0	27.3
2100+	(converged)	10.0	3.0	30.0
	Compare:	12.5 billion ×	3 kW	= 37.5 TW
		10 billion ×	5 kW	= 50 TW
		12.5 billion ×	5 kW	= 62.5 TW

* Includes industrial and traditional energy forms.

Box C: The meaning of sustainability

Sustainability means, most fundamentally, that the provision of environmental, economic, and social goods and services to the human population is done in ways that do not reduce, over time, the quantity and quality of goods and services that the planet's environmental, economic, and social systems are able to provide.

The size of the human population, the magnitude of its material appetites measured in use of physical resources per person, and the sophistication of the technology with which those resources are obtained and transformed are the three most basic determinants of civilization's impact on the planet's environmental systems. Technology can be improved, but not without limit. For any given state of technology there is a maximum product of population and resource use per person consistent with sustainability.

An energy system that depends on nonrenewable resources is sustainable for only as long as technological improvements that increase the efficiency of use of these resources can offset declining quality of the resource base. Not only fuels but also construction materials for energy systems must be considered.

The flow of sunlight and the derivative renewable energy flows in the forms of hydropower potential and wind cannot readily be depleted by human action, but the sites for capturing these flows and hence the rates at which they can be captured are limited in extent. Since the sites vary in quality, moreover, expression of the rate of use must eventually encounter rising marginal costs. Thus, while a particular fixed rate of energy use based on such renewables may be sustainable, unending growth of energy use based on them is not.

Biomass energy – solar energy stored by photosynthesis in organic matter – is renewable in principle but sustainable at a given use rate only if the associated cultivation, harvesting, conversion, and end-use practices (and other contemporaneous human activities) do not consume the bases of renewability in the form of soil fertility, water availability, and stability of climate.

Some fuels that are nonrenewable in principle are so abundant in practice that significant depletion is hardly plausible on any time scale of conceivable policy interest. Uranium for use in fission reactors and lithium and deuterium for use in fusion reactors are good examples. Even the fossil fuels, if we are speaking not only for petroleum and natural gas but also of coal and oil shale, would last for centuries at almost any imaginable use rate. For at least hundreds of years, then, sustainability will be a question of the combination of economic, environmental, and sociopolitical costs more than a question of depletion.

An energy system that claims a continuously rising share of economic product is not sustainable.

No global energy and economic system that constrains human well-being in the less-developed countries to levels greatly below those enjoyed in the more-developed countries will prove to be politically sustainable. Another dimension of sociopolitical sustainability is whether the energy system presents opportunities for disruption of society by the discontented that are too large in relation to society's capacity to control the discontented or shrink the sources of discontent.

Acknowledgements

This chapter is an updated and greatly expanded version of the author's article, "Energy in transition", which appeared in the September 1990 issue of *Scientific American*. I thank Mark Christensen, Paul Ehrlich, John Harte, Gene Rochlin, Lee Schipper, and Bob Williams for helpful suggestions.

References

Ahuja, D. R. 1990. Research needs for improving biofuel burning cookstove technologies. *Natural Resources Forum*. May, pp. 125–34.

Albright, D. & Feiveson, H. A. 1988. Plutonium recycling and the problem of nuclear proliferation. *Annual Review of Energy*, **13**, 239–65.

Anderson, K. 1987. Conservation versus energy supply: an economic and environmental comparison of alternatives for space conditioning of new residences. PhD Dissertation. Energy and Resources Group, University of California, Berkeley. November.

Bogue, D. J. 1969. *Principles of Demography*. New York: Wiley.

British Petroleum. 1990. BP *Statistical Review of World Energy*. London.

Brobst, D. A. & Pratt, W. P. (eds.). 1973. U.S. Mineral Resources. Washington, DC: Government Printing Office.

Brower, M. 1990. *Cool Energy: The Renewable Solution to Global Warming*. Cambridge, MA: Union of Concerned Scientists.

Brown, H. 1954. *The Challenge of Man's Future*. New York: Viking. (Reprinted by Westview Press, Boulder, CO, 1984.)

Brown, L. R., Durning, A., Flavin, C., French, H., Jacobson, J., Lowe, M., Postel, S., Renner, M., Starke, L. & Young, J. 1990. *State of the World 1990*. New York: Norton.

Carter, L. J. 1987. *Nuclear Imperatives and Public Trust: Dealing With Radioactive Waste*. Washington, DC: Resources for the Future.

Colglazier, E. W. & Langum, R. B. 1988. Policy conflicts in the process for siting nuclear waste repositories. *Annual Review of Energy*, **13**, 317–57.

Cook, E. 1976. *Man, Energy, Society*. San Francisco: W. H. Freeman.

Daly, H. E. 1974. The economics of the steady state. *American Economic Review*, **64**, 15–21.

1988. Moving to a steady-state economy. In P. R. Ehrlich and J. P. Holdren. (eds.). *The Cassandra Conference: Resources and the Human Predicament*, pp. 271–285. College Station, TX: Texas A&M University Press.

Darmstadter, J. 1968. *Energy in the World Economy*. Baltimore: Johns Hopkins University Press.

Deese, D. A. & Nye, J. S. 1981. *Energy and Security*. Cambridge, MA: Ballinger.

Douglas, J. 1990. Beyond steam: breaking through performance limits. EPRI Journal. December, 4–11.

Ehrlich, P. R. 1986. *The Machinery of Nature*. New York: Simon & Schuster.

Ehrlich, P. R. & Ehrlich, A. H. 1990. *The Population Explosion*. New York: Simon & Schuster.

Ehrlich, P. R., Ehrlich, A. H. & Holdren, J. P. 1977. *Ecoscience: Population, Resources, Environment*. San Francisco: W. H. Freeman.

Feiveson, H., Williams, R. & von Hippel, F. 1979. Fission power: an evolutionary strategy. *Science*, (January).

Flavin, C. & Lennsen, N. 1990. *Beyond the Petroleum Age: Designing a Solar Economy* (Worldwatch Paper 100). Washington, DC: Worldwatch, December.

Forsberg, C. W. & Weinberg, A. M. 1990. Advanced reactors, passive safety, and acceptance of nuclear energy. *Annual Review of Energy*, **15**, 133–52.

Fulkerson, W., Judkins, R. R. & Sanghvi, M. K. 1990. Energy from fossil fuels. *Scientific American*, **263** (3), 129–35.

Gilinsky, V. 1977. Plutonium, proliferation, and policy. *Technology Review*, **79** (4), 58–65.

Gleick, P. H. 1989. The implications of global climatic changes for international security. *Climatic Change*, **15**, 309–25.

1990. Climate change and international politics. *Ambio*, **18**, 333–39.

Goldemberg, J., Johansson, T. B., Reddy, A. K. N. & Williams, R. H. 1987. *Energy for a Sustainable World*. Washington, DC.: World Resources Institute.

Graedel, T. E., & Crutzen, P. J. 1989. The changing atmosphere. *Scientific American*, September, pp. 58–68.

Haefele, W. 1981. *Energy in a Finite World: A Global Systems Analysis*. Cambridge, MA: Ballinger.

1990. Energy from nuclear power. *Scientific American*, September, pp. 136–44.

Hall, C. A. S., Cleveland, C. J. & Kaufmann, R. 1986. *Energy and Resource Quality: The Ecology of the Economic Process*. New York: Wiley.

Hall, D. O., Barnard, G. W. & Moss, P. A. 1982. *Biomass for Energy in Developing Countries*. Oxford: Pergamon.

Harte, J. 1985. *Consider a Spherical Cow: A Course in Environmental Problem Solving*. Los Altos, CA: Kaufmann.

1988. Acid rain. In P. R. Ehrlich and J. P. Holdren. (eds.). *The Cassandra Conference: Resources and the Human Predicament*, pp. 125–146. College Station, TX: Texas A&M University Press.

Holdren, J. P. 1976. The nuclear controversy and the limitations of decision-making by experts. *Bulletin of the Atomic Scientists*, **32** (3), 20–2.

1983. Nuclear power and nuclear weapons: the connection is dangerous. *Bulletin of the Atomic Scientists*. January.

1986. Energy and the human predicament. In K. R. Smith, F. Fesharaki & J. P. Holdren (eds.). *Earth and the human future: essays in honor of Harrison Brown*, pp. 124–160. Boulder, CO: Westview.

1987. Global environmental issues related to energy supply. *Energy*, **12**, 975–92.

1989. Civilian nuclear technologies and nuclear weapons proliferation. In C. Schaerf, B. Holden-Reid, & D. Carlton (eds.). *New Technologies and the Arms Race*, pp. 161–198. London: Macmillan.

Holdren, J. P. & Ehrlich, P. R. 1974. Human population and the global environment. *American Scientist*, **62**, 282–92.

Holdren, J. P., Anderson, K. B., Deibler, P. M., Gleick. P. H., Mintzer, I. M. & Morris, G. P. 1983. Health and safety impacts of renewable, geo- thermal, and fusion energy. In C. C. Travis & E. L. Etnier (eds.) *Health Risks of Energy Technologies*, pp. 141–208. Boulder, CO: Westview.

Holdren, J. P., Berwald, D., Budnitz, R., Crocker, J., Delene, J. G., Endicott, R., Kazimi, M., Krakowski, R., Logan, G. & Schultz, K. 1988. Exploring the competitive potential of magnetic fusion energy: The interaction of economics with safety and environmental characteristics. *Fusion Technology*, **13**, 7–56.

Houghton, J. T., Jenkins, G. J. & Ephrams, J. J. (eds.). 1990. *Climate Change: the IPCC Scientific Assessment*. Intergovernmental Panel on Climate Change. Cambridge, UK: Cambridge University Press.

Hubbert, M. K. 1969. Energy resources. In National Research Council, *Resources and Man*, pp. 157–241. San Francisco: W. H. Freeman.

Hueting, R. 1980. *New Scarcity and Economic Growth*. New York: North Holland Publishing Co.

Hughart, D. 1979. *Prospects for Traditional and Non-Conventional Energy Sources in Developing Countries*. Washington, DC: World Bank.

Lashof, D. A. & Tirpak, D. A. (eds.). 1989. *Policy Options for Stabilizing Global Climate*. Washington, DC: Environmental Protection Agency.

Lipschutz, R. D. 1989. *When Nations Clash: Raw Materials, Ideology, and Foreign Policy*. Cambridge, MA: Ballinger.

MHB Associates. 1990. *Advanced Reactor Study*. Union of Concerned Scientists, Cambridge, MA.

Miller, A. S., Mintzer, I. M. & Hoaglund, S. H. 1986. *Growing Power: Bioenergy for Development and Industry*. (WRI Study No. 5.) Washington, DC: World Resources Institute.

Morgan, M. G., Morris, S. C., Meier, A. D. & Shenk, D. L. 1978. A probabilistic method for estimating air-pollution health effects from coal-fired power plants. *Energy Systems and Policy*, **2**, 287–310.

Myers, N. (ed.). 1984. *Gaia: An Atlas of Planetary Management*. London: Gaia Books.

National Academy of Sciences, Committee on Science and Public Policy. 1979. *Risks Associated with Nuclear Power: A Critical Review of the Literature*. Washington, DC: National Academy of Sciences.

National Research Council, Commission on Natural Resources. 1975. *Air Quality and Stationary Source Air Pollution Control* (Report for the Committee on Public Works of the U.S. Senate). Washington, DC: Government Printing Office.

National Research Council, Committee on Nuclear and Alternative Energy Systems. 1980. *Energy in Transition 1985–2010*. San Francisco: W. H. Freeman.

Ogden, J. M. & Williams, R. H. 1989. *Solar Hydrogen: Moving Beyond Fossil Fuels*. Washington, DC: World Resources Institute. October.

Okken, P., Swart, R. & Zwerver, S. (eds.) 1989. *Climate and Energy: The Feasibility of Controlling CO_2 Emissions*. Dorchtecht, Holland: Kluwer Academic Publishers.

Page, T. 1980. *Conservation and Economic Efficiency*. Baltimore: Johns Hopkins.

Pandey, M. R., Smith, K. R., Boleij, J. S. M. & Wafula, E. M. 1989. Indoor air pollution in developing countries and acute respiratory infection in children. *Lancet*, 25 February, 427–9.

Population Reference Bureau. 1990. *1990 World Population Data Sheet*. New York.

Ramakrishna, J., Durgaprasad, M. B. & Smith, K. R. 1989. Cooking in India: The impact of improved stoves on indoor air quality. *Environment International*, **15**, 341–52.

Rubin, E. S. 1989. Implications of future environmental regulation of coal-based electric power. *Annual Review of Energy*, **14**, 19–46.

Sathaye, J. & A. Ketoff. 1991. CO_2 emissions from major developing countries:

Better understanding the role of energy in the long term. *The Energy Journal*, **12** (1), 161–96.

SCEP (Study of Critical Environmental Problems). 1970. *Man's Impact on the Global Environment*. Cambridge, MA: MIT Press.

Schneider, S. H. 1989. *Global Warming*. San Francisco: Sierra Club Books.

Schneider, S. H. & Londer, R. 1986. *The Coevolution of Climate and Life*. San Francisco: Sierra Club Books.

Schneider, H. K. & Schulz, W. 1987. *Investment Requirements of the World Energy Industries 1980–2000*. London: World Energy Conference.

Seiler, W. & Crutzen, P. J. 1980. Estimates of gross and net fluxes of carbon between the biosphere and the atmosphere from biomass burning. *Climatic Change*, **2**, 207–47.

Smith, K. R. 1987. *Biofuels, Air Pollution, and Health*. New York: Plenum. 1988. Rural air pollution. *Environment*, **30** (10), 17–34.

Smith, K. R., Aggarwal, A. L. & Dave, R. M. 1983. Air pollution and rural biomass fuels in developing countries. *Atmospheric Environment*, **17**, 2343–62.

Solar Energy Research Institute. 1989. *The Potential of Renewable Energy*. (Prepared jointly with the Idaho National Engineering Laboratory, the Los Alamos National Laboratory, the Oak Ridge National Laboratory, and the Sandia National Laboratories). Golden, CO.

Spector, L. S. & Smith, J. R. 1990. Deadlock damages nonproliferation. *Bulletin of the Atomic Scientists*, **46** (10), 39–44.

Spiewack and Weinberg 1985. Inherently safe reactors. *Annual Review of Energy*, **10**, 431–62.

Spinrad, B. 1983. Nuclear power and nuclear weapons: the connection is tenuous. *Bulletin of the Atomic Scientists*. February.

Stigliani, W. M. & Shaw, R. W. 1990. Energy use and acid deposition: the view from Europe. *Annual Review of Energy*, **15**, 210–16.

Sweet, W. 1984. The nuclear age: power, proliferation, and the arms race. *Congressional Quarterly*. Washington, DC: US Congress.

UN Environment Programme. 1987. *Environmental Data Report*. Oxford: Blackwell.

US DOE (US Department of Energy). 1980. *Nuclear Proliferation and Civilian Nuclear Power: Report of the Non-Proliferation Alternative Systems Assessment Program*. Washington, DC: Government Printing Office.

US EIA (Energy Information Administration). 1989. *International Energy Annual 1988*. Washington, DC: Government Printing Office. 1990. *Monthly Energy Review* (January). Washington, DC: Government Printing Office.

US Nuclear Regulatory Commission. 1975. *Reactor Safety Study*. Report WASH-1400 NUREG-75/014. Washington, DC: Government Printing Office.

US OMB (Office of Management and Budget). (various years). *Budget of the U.S. Government*. Washington, DC: Government Printing Office.

US OTA (Office of Technology Assessment). 1977. *Nuclear Proliferation and Safeguards*. Washington, DC: Government Printing Office. 1984. *Nuclear Power in an Age of Uncertainty*. Washington, DC: Government Printing Office.

Williams, R. H. 1990. Hydrogen from coal with gas and oil well sequestering of the recovered CO_2. Unpublished manuscript. Princeton University, Center for Energy and Environmental Studies.

Willrich, M. & Taylor, T. 1974. *Nuclear Theft: Risks and Safeguards.* Cambridge, MA: Ballinger.

Wilson, R., Colome, S. D., Spengler, J. D. & Wilson, D. G. 1980. *Health Effects of Fossil Fuel Burning.* Cambridge, MA: Ballinger.

World Bank. 1983. *The Energy Transition in Developing Countries.* Washington, DC.

1990. *World Development Report 1990.* New York: Oxford University Press.

World Energy Conference. 1983. *Energy 2000–2020: World Prospects and Regional Stresses.* London: Graham and Trotman.

World Resources Institute 1990. *World Resources 1990–91: A Guide to the Global Environment.* (In collaboration with the United Nations Environment Programme and the United Nations Development Programme). New York: Oxford University Press.

Yergin, D. 1988. Energy security in the 1990s. *Foreign Affairs*, **67** (1), 110–32.

About the author

John P. Holdren is Professor of Energy and Resources and Chair of Graduate Advisors in the interdisciplinary Energy and Resources Group at the University of California, Berkeley, where his research interests include global environmental problems, comparative environmental assessment of energy options, fusion reactor design to minimize radiological hazards, energy policy, and international security implications of energy and environmental issues. He is also Faculty Consultant in Magnetic Fusion Energy at the Lawrence Livermore National Laboratory, Chairman of the Steering Committee of the University of California's system-wide Institute on Global Conflict and Cooperation, Chairman of the International Executive Committee of the Pugwash Conferences on Science and World Affairs, and a former Chairman of the Federation of American Scientists. In 1981 he received one of the first MacArthur Foundation Prize Fellowships. Professor Holdren was elected to the National Academy of Sciences in 1991.

1

Looking at energy use: an introduction

In the 1950s and 1960s, questions about energy use were mainly of interest to those people whose job it was to provide energy in its various forms. In order to plan their operations, they needed to project how much oil, gas, coal, or electricity might be demanded over a given period of time. For most energy users, energy was something that worked when they turned on the lights or cooked their food. For most governments, energy was not something they paid a great deal of attention to, except in some cases as a source of revenue, or as an area for public investment. There seemed to be plenty of fossil fuel available at low cost, and the future of nuclear energy looked bright.

With the 1973 oil crisis, energy suddenly took center stage, and confidence about the future gave way to anxiety. Sharply higher prices and gasoline lines got the attention of users, while the cost and security of oil supplies roused governments to action. At the same time, many people were becoming more aware of the environmental impacts connected with energy supply and use. The situation calmed down somewhat for a few years, but the jump in world oil prices in 1979–80, combined with the accident at the Three Mile Island nuclear power plant, brought public and political attention on energy once again. In much of the developing world, the high cost of importing oil gave rise to efforts to develop alternatives to oil, but modifying energy use was given much less attention.

The atmosphere of an energy crisis did not last long, however, at least not in the industrial countries. The rise in energy prices led to production of new energy supplies and increased exploration, and estimates of world oil and gas resources were raised. With the collapse of world oil prices in 1986, there was almost a sense of having returned to the period before 1973 when there was no "energy problem".

In fact, the world has changed in significant ways. The scale of human

52

energy use today is considerably larger than in 1970. Demand is growing rapidly in the developing world, raising concerns that the energy sector will claim an unsustainable share of financial resources in these countries. The previous confidence in nuclear power has disappeared in much of the world. Perhaps most significant, the environmental impacts of energy use are being increasingly perceived at local, regional, and global levels. As human civilization approaches the twenty-first century, with the number of people and volume of activities expanding greatly, there appears to be a significant potential that impacts associated with energy supply and use will seriously harm human and ecological well-being.

The Prologue of this book presents a strong case that the world is facing a future in which the overall costs per unit of energy delivered are rising. Since energy use is also rising, the possibility that the environmental, sociopolitical, and economic costs associated with that use will place a growing burden on the lives of future generations cannot be taken lightly. Given this troubling prospect, gaining a better understanding of energy use is very important for two key reasons. First, understanding how and why energy use is evolving provides important insights about the range of problems the world may face. Secondly, improving knowledge about how and why energy is used, and especially about the economic and social forces that underlie energy use, is essential to efforts to shape energy use in a sustainable direction.

1.1 Looking at energy use: a metaphor

Energy use in a society is connected to a large number of diverse activities undertaken by many different types of actors. To understand energy use, one must look at these activities, and at how energy is used in relation to them.

A useful metaphor for thinking about energy and human activities is a forest. From high above, we can easily see that a forest is growing. We can perhaps observe that the number of trees is increasing, and presume that many of the trees are themselves growing. From on high, however, we can learn very little about what is really going on in the forest. We cannot know very much about what is causing it to grow, or how its growth might change.

The situation is similar for energy use. Knowing how much energy is used in any given country, or even in the world, is not very difficult – at least not for the traded energy forms for which records are kept. We can easily observe growth in such energy use, but understanding why a certain amount of energy is used, and why the amount is changing, is not so simple. As with the trees in the forest, we can perhaps observe that the overall magnitude of activities for which energy is used is increasing. From on high, however, we might not know which activities are increasing and decreasing. Moreover, we would not know how much energy is being used to support various activities, how and why that quantity is changing, or how it could be different.

1.1.1 The macro view: losing sight of the trees?

One common way of describing the energy forest from above is the ratio of energy use to Gross Domestic Product (GDP), which is a measure of the aggregate economic activity in a given society. One often finds the energy/GDP ratio presented as an indicator of national energy efficiency. In fact, however, this indicator obscures far more than it reveals. To be sure, change in the ratio suggests that something is happening in the forest, but says little about just what that might be.

Part of the problem lies with GDP itself, which is not really an aggregate measure of all activities that use energy, but rather an attempt to measure the exchange value of economic activities. Many activities that use energy – including illegal and unreported work, private travel, and various home activities – are not counted in GDP or are only partially counted. In some countries, the system for collecting the data that go into national accounts may be weak. Where inflation has been very high, expressing GDP in "real" terms is problematic. In the former centrally-planned economies, the framework of GDP was not used. The so-called Net Material Product (NMP) did not include economic output in services (in part because many services that are delivered by the private sector in market economies were delivered by the public sector or were highly subsidized by the government). In addition, since prices were administered rather than set by the market, expressing the value of production of goods is difficult.[1]

The composition of a country's GDP changes over time as its economy evolves. Since certain activities tend to be more energy-intensive than others, such structural change may significantly affect the energy/GDP ratio. For example, the shift in the structure of GDP from agriculture to

[1] Official Soviet statistics show an average growth in NMP of 3.4% per year between 1975 and 1985, while the US Central Intelligence Agency's estimate of growth in Soviet GNP is 2.3%. Using Soviet statistics, the energy/NMP ratio decreased by 9% between these years; using the CIA's estimate, the energy/GNP ratio increased by 8% (Schipper & Cooper, 1991).

manufacturing increases the energy/GDP ratio, while the shift from manufacturing to services depresses it.

Change in the energy/GDP ratio is a very uncertain measure of the improvement in energy efficiencies in an economy. The ratio is also a misleading indicator of the relative energy efficiency of countries, as we illustrate later in this chapter in a comparison of the United States and Japan. Structural factors are often as important as energy efficiencies in determining the ratio. Asking whether aggregate energy use is "efficient" is the wrong question. Instead, one should examine the energy efficiency of well-defined sectors or end uses.

The macro view is also found in much modeling of energy–economy interactions at an aggregate level. While such modeling has become more sophisticated and detailed over the years, it often still relies on aggregate energy statistics, and tends to have limited data on the activities for which energy is used. The relationships that are found between energy use and energy prices, for example, usually reveal little about the underlying dynamics of change in energy use. The physical nature of energy use – the interaction between people and a diverse set of changing technologies – is not well captured by most macroeconomic modeling. Moreover, one does not get a sense for how much the energy efficiency of particular end uses might change, or what the impact of specific policies might be.

1.1.2 The micro view: losing sight of the forest?

While economists have focused on their models, researchers grounded in the physical sciences or engineering have studied the technical efficiency of energy end uses, usually with an eye to describing how the efficiencies might be improved. Over the past 15 years, studies conducted throughout the world have been extremely valuable in raising awareness of the potential for improving energy efficiency. Most studies estimate the cost of raising the energy efficiency of various end uses to a particular level, often showing that considerable improvement over average practice appears to be cost-effective.

The validity of efficiency potential studies depends on how well the characteristics of the existing and new stocks of energy-using technologies are understood, and on the accuracy of the estimates of the cost of efficiency improvement. Research on these factors has improved the validity of efficiency potential studies in many aspects. Yet, most of these studies are static, describing the energy savings arising if technology "A" is employed instead of technology "B," where "B" usually represents the present average efficiency. The studies are often not well grounded in how "B" has been evolving, and they may not capture changes in "B" that interact with the effects of using "A." In addition, it is difficult to incorporate the behavior of people that shapes the real-world performance of "B."

Put simply, studies of potential end-use efficiency may "describe the trees" in important ways, but can easily lose sight of the forest. It is important that such studies reflect an awareness of how the end uses are evolving in the real world. To understand the cumulative potential impact of changes in end-use efficiencies, it is also necessary to account for change in the magnitude and structure of activities in each sector.

1.2 Our approach to studying energy use

Our approach to understanding energy use had its genesis in the mid-1970s in a study that explored why Sweden used so much less energy per person than the United States (Schipper & Lichtenberg, 1976). At that time, most energy studies did not look very deeply at the structure of energy use, and future energy use was often projected by extrapolating simple relationships from the past. The guiding principle behind the US/Sweden study, which also underlies the work presented in this book, is that understanding trends in energy use, or differences among countries, requires study of the activities for which energy is used, and of how people and technology interact to provide the services that are desired.

An important consequence of linking energy use with activities is that statistics on energy use are only one side of the story. Such statistics must be coupled with corresponding data on the activities for which energy is used. The story of energy use is contained in industrial, transport, and housing statistics as well as in energy statistics.

An example of the need to "dig" below the surface of aggregate energy statistics to understand trends in energy use is the case of oil heating in West German homes. Average oil use in homes using oil did not fall very much between 1973 and 1988. Does this mean there was no conservation? Not at all. In fact, average oil use in detached homes with modern central heating systems fell by almost 40%; a similar decline occurred in homes with other types of systems. However, there was also a large shift to central heating systems, which increased energy use; in addition, the average size of oil-heated homes rose. The reason why use averaged over all homes did not decrease is that these "structural" changes counteracted the physical and behavioral changes that took place.

Similar examples of the need to look beneath the surface exist in all sectors and countries. Over the past decade, we have worked with our associates at Lawrence Berkeley Laboratory and with colleagues in many countries to assemble and analyze historical portraits of energy use and the structural factors that shape it. An important goal of our work is to organize information on individual countries into larger groupings, and thus to analyze trends that are relevant at a global level. It also allows us to make meaningful comparisons among countries. As one would expect, both the style of disaggregation and the historical dimension of the available data vary among countries. Far more information is available for the OECD countries than for most developing countries. While gathering of energy and related data is improving throughout the developing world, assembling consistent historical portraits is often difficult. Thus, most of the discussion and quantitative analysis in this book centers on the OECD countries, which have also been the primary focus of our work over the past decade. We discuss trends in non-OECD countries, but the available data do not allow the same depth of analysis. Despite this shortcoming, understanding the relationships between energy use, income, prices, and other factors in the OECD countries shed some light on how the situation may evolve in other parts of the world.

We view our approach to understanding energy use as a "middle way" between the macro and micro approaches. We endeavor to penetrate to a level of detail at which one gains insight into the role of underlying forces in shaping energy demand, but not so deep that one gets "lost in the forest." It is important, after all, to be able to put the pieces back together into an overall picture that helps to understand past trends, future prospects, and how to most effectively change course.

1.2.1 Activity, structure, and energy intensity

Total energy consumption in a society is comprised of the energy used in a great variety of activities. Broadly speaking, there are two large classes of energy users: (1) enterprises involved in the transformation of primary energy sources into fuels, electricity, and heat; and (2) all other users, which are called "end" or "final" users. Total energy use ("primary energy") is the sum of the energy consumed directly by end-users ("final energy") and the energy "lost" in the production and delivery of energy products (mainly in the production and delivery of electricity). We focus on final energy because it provides the energy services that users want. However, it is important to be aware that final use of electricity entails loss of considerably more energy in the production and delivery (approximately twice as much if electricity is produced from fossil fuels).

The many types of activities in a society may be aggregated in different ways. Wherever possible, we examine five sectors: manufacturing, passenger travel, freight transportation, residential, and services (often called the "commercial" sector). We divide transport into passenger and freight sectors because we view these activities as fundamentally different. We have not analyzed energy use in agriculture, mining, and construction, the other main end-use sectors, as energy-use data for these sectors are often uncertain. These sectors account for a small fraction (about 10%) of total final energy use in the industrial countries; they are more important in the developing countries, especially agriculture.

The purposes for energy use are different in each sector, but in each it is possible to examine three basic elements: (1) the level of aggregate *activity*, (2) the *structure* of activities, and (3) the *energy intensities* of specific types of activities. We describe these three elements below; particular issues relevant to each sector are further discussed in Chapters 3 through 6.

Activity expresses the main enterprise in each sector, the underlying purpose for which energy is used. In manufacturing, it is production of goods; in passenger travel, movement of people; and so on in other sectors. In each sector, of course, there are a large number of different activities and purposes, which change over time and differ among countries. It is impossible to truly aggregate across the many energy-using activities in each sector, but one can use various measures to provide an indication of total activity. In general, these are either economic or physical measures. Both have their merits, but it is not possible to use physical measures at an aggregate level in some sectors (e.g., manufacturing), and economic measures are difficult to apply in passenger travel, for example.

We use particular measures based on their logical coherence and empirical tractability. For the manufacturing and service sectors, we use value-added, an economic measure of output (we also consider measures of physical production in manufacturing). For the service sector, a physical indicator of activity is floor area, but historical data on service-sector floor area are often not available. For passenger travel and freight transport, we use physical measures of activity: passenger – kilometers (p – km) and tonne – kilometers (t – km), respectively. These units measure both the quantities that are transported and the distance. (See Chapter 4 for definitions of these measures.) We use vehicle – km in specific cases. For the residential sector, measuring activity is difficult since there are many different energy-using activities that take place in homes but no single measure of "output." We use population as an activity indicator, since it is people who engage in residential activities that use energy. The number of households is another possible measure. The level of activity per person or household typically changes over time, often with important implications for energy use.

Growth in sectoral activity increases energy use. The level of activity is shaped by many factors; these differ among sectors, but population growth underlies increase in activity in all of them. Activity in one sector often affects activity in others. For example, increase in manufacturing production contributes to growth in freight transport; but the magnitude of the effect depends on the type of goods whose production is increasing, as well as the distance they are shipped.

Structure generally refers to the mix of different activities within a sector. In manufacturing, we define structure in terms of the shares of total value-added accounted for by the major sub-sectors. In passenger and freight transportation, it refers to the shares of total activity accounted for by each mode of transport (i.e., automobiles, rail, etc.). In the service sector, we view structure in terms of the shares of total sector floor area accounted for by different types of activities (where data permit). In the residential sector, structure is defined somewhat differently, since there is no measure of the activities themselves. Instead, we consider household size, dwelling area per person, and stocks of energy-using equipment.

Structural change may increase or decrease energy use. In manufacturing, for example, the production of paper and pulp, chemicals, cement, steel, and nonferrous metals is on average much more energy-intensive than other manufacturing activities. Thus, a shift toward the production of less energy-intensive commodities decreases energy use relative to aggregate manufacturing activity. In freight and passenger transport, decline in the shares of relatively energy-efficient modes, such as water and rail transport, relative to energy-intensive modes, such as trucks and automobiles, implies increasing energy use.

We generally use the term "structure" or "structural change" as defined above. One can also consider structure at more macro and micro levels. An example of the former is the mix of basic economic activities such as agriculture, manufacturing, and services in the total economy. An example of micro structure is the share of specific types of industries within any manufacturing sub-sector. Micro structural changes are less important to aggregate sectoral energy use than the changes at a higher level, but they may impact energy intensities.

Energy intensity refers to the amount of energy used per unit of activity or service. Energy intensity by itself does not have a distinct meaning; rather, one must speak of the energy intensity of some particular activity. We generally refer to energy intensity at either the aggregate sectoral level (e.g., total manufacturing energy use per unit of value-added) or at the next level of disaggregation (e.g., energy use per value-added in the ferrous metals industry). Aggregate sectoral energy intensity is determined by the sub-sectoral energy intensities, so it is affected by change in the relative importance of different sub-sectors. In the manufacturing sub-sector paper and pulp, for example, a shift from production of energy-intensive pulp to finished paper goods lowers the energy intensity of the sub-sector, even if the energy intensities in each area remain unchanged. Similarly, when comparing two countries with equal output in paper and pulp, one may consume considerably more energy per unit of output if it produces more pulp than paper goods.

Energy intensities depend on *operation* of equipment or buildings as well as *technical energy efficiency*, by which we mean energy use of equipment or processes under uniform operating conditions. In transport, for example, the energy intensities of each mode are affected by how much of the vehicle capacity is utilized (load factor) and by operator behavior. Falling load factors (i.e., decline in the ratio of passengers to available seats) raise energy intensity even if the technical energy efficiency of the vehicle stock is constant. In the residential sector, the energy intensity of space heating (energy use per unit of floor area) depends not only on the thermal characteristics of building shells and the efficiency of heating systems, but also on the heating habits of households.

The technical energy efficiency of any given process, building, or equipment depends on the characteristics of various components, and how they interact. Measures of technical energy efficiency are difficult to estimate for an entire national stock of equipment or buildings, since real-world energy use depends on operation. They can sometimes be estimated for the stock of new equipment (such as cars or appliances) purchased in a given year. The energy-intensity indicators that we use reflect what could be called "system energy efficiency" in that energy efficiency is seen as part of a system of equipment and operation.[2] The room for improvement in each of these areas varies among activities. For example, lighting energy intensity in a building can be reduced through more efficient end-use equipment as well as changes in operation. Similarly, the intensity of truck

---

[2] Properly speaking, energy intensity and energy efficiency are inversely related. Energy intensity measures energy use per unit of output or service, while energy efficiency refers to amount of service delivered per unit of energy.

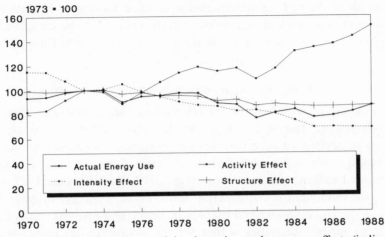

Fig. 1.1. US Manufacturing: activity, intensity, and structure effects (indices)

freight transport can be reduced through change in the technical characteristics of the vehicle stock and in operation, which includes driver behavior and how well vehicle capacity is utilized.

1.2.2 Impact of changing activity, structure, and energy intensities

Total energy use in each sector is the sum of activity times energy intensity for each type of activity considered. Thus, energy use may be expressed by the identity:

$$E = \sum A_k I_k.$$

where k refers to each type of activity.

We illustrate the roles of changing activity, structure, and energy intensities with the example of manufacturing in the United States. Total manufacturing energy use declined by 13% between 1973 and 1988 (Fig. 1.1). Aggregate activity (value-added) grew considerably (52%). If manufacturing structure and energy intensities had remained constant, energy use would have increased by 52%. However, the structure of the sector changed in a way that dampened growth in energy use slightly. Decline in sub-sectoral energy intensities had an even more powerful downward effect. The combined result of these three forces was the 13% decline in energy use.

One can measure the impact of any one of the three factors by holding the other two constant. For the OECD countries that we have studied, we quantitatively describe the relative impact of each factor on historic energy use in all of the end-use sectors except services, for which the data are insufficient. The method that we use, which is rooted in the use of fixed-weight or Laspeyres indices, is described in the Appendix to Chapter 3.

1.3 Using the activity-structure-intensity approach

The example presented above for US manufacturing illustrates how we analyze sectoral trends in energy use. We also combine results across sectors to develop an aggregate analysis of trends in a country (or set of countries). This approach provides far greater insight than a highly aggregated indicator such as the energy/GDP ratio. Between 1973 and 1988, for example, the energy/GDP ratio (primary energy) dropped by 26% in the United States and by 34% in Japan. Our analysis shows that declining energy intensities within sectors accounted for about three-quarters of the decrease in the United States, but for only half of it in Japan (see Chapter 7). Part of the reason for this difference is that energy intensity in passenger travel and the residential sector increased in Japan, but these values decreased in the United States. Macro-level structural change (i.e., faster growth in services than in manufacturing) was also somewhat stronger in Japan, which contributed to the decline in energy/GDP.

Looking at differences in levels of sectoral activity, structure, and energy intensities is also very useful in comparing energy use among countries. For example, in 1987 the United States used over twice as much primary energy per person as Japan. What accounts for this large difference? Higher levels of per capita activity in different sectors explain part of it. Manufacturing output per capita (measured in purchasing power parity units) is about the same in both countries, but in other sectors per capita activity is significantly higher in the United States. The average Japanese travels (domestically) around 10,000 kilometers per year, while the US average is around 22,000. Per capita freight transport is also much higher in the US.

Differences in the structure of each sector play an important role. In manufacturing, energy-intensive industries comprise a larger share of output in Japan than in the United States. Japan ships far more of its domestic freight by truck, the most energy intensive mode, than does the United States. In consumer-oriented sectors, on the other hand, Japan's structure is far less energy-intensive than that of the United States. Only 55% of Japan's total domestic travel is by car, vs. 85% in the United States. Japanese homes average only 27 square meters per person vs. over 50 in the United States. Only 5% of Japanese homes had central heating, vs. over

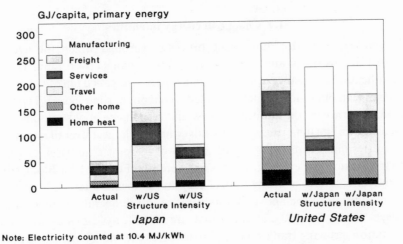

GJ/capita, primary energy

Manufacturing
Freight
Services
Travel
Other home
Home heat

Actual   w/US   w/US

Structure   Intensity

Japan

Actual   w/Japan   w/Japan

Structure   Intensity

United States

Note: Electricity counted at 10.4 MJ/kWh

Fig. 1.2. Japanese and US energy use in 1987: effects of structure and intensity differences

80% in the United States, and Japanese had few clothes dryers, dishwashers, or freezers. If the United States had the same sectoral structure as Japan (but its own energy intensities), its energy use would have been about 20% less than it actually was (Fig. 1.2). If Japan had US sectoral structure (but its own energy intensities), its energy use would have been 75% higher than it actually was.

Did Japan use energy more efficiently for particular activities than the United States? When allowance is made for differences in the structure of manufacturing, Japan used approximately 30% less energy per unit of output than the United States. Japanese trucks used less energy per tonne-km than American ones, and Japan's auto fleet used about 30% less energy per kilometer driven. Japanese appliances – refrigerators and air conditioners – appear to have used 25% less energy than comparable American ones. Japanese households only used 40% as much heat per unit of floor area as American ones (adjusted for climate), but this difference is caused almost entirely by the spartan heating habits of the Japanese, who only heat to 14–15°C in rooms that are occupied. In all, if Americans used energy with the same intensities as the Japanese (but with US sectoral structure), US primary energy use would have been about 20% less than it actually was. Leaving out Japanese heating practices, it would have been 15% less. Lower energy intensities in Japan account for less than half of the difference in per capita energy use. Differences in the level of activities and structure account for the rest.

1.4 Change in energy intensities

In general, there are three basic processes through which the average energy intensity of any given activity may change over time. One is modification of existing equipment or facilities – so-called retrofit. Another is change in operation (which is related to retrofit in some cases). The third is stock turnover, the addition of new and retirement of old equipment or facilities. In the long run, stock turnover has the largest overall impact. In countries that are growing rapidly (many developing countries), or where a large part of the existing stock is obsolete (the Former East Bloc), stock turnover can have a strong impact even in the medium term.

In addition to the above factors, change in the operating environment affects energy intensity in ways that are not always obvious. In transportation, growing traffic congestion increases the energy intensity of road transport modes, and affects air travel as well. The outdoor environment is a key factor for buildings, and is influenced by the nature of the built environment. Akbari et al. (1990) have estimated that about 5–10% of the current urban electricity demand in major US cities is spent to cool buildings just to compensate for the "heat island effect".

1.4.1 Retrofit, operation, and maintenance

Retrofit mainly effects buildings and industry. Buildings have very long lives, so the turnover of the stock tends to be relatively slow (more so in the OECD countries than in the developing countries, however). Improvement in the thermal properties of buildings or replacement of equipment may result from general-purpose renovation as well as from modifications made explicitly to conserve energy. Similarly, modifications to industrial facilities to improve productivity generally may have a strong impact on energy intensity. In transportation, and for home appliances, there is little practical scope for retrofit.

Change in operation and maintenance of equipment and buildings may be short-term in nature, but some changes endure over time, or are only partly reversed. Procedures that lower operating costs without hindering productivity are likely to endure in competitive sectors like manufacturing, air travel, and freight transport. This is less the case in homes and personal travel, where, for example, the reductions in indoor temperature and driving speed (in the United States) that occurred in response to increased energy prices partially or entirely reversed over time.

The way in which equipment and buildings are operated changes with the characteristics of the technologies themselves. For example, more powerful cars allow people to drive faster. In addition, technologies may be incorporated that facilitate more energy-efficient operation. Computerized control systems for buildings, industrial processes, and freight transport are increasingly being used to better manage operations. In many cases, decisions that were formerly made (or not made) by people are now made by control technologies. For example, sensors that turn lights off when a room is unoccupied remove the behavior component (although users may sometimes dismantle control devices that are unwanted). The penetration of control technology into consumer goods is slower, but is nonetheless apparent.

1.4.2 Stock turnover and new capital stock

The impact of stock turnover on energy intensities may be positive or negative depending on the characteristics of existing and new equipment and facilities. In productive sectors, the energy efficiency of new equipment or processes often improves over time – irrespective of energy prices – due to technological innovation, which may increase energy efficiency without that being a primary purpose of the change. In some cases, however, technological change may improve technical energy efficiency even as change in features increases energy intensity. This phenomenon has occurred for automobiles and home appliances in the OECD countries. In both cases, increase in size and performance has counteracted the impact of improved technical energy efficiency.

The rate of stock growth and turnover are key factors. In mature economies, the growth of the overall stock is relatively slow, but retirement of old equipment is often relatively rapid. In developing economies, the addition of new equipment and facilities tends to be rapid, but retirement of the old is usually slow. Because of shortage of capital, equipment tends to be kept in operation rather than scrapped. Generally speaking, where there are clear incentives to reduce production costs per unit of output, faster economic growth results in more rapid technological innovation as well as more rapid incorporation of technological advances.[3] Faster growth implies expansion of production, which involves addition of new equipment or facilities. If businesses successfully take advantage of opportunities during periods of higher economic growth, profits rise,

---

[3] Where incentives to innovate are weak, output may expand rapidly without there being a reduction in unit costs. Further, if energy is priced low, the innovation that does take place may seek to conserve capital at the expense of energy. These two factors explain the relative lack of decline in manufacturing energy intensities in the Former East Bloc.

leaving more resources for R&D to develop better technologies. If markets are functioning reasonably well, firms have incentive to incorporate the fruits of R&D if doing so improves their competitive position.

For households, however, faster economic growth has a different effect. Increase in disposable income allows consumers to acquire new energy-using goods, or to replace old equipment with models that offer more service. It also allows purchase of more living area per person and attainment of a more comfortable indoor environment, both of which increase energy use. Saturation in equipment ownership appears at some point, but the market penetration of a device only tells part of the story. In the former Soviet Union, for example, the ownership level of refrigerators is quite high, but there is considerable room for growth in size and features. As the market for particular goods expands, innovations in production often result in lower real prices for goods, making them affordable to a greater portion of the population. This kind of positive feedback between producers and consumers characterized much of the post-World War II period in the OECD countries, resulting in rapid buildup of stocks of energy-using consumer goods, and is seen today in many developing countries.

How changes balance out during periods of faster economic growth depends on the actual circumstances at a given time and location. For example, the wealthy OECD countries are entering a phase in which marginal consumer expenditures in the home are not increasing household energy intensity by very much, since ownership of the main energy-intensive equipment is approaching saturation. In contrast, many middle-income developing countries are now in a period of rapid growth in ownership of appliances and motor vehicles similar to what occurred in Western Europe and Japan in the 1960s. (Ownership of automobiles and central heating in West Germany increased at more than double the rate of disposable income during the 1960s.) If reform is successful, the Former East Bloc could enter a similar phase in the not-too-distant future.

1.5 Looking at the world

Just as one must add up the end-use sectors to arrive at total final energy use in a country, one must add up countries to get a broader view of what is happening in the world. In this book, we divide the world into three large groups of countries: (1) the OECD countries, (which we sometimes call the "industrial countries"). (2) the Soviet Union and Eastern Europe (which we call the "Former East Bloc"),[4] and (3) the developing countries (which we sometimes call the "LDCs"). We include all countries not in the first two groups with the developing countries. In reality, the lines between the three groups are somewhat indistinct. China has characteristics of both the Former East Bloc and the developing world. South Africa is sometimes included among the "industrial" countries. Many of the countries of Eastern Europe are relatively urban and industrial, but are poorer than some of the more industrialized LDCs. In terms of GDP per capita, a few countries grouped with the LDCs rank with (or even above) the lower-income OECD countries.

Among the 24 member countries of the OECD, the so-called "G-7" – the United States, Japan, West Germany, the United Kingdom, France, Canada, and Italy – account for about 85% of total energy use. The United States alone accounts for half of the total. In the Former East Bloc, the Soviet Union accounted for about three-fourths of 1989 energy use. Poland and the former East Germany accounted for 6% and 5%, respectively. Among the LDCs, ten countries account for about 70% of total commercial energy use. China alone accounts for 32%, followed by India with 9%, Mexico with 6%, and Brazil and South Africa with 5% each.

Where generalizations are made, it should be clear that they apply to varying degrees across the countries in each group. Among the OECD countries, most of what we say applies more to the higher-income countries than to countries like Portugal, Greece, and Turkey. In the Former East Bloc, we primarily discuss the Soviet Union. While the countries of Eastern Europe have many similarities with the Soviet Union, there are differences as well. The LDC group is especially heterogeneous, ranging from very poor, primarily rural nations to relatively wealthy and rapidly modernizing ones. We often discuss LDC regions separately (i.e., Latin America, Asia, Africa), but differences also exist within these regions.

1.6 Overview of the book

This book is divided into three parts. Part I describes how and why energy use has changed in the world since 1970, using the framework discussed above. Chapter 2 presents an overview of energy-use trends for the world and for the three country groups. It illustrates the rising importance of

---

[4] Throughout the book, we include the former East Germany within the Former East Bloc. The Soviet Union includes all of the republics that were part of the Union prior to the failed coup of August 1991.

energy use in the LDCs, and shows how the role of the end-use sectors has changed in each group. Chapter 3 covers manufacturing, which at a global level accounts for more energy use than any other sector. We show how energy intensities in manufacturing have declined continuously in the OECD countries, contributing to a decline in energy use despite considerable growth in activity. Chapter 4 discusses trends in passenger travel and freight transport, sectors that account for the largest share of world oil use. We highlight the importance of structural change – shifts toward more energy-intensive modes such as automobiles, air travel, and trucks – in pushing energy use upward. Chapter 5 covers the residential sector. Here too we illustrate how structural change – especially increase in appliance ownership – has caused energy use to grow in much of the world. Chapter 6 describes trends in the service sector, in which electricity plays an important and growing role. We show that energy use for heating has declined greatly in the OECD countries, but electricity intensity has risen. In Chapter 7, we summarize the changes in sectoral energy use and discuss several cross-cutting issues that emerge from the OECD sectoral analyses: What was the role of higher energy prices? Of energy-efficiency policies? Did energy savings represent breaks in long-term trends?

Part II looks into the future. We use the historical analysis as a guide, supplementing it with judgment about the possible evolution of key trends. Chapter 8 discusses prospects for growth in activity and change in structure in each sector and country group. Growth in activity will be especially strong in the developing countries, and structural change is likely to increase energy use considerably as well. The greatest potential for structural change is in the Former East Bloc, but the pace of change is very uncertain. Structural change is occurring much more slowly in the OECD countries. Chapter 9 considers future energy intensities in each sector. We discuss both the direction in which trends seem to be pointing, and the potential for reducing energy intensities. Drawing on studies from many countries, we show that there is considerable potential for lowering intensities in all sectors and country groups. However, the direction of current trends makes it unlikely that most of the potential will be realized without greater action than is now occurring. In Chapter 10, we illustrate the extent to which energy intensities might be reduced by presenting scenarios for the OECD countries and the Soviet Union. The scenarios show that intensities in the OECD countries in 2010 could be nearly 50% less on average than the level to which trends seem to be pointing. The Soviet scenarios suggest that intensities could decline by almost that much (relative to 1985 levels) if economic reform is successful and technology levels are upgraded to Western standards. In both cases, however, achieving reductions on this magnitude will take considerable policy effort.

Part III presents an overview of policies and programs that can help to change the course of energy use. Chapter 11 presents an overview of policies and programs that can increase the rate of energy-efficiency improvement. These include reform of energy pricing, efficiency standards, financial incentives, as well as innovative methods and policies outside the sphere of energy. In Chapter 12, we summarize the reasons why restraint in energy use is an important strategy for reducing environmental problems and supporting sustainable development world-wide, and outline key steps for achieving such restraint. We call for strong action by governments at all levels, but stress the importance of considering the full range of factors that will shape realization of the energy-efficiency potential around the world, and the need to look beyond energy policy for solutions. We suggest how the emergence of a global market economy with relatively open borders could enhance efforts to improve energy efficiency and contend that strong policies in the OECD countries to encourage greater energy efficiency would have profound effect world-wide.

In this book, we seek to illuminate the complex relationship between energy and human activity: how energy use is linked to what people do, and to how they do things. We do not claim that solutions to the problems connected to energy use will be easy to put into effect. Rather, we seek to provide a coherent analysis of what has occurred over the past 20 years, where trends are pointing, and of the means for affecting change. We hope that it contributes to better understanding of global prospects and to responsible actions to lay the foundations for a sustainable future.

References

Akbari, H., Rosenfeld, A. H. & Taha, H. 1990. Summer heat islands, urban trees, and white surfaces, ASHRAE *Transactions*, **96**, (1), 1381–88.

Schipper, L. & Cooper, R. C. 1991. Energy Use and Conservation in the USSR: *Patterns, Prospects, and Problems*, Berkeley, CA: Lawrence Berkeley Laboratory Report LBL-29831.

Schipper, L. & Lichtenberg, A. 1976. Efficient Energy Use and Well Being: The Swedish Example. *Science*, **194**, 1001.

Part I
Past trends

2

World energy use since 1970: an overview

World primary energy use, including biomass, grew by over one-third between 1970 and 1990. The average increase over these two decades of 2.3% per year hides periods of growth and stagnation. World energy use was steady in the 1974–1975 and 1980–1983 periods as a result of response to higher energy prices and slowing of economic activity; but growth continued after each period of stagnation as real oil prices declined and economic growth resumed (Fig. 2.1). Between 1983 and 1989, growth in world energy use was rather steady, averaging 2.8% per year. There was little increase in 1990, however, due to a slowing in the OECD countries and a sharp drop in the Former East Bloc.[1]

Most of the increase in energy use since 1970, and especially since 1973, has been in the LDCs. Their share of the total grew from 20% in 1970 to 31% in 1990. (Not counting biofuels, the LDC share would be about 27%.) The share of the OECD countries fell from 60% to 48%, while the share of the Former East Bloc rose slightly from 20% to 21% (in part because these countries were insulated from the effect of higher oil prices). The average annual growth in energy use between 1970 and 1990 was 1.3% in the OECD countries, 2.4% in the Former East Bloc, and 4.5% in the LDCs.

The reasons for these changes in energy use are the subject of the next few chapters of this book, but a few basic factors are worth mentioning at this stage. First, population grew much faster in the LDCs than in the other groups, as did economic activity. Further, many LDCs have been in a stage

---

[1] The data on commercial energy use in this section are from BP (1991). BP presents hydro and nuclear electricity generation in terms of the amount of oil required to fuel an oil-fired plant in order to generate the same amount of energy. These data include use of fuels for non-energy purposes (chemical feedstocks, lubricants, etc.). The data on biomass use are from the International Energy Agency (IEA 1989, 1990). These data are very uncertain, and probably underestimate actual use. We have modified the IEA data on biomass use in China to reflect recent Chinese estimates, which suggest that use is over four times as large as the value given by the IEA.

73

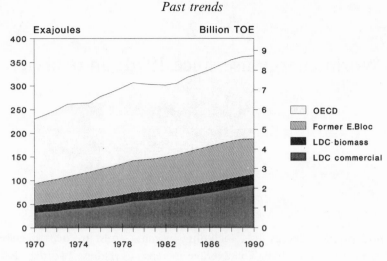

Sources: BP, IEA, & LBL

Fig. 2.1. World primary energy use, 1970–1990

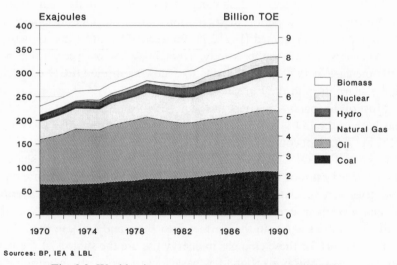

Sources: BP, IEA & LBL

Fig. 2.2. World primary energy use by source, 1970–1990

of economic development in which use of commercial energy tends to grow
faster than GDP due to development of manufacturing and basic
infrastructure. In the OECD countries, on the other hand, structural
change in economic activity and maturation of physical infrastructure has
contributed to decline in the energy/GDP ratio. This "structural" decline
was complemented by response to higher energy prices, as well as ongoing
technological innovation. In many LDCs, energy users have been

somewhat insulated from higher energy prices. In addition, most LDC energy users have had far less information, technology, and capital to improve energy efficiency than have users in the OECD countries.

The evolution of world energy use by type is shown in Fig. 2.2. The share of oil has declined from 41% to 36%, but total oil consumption in 1990 was over one-third higher than in 1970. The share of natural gas has grown from 17% to 20%. Coal has fallen from 28% to 25%, but total consumption has grown slightly. Nuclear energy, which accounted for 5% of total energy use in 1990, has seen the most rapid growth. The share of biomass has remained at about 8%, as an increase in share in the OECD countries has been balanced by a decline in the LDCs (from 32% to 20%).

2.1 Regional overview

Primary energy use declined sharply in the OECD countries in 1974–1975 and 1980–1982 (Fig. 2.3). Users responded to higher prices in the short run, but slowing of economic activity played the main role. The increase in prices accelerated the historic decline in the energy/GDP ratio resulting from structural and technological change. Between 1983 and 1990, however, energy demand rose at an average rate of 2.7% per year. GDP growth was strong in this period, and the collapse of world oil prices in 1986 helped to diminish interest in energy saving. Electricity use increased faster than use of fuels, contributing to growth in primary energy use.[2] The United States accounts for nearly half of total OECD energy use, though its share has fallen somewhat since 1970. The share of Japan has grown slightly (from 9% to 11%), while the share of Western Europe has remained at about 33%.

In the Former East Bloc, primary energy use grew at a rapid rate in the 1970s as industrial output expanded (Fig. 2.4). Considerable increase in transport of Soviet energy products (coal, oil, and natural gas) for domestic use and export also played a role. The pace slowed considerably in the early 1980s, then grew faster again for several years. Energy use declined slightly in 1989 as the old economic and political order began to change, and fell even more sharply in 1990 as economic recession took hold. The Soviet Union accounts for 75% of total energy use in the Former East Bloc, followed by Poland (6%) and the former East Germany (5%).

Primary energy use in the LDCs increased over two-fold between 1970 and 1990 (Fig. 2.5). There was some reduction in the pace of growth after

---

[2] Electricity, though convenient, is not as efficient as the direct use of fuels because of losses in conversion, transmission, and distribution.

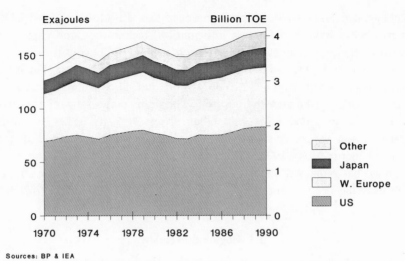

Sources: BP & IEA

Fig. 2.3. OECD primary energy use, 1970–1990

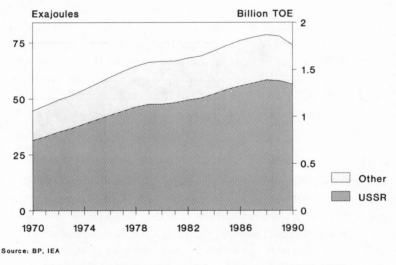

Source: BP, IEA

Fig. 2.4. Former East Bloc primary energy use, 1970–1990

the two oil-price shocks, but slower growth in oil-importing countries was partly balanced by rapidly increasing use in the oil exporters. Between 1982 and 1990, LDC energy use grew at an average of 5.1%/year. Growth has varied among regions, however. It averaged 6.5%/year in the Middle East, which developed considerably based on oil-export earnings. The pace of increase was also high in Asia: 5.7%/year in China, and 6.7%/year in the

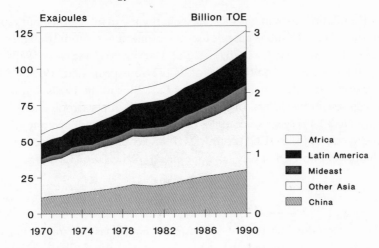

Source: BP, IEA & LBL

Fig. 2.5. LDC primary energy use, 1970–1990

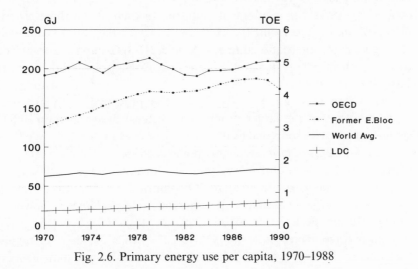

Fig. 2.6. Primary energy use per capita, 1970–1988

rest of developing Asia, which experienced considerable economic growth in the latter half of the 1980s. Growth was much less in Latin America (3.1%/year) and Africa (2.6%/year) due to slower economic growth in those regions. China accounted for 31% of total LDC energy use in 1990, the rest-of-Asia for 26%, Latin America for 22%, Africa for 10%, and the Middle East for 11%. Overall, growth in commercial energy use in the LDCs was faster than that of total energy use due to the transition away from biomass fuels.

Population growth has increased energy use in all regions, but especially in the LDCs. Primary energy use per capita has risen only slightly in the OECD countries since 1970 (Fig. 2.6). It declined in the early 1980s, but by 1990 had increased to the level of its previous peak year, 1979. Per capita use grew considerably in the Former East Bloc in the 1970s, but increased much less in the 1980s. In the LDCs, growth in per capita energy use has averaged 2.4%/year since 1970, but in 1990 it was still just one-seventh of the average of the OECD countries. If one counts only commercial energy, the per capita LDC average is one-ninth of the OECD average.

2.2 Trends in final energy use by sector

The energy use described above (primary energy) is the sum of final energy use in each sector and the losses in energy transformation (mainly in electricity production) and delivery (mainly electricity transmission and distribution losses). In this section, we show how final energy use has evolved by sector for a subset of countries in each of the three groups discussed above. The countries are those for which the authors and their colleagues in the International Energy Studies (IES) Group have organized data on sectoral energy demand. For the OECD countries and the Former East Bloc, the selected countries account for 75% or more of the total energy use in each group, so the portraits presented are reasonable representations of the larger group. The sectoral disaggregation differs somewhat in each group due to the nature of the available data. We have included estimates of biomass energy use in each sector, but have excluded or removed data on use of energy products for non-energy purposes.[3]

Total final energy use in nine OECD countries[4] was about the same in 1988 as in 1973. The shares of different sectors changed, however. Manufacturing declined from 36% of the total to 27% (Fig. 2.7). Passenger travel increased from 19% to 22%, while the share of freight transport rose from 7% to 10%. In each case, a transition to more energy-intensive modes (automobiles and trucks, respectively) played an important role in boosting energy use. The residential sector's share grew slightly from 20% to 21%, and the share of the service sector rose from 10% to 11%.

[3] Energy use as given in most national and international statistics includes the use of fossil fuels for non-energy purposes. While such data are useful for understanding the demand for fossil fuels, the energy content of the fuels is not utilized. Thus, such use is actually not energy use; where possible, we exclude it from our analysis.
[4] The United States, Japan, West Germany, France, Italy, the United Kingdom, Sweden, and Denmark. Canada is the only major consumer left out.

Exajoules

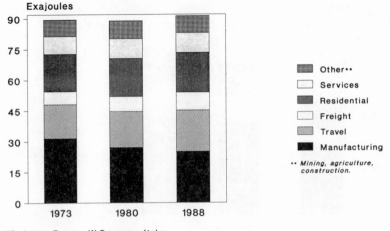

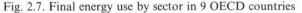

* US, Japan, France, W.Germany, Italy,
UK, Norway, Denmark, Sweden

Fig. 2.7. Final energy use by sector in 9 OECD countries

The above picture of final energy use obscures changes in the role of different energy sources (which we describe in Chapter 7). The share of electricity increased in manufacturing and in the residential and service sectors. Thus, the sectoral shares would be slightly different than described above if losses in electricity supply were allocated among the sectors. The share of the residential and services sectors would increase the most, since electricity's share of final energy use has grown considerably in those sectors.

The sectoral picture of final energy use in the USSR is quite different from that in the OECD countries. The share of industry has fallen somewhat since 1970, but at 55% it still dominated final energy use in 1985 (Fig. 2.8).[5] The residential sector is the next largest, which reflects the cold Soviet climate, but its share fell slightly from 23% to 20%. The share of passenger travel has increased substantially, but it was still only 5% in 1985, reflecting the low level of automobile use.

For the LDCs, we present sectoral data for 12 major countries which account for approximately 35% of total LDC final energy use.[6] As with the Soviet Union, the data do not allow disaggregation of industrial energy use

[5] The Soviet data do not allow disaggregation of manufacturing from total industrial energy use; the latter includes mining and construction, as well as agriculture. We present 1985 as the final year because later data on sectoral energy use are of somewhat questionable accuracy.

[6] The 12 are India, Indonesia, Malaysia, Pakistan, the Philippines, South Korea, Taiwan, Thailand, Argentina, Brazil, Mexico, and Venezuela. China is not included because sectoral data are only available from 1980 onwards. Large users left out include South Africa and Egypt.

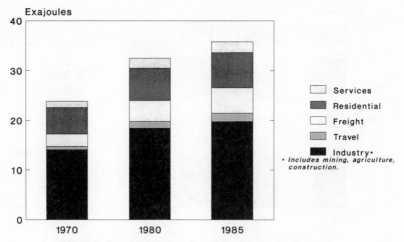

Fig. 2.8. Final energy use by sector in the Soviet Union

into its components, though agricultural use is given for most countries. In addition, few developing countries have an accurate disaggregation of residential and service sector energy use, or of passenger travel and freight transport.

The picture of sectoral energy use in the 12 LDCs is strongly affected by including estimates of biomass use, which plays a major role in the residential sector. The residential and commercial sectors together accounted for 38% of final energy use in 1988, about the same share as industry (Fig. 2.9). Their share has declined since 1970, however. The transportation sector has seen considerable growth in share (from 18% to 22%), as automobiles and trucks have come into greater use. The share of industry has also grown, reflecting structural change in some of the economies. If only commercial energy were counted, the residential/commercial share would be much smaller.

China alone accounts for as much energy use as the above 12 LDCs together. The share of industry (48%) is higher, however, reflecting the large presence of heavy industry in China (Fig. 2.10). The combined residential/commercial sector is the next largest user at 42%. Similar to the Soviet Union, transportation accounts for a relatively small share (6%) of final energy use due in part to the very low level of automobile use. Since 1980, the share of industry has grown slightly.

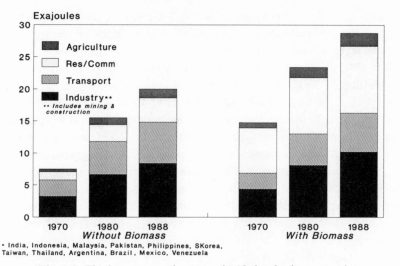

Fig. 2.9. Final energy use by sector in 12 developing countries

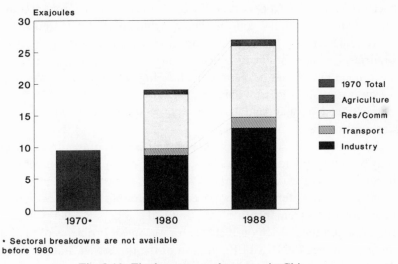

• Sectoral breakdowns are not available before 1980

Fig. 2.10. Final energy use by sector in China

2.3 Introduction to the sectoral chapters

The reasons for the changes in sectoral energy use described in the preceding section are the subject of Chapters 3 through 6, which cover the manufacturing, transportation, residential, and service sectors in each of the country groups. The analysis of the OECD countries presented in Chapters 3–6 is primarily based on work done over the past decade by the authors and their colleagues in the IES Group. A number of ongoing studies have assembled detailed time-series data bases on energy use and related factors in the nine countries listed above. An important focus of our effort are data on the characteristics and structure of each end-use sector. These include the types of homes, heating systems, and electric appliances; the types of motor vehicles and their usage; the levels of passenger travel and freight activity; and levels of monetary and physical output in the major sectors of manufacturing. To these "structural" data we join energy use by fuel for the main subsectors or purposes. In disaggregating energy use, assumptions are required to separate household energy consumption into end-uses, or to separate the use of road, rail, and marine fuels into the various travel and freight modes. We often use a "bottom-up" approach, by which we mean combining data on energy-using stocks with estimates of average energy use for various purposes (either our own or those of in-country experts). The assumptions we have used are described in the relevant reports cited in Chapter 3–6.

The main data sources for the nine OECD countries that we refer to are described in detail in Appendix A. We use data gathered from each country rather than international sources. These include official data and surveys, private data and surveys, and authoritative reports from the individual countries we have studied. Official data include figures published by government ministeries, transport and housing authorities, and the central bureau of statistics. Private data include surveys of household equipment commissioned on a regular basis by energy suppliers, as well as energy consumption data from suppliers.

The analysis of sectoral energy use in the USSR is drawn from a recent study by Schipper and Cooper (1991). Data are mostly from sources within the Soviet Union: published data from various ministeries; energy studies in the open Soviet literature; and a wealth of unpublished data provided by the Energy Research Institute (Moscow), the Siberian Energy Institute (Irkutsk), and the All-Union Institute for Technical Problems of the Fuel-Energy Complex (VNIIKTEP), an arm of the the Soviet State Economic Planning Agency (GOSPLAN). Additionally, we used information from a series of studies led by M. Sagers of the US Bureau of the Census. The methods used in disaggregation are described in the report cited above.

For the LDC analysis, a key source is a time-series data base on sectoral energy use and related factors for 13 major LDCs (including China) that has been organized by the IES Group (see list above). Together, these 13 countries account for about 70% of total LDC energy use (and a somewhat higher share of total commercial energy use). These data are not nearly as detailed as those for the OECD countries, but much of the description and analysis of energy use in the LDCs derive from this and related work. An overview is given in Sathaye et al. (1987). We also use results from studies conducted in various LDCs, as cited in each chapter.

Unless stated otherwise, the data presented in the text, tables, and figures in Chapters 3–7 come from the OECD, Soviet, and LDC data bases discussed above.

References

BP (British Petroleum Company). 1991. *BP Statistical Review of World Energy*, London, UK.

IEA (International Energy Agency). 1989. *World Energy Statistics and Balances 1971–1987*, Paris, France.

IEA (International Energy Agency). 1990. *World Energy Statistics and Balances 1985–1988*, Paris, France.

Sathaye, J., Ghirardi, A. & Schipper, L. 1987. Energy demand in developing countries: a sectoral analysis of recent trends, *Annual Review of Energy*, **12**, 253–81.

Schipper, L. & Cooper, R. C. 1991. *Energy Use and Conservation in the U.S.S.R.: Patterns, Prospects, and Problems*, Berkeley, CA: Lawrence Berkeley Laboratory Report LBL-29831.

3

Historic trends in manufacturing

Manufacturing accounts for more of the world's energy use – around 40% of final energy – than any other major sector. In the OECD countries, its share of final energy use declined from about 35% to 30% between 1973 and 1988 due to rising consumption in other sectors and decline in manufacturing. It plays a more important role outside the OECD countries, however, accounting for approximately 40% of energy use in the Soviet Union, 60% in China, and 35–45% in other LDCs.

At a global level, it is difficult to separate manufacturing from total industrial energy use, but trends in the latter give a reasonably good picture of those in manufacturing, since the latter dominates industrial activity and energy use. The other main industrial activities – mining and construction – account for 10–20% of total industrial energy use in the OECD countries, about 20% in the former East Bloc, and around 20–25% in the LDCs. As shown in Fig. 3.1, total OECD industrial energy use has fluctuated since the early 1970s, but was less in 1988 than in 1973. In the LDCs and former East Bloc, growth has been more steady, though the growth trend for the latter has levelled off since 1988.

3.1 Analyzing manufacturing energy use

Disaggregated analysis of manufacturing energy use is often based on classifications of manufacturing activities such as the International Standard Industrial Classification (ISIC), which groups industries in nine broad "2-digit" sectors, which in turn are divided into "3-digit" sectors. Production within any given 2-digit sector usually consists of a large number of different commodities. For particular products, such as steel, cement, and aluminum, one can measure physical output in tonnes, but it is difficult to aggregate across different physical products within most

84

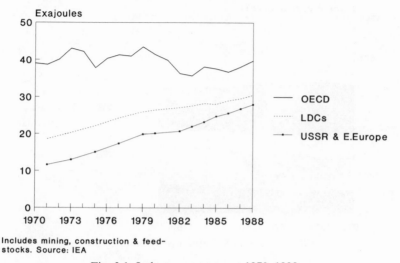

Includes mining, construction & feedstocks. Source: IEA

Fig. 3.1. Industry energy use, 1970–1988

2-digit sectors. Thus, it is necessary to value different commodities in a common unit. A standard approach in market economies is measurement of monetary value-added, which refers to the value of sectoral production minus purchased material inputs. This method has not been used in the centrally planned economies, however, which makes international comparisons difficult.

If data on manufacturing energy use can be disaggregated into the same categories as manufacturing activities, one can divide energy use by production to calculate the energy intensity of a given sector. Energy intensity at this level is affected not only by energy management practices and the types of processes used, but also by the type of goods produced within each industry group. A few industry sectors are much more energy intensive – in terms of energy use per unit of value-added – than the others, and usually account for a large share of total manufacturing energy use. Generally, the most energy-intensive industries are paper and pulp (ISIC 341), chemicals (ISIC 351–352); stone, clay and glass (ISIC 36); iron and steel (ISIC 371); and nonferrous metals (ISIC 372). The common feature of these industries is that their output is dominated by products that one can view as "building blocks" – either for production of capital goods and consumer products, or for construction of buildings and transport infrastructure. For the eight OECD countries that we studied (see below), the average energy intensity of these five sectors in 1988 ranged from 31 MJ/1980$ of value-added for chemicals to 97 MJ for iron and steel. The

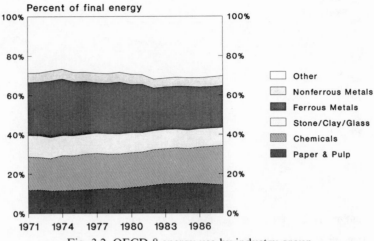

Fig. 3.2. OECD-8 energy use by industry group

average energy intensity of all other manufacturing was only 6 MJ/1980$. The five energy-intensive sectors accounted for only around 20% of manufacturing value-added, but claimed 70% of energy use (Fig. 3.2). Because of the great difference in energy intensity between the energy-intensive industries and all others, changes in the output shares of these industries have a major impact on aggregate manufacturing energy use.

3.2 OECD countries

For eight OECD countries – the United States, Japan, France, the United Kingdom, West Germany, Norway, Sweden, and Denmark – we assembled data on value-added and energy use for the five energy-intensive industries listed above, and placed all other manufacturing activities (excluding petroleum refining) in a residual "Other" category.[1] These eight countries account for around three-quarters of total OECD manufacturing energy use.[2] The level of aggregation chosen ensures that the data are fairly comparable among countries. While more detailed data are

---

[1] Data on refining energy use are inconsistent among countries and are often not available at all, so we excluded this sector in our analysis. Energy used in petroleum refining is usually not included in manufacturing in national energy balances, as it is considered an energy transformation industry (like electric utilities).

[2] We have not included Italy in the analysis due to data uncertainties. The data on real value-added from the national accounts and the industrial production index show very different trends in sectoral production. The value-added data show growth in sectoral output of 3.7%/year between 1973 and 1988, while the industrial production index indicates that manufacturing output grew by only 1.75%/year. In addition, the national data do not allow a sectoral disaggregation that matches the one we used for other countries.

Table 3.1. *OECD Manufacturing energy use and value-added in 1988*

	Energy use		Value-added	
	exajoules	%	bn 1980 $	%
United States	13.0	54	767	48
Japan	4.9	20	375	24
Europe-6	6.2	26	445	28
West Germany	2.2	9	181	11
United Kingdom	1.8	7	120	8
France	1.4	6	110	7
Sweden	0.5	2	18	1
Norway	0.3	1	6	0.4
Denmark	0.1	0.4	8	0.5
OECD-8	24.2	100	1587	100

available for some countries, studies have shown that the level of aggregation that we use captures the major effects of changes in energy intensity and structure on manufacturing energy use (Boyd et al., 1987; Jochem, 1990).

The energy data used in our analysis reflect all total fuels and electricity consumed for heat and power, including biomass used in the paper and pulp sector wherever possible. We subtracted chemical industry feedstocks from energy use statistics, since these are properly viewed as material rather than energy inputs. Energy is counted in terms of its thermal equivalent at the point of utilization (final energy). An examination of manufacturing energy use trends in the United States found that similar results are obtained for indicators based on final energy, primary energy, and price-weighted measures of energy use (Howarth, 1991).

In the sections below, we report data for the OECD-8 aggregate, as well as for the United States, Japan, and a European aggregate of six countries ("Europe-6"). Within the OECD-8 aggregate, the United States accounted for 54% of total manufacturing energy use in 1988, while Japan and Europe-6 accounted for 20% and 26%, respectively (Table 3.1). Changes in the United States obviously have a large effect on the OECD-8 aggregate. Within the Europe-6 aggregate, the Scandinavian countries comprise only 15% of energy use.

Table 3.2. *Changes in OECD manufacturing energy use, value added, and energy intensity, 1973–1988 (total % change)*

	Energy use	Value added	Aggregate intensity
United States	−13	52	−43
Japan	−10	64	−45
Europe-6	−25	14	−34
West Germany	−20	18	−32
United Kingdom	−36	2	−37
France	−25	18	−36
Norway	0	6	−5
Sweden	−11	20	−26
Denmark	−20	25	−36
OECD-8	−16	41	−40

3.2.1 Overview of trends

Manufacturing value-added in the OECD-8 rose at an average rate of 2.3% per year between 1973 and 1988.[3] Despite the growth in activity, energy use fell by 1.2% per year, although not smoothly over time. Energy use fell during the recessions that followed the 1973 and 1979 energy shocks but remained relatively constant during the 1976–79 and 1982–88 periods. Energy use declined by roughly similar proportions between 1973 and 1988 in the United States, Japan, and Europe-6 (10%-25%), but value-added grew much more in the United States and Japan than in Europe. The largest decline in energy use (36%) occurred in the United Kingdom; sectoral production fell between 1973 and 1982 and returned to its 1973 value only in 1988, but aggregate energy intensity declined considerably. The only country with no decline was Norway, where there was also little growth in value-added and also only a small reduction in aggregate energy intensity.[4] Aggregate manufacturing energy intensity in the OECD-8 fell by 3.7% per year, or 40% overall. The decline was greater in the United States and Japan (43% and 45%) than in Europe (34%) (see Table 3.2).

The changes in aggregate energy intensity have been shaped by two

[3] Value-added data were gathered from national sources in real 1980 currency and converted to 1980 U.S. dollars using the purchasing power parities published by the OECD. The use of purchasing power parities for this purpose is superior to the use of market exchange rates, which are affected by international monetary imbalances and are often poor measures of the relative purchasing power of different currencies.

[4] In Norway and the United Kingdom, the development of offshore oil resources in the 1970s drew capital away from manufacturing and thus restrained sectoral growth.

Table 3.3. *Decomposition of change in aggregate manufacturing energy intensity in OECD countries, 1973–1988 (total % change)*

	Change in aggregate intensity	Effect of:		
		Structure	Intensities	Interaction[a]
United States	−43	−13	−32	2
Japan	−45	−16	−34	5
Europe-6	−34	−5	−30	1
West Germany	−32	−12	−23	3
United Kingdom	−37	−3	−35	1
France	−36	2	−37	−1
Norway	−5	26	−20	−11
Sweden	−26	2	−28	0
Denmark	−36	−4	−29	−3
OECD-8	−40	−11	−32	3

[a] Because the structural and intensity variables interact in a nonlinear fashion, the two effects do not sum to the total change in aggregate intensity.

broad forces: (1) structural change within the manufacturing sector, and (2) change in the energy intensities of specific subsectors. We analyzed the relative contributions of these forces using the method described in Chapter 1; for further description of the methodology, see the Appendix of this chapter. To gauge the effects of structural change, we estimated the trends in energy use that would have developed if the energy intensities of each sector had been constant at their 1973 values. Structural change is defined as change in the output shares of the six sectors listed above. Structural change within these sectors is not considered in these calculations. The effect of change in energy intensities is calculated in a similar fashion, holding sectoral structure constant in its 1973 configuration.

The results show that structural change – decline in the output share of energy-intensive industries – accounted for about one-fourth of the decrease in aggregate intensity for the OECD-8 (Table 3.3). It played a more important role in the United States and Japan than in Europe-6, which largely explains why the decline in aggregate intensity was less in Europe-6 than in the United States and Japan. The only one of the eight countries in which structural change pushed significantly upwards on aggregate intensity was Norway.

Decline in subsectoral energy intensities was the major force driving aggregate intensity. For the OECD-8, declining energy intensities reduced energy use by 32% between 1973 and 1988. Expressed differently, the

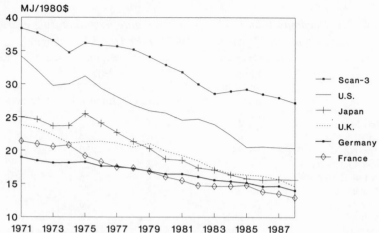

Fig. 3.3. OECD manufacturing energy intensities: constant 1973 industry structure

"structure-adjusted" energy intensity declined by 32%. The effect in the three regions was rather similar: 32% in the United States, 34% in Japan, and 30% in Europe-6. Within Europe-6, large declines in "structure-adjusted" energy intensity in the United Kingdom and France were partly balanced by smaller decline in West Germany and Scandinavia. In several countries, there was an increase in "structure-adjusted" intensity in the recession year of 1975. Decline was relatively smooth thereafter, with a plateau evident in the United States and Japan after 1985 (Fig. 3.3).[5]

The trends in *energy intensity* mask very different trends for *electricity intensity* and *fuel intensity*. Structure-adjusted fuel intensity declined considerably in all countries (Fig. 3.4). Electricity intensity declined in the United States and Japan, but much less than fuel intensity, and was fairly steady in West Germany and France, with some increase since 1984 in the latter (Fig. 3.5). Since there was growth in industrial electricity cogeneration in the study period, and the energy used in cogeneration is counted as fuel use, the true increase in electricity intensity was somewhat higher than is the case if one counts only purchased electricity (as we have done). Still, the trends in electricity intensity suggest that there has been efficiency improvement in electrical processes at the same time that reliance on them has grown. In addition, application of electronic control and automation has enhanced the productivity of many processes, thereby contributing to growth in value-added.

[5] The 1985 and 1988 values for the United States are not entirely comparable, as the survey frame on which the data are based was enlarged in 1988. This change means that the 1988 value may be too high relative to the earlier series.

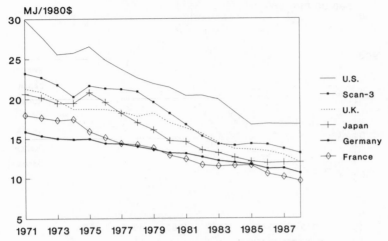

Fig. 3.4. OECD manufacturing fuel intensities: constant 1973 industry structure

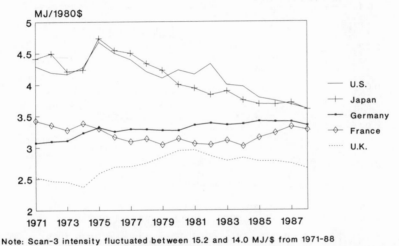

Note: Scan-3 intensity fluctuated between 15.2 and 14.0 MJ/$ from 1971-88

Fig. 3.5. OECD manufacturing electricity intensities: constant 1973 industry structure

3.2.2 Change in energy prices

The decline in fuel intensities was partially a result of increase in fuel prices, particularly oil prices. While an oil price level of around 75 1980$/ton prevailed in most nations in 1973, differences in government policies led to a significant divergence in later years (Fig. 3.6). In the United States, prices peaked in 1981 at about 270 1980$/mt. In Sweden, on the other hand, heavy taxation caused the price to rise continuously to around 360

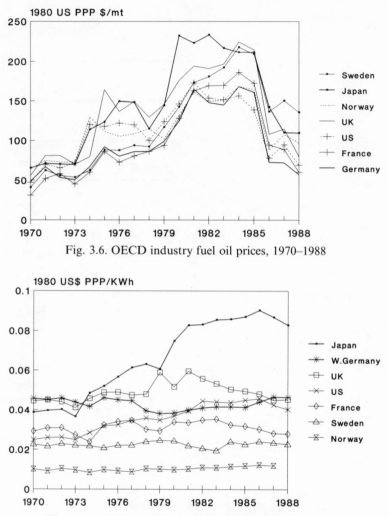

Fig. 3.6. OECD industry fuel oil prices, 1970–1988

Fig. 3.7. OECD industry fuel electricity prices, 1970–1988

1980$/mt in 1985. With these exceptions, however, the trend was broadly similar among countries. Not surprisingly, these price increases led manufacturers to reduce their reliance on oil in favor of other fuels and, in some cases, electricity. A substantial decline in oil prices occurred in 1986, largely offsetting the earlier price rises. While this development did not lead to increasing oil intensity, the declining trend in oil intensity came to a halt.

While industrial electricity prices remained stable in most European nations over the period, sharp increases occurred in Japan and the United States (Fig. 3.7). This may partly explain why electricity intensity, adjusted

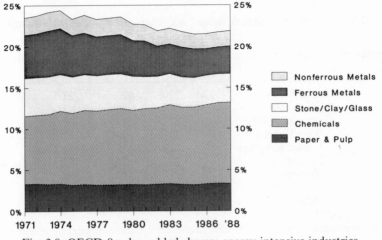

Fig. 3.8. OECD-8 value-added shares: energy-intensive industries

for structural change, increased in Europe-6 but decreased in the United States and Japan. Substantial price differences have existed among countries, driven mainly by differences in the forms of energy used to generate electricity. Electricity prices in Norway are far lower than in other countries due to the availability of low-price hydropower. This situation led to strong growth in electricity-intensive industries and retarded progress towards higher energy efficiency.

Gas prices tracked oil prices where this fuel was widely available. Because gas has long been relatively inexpensive in the United States, its share of total energy use is high in comparison with other nations. The disparity, however, has narrowed significantly in recent years. Coal has also been relatively inexpensive in the United States. While the price of coal increased in most countries during the 1970s, it declined substantially after 1982.

Disentangling the role of price changes in reducing intensities relative to other factors is difficult, especially since "autonomous" technological change has been so important.[6] We discuss this issue later in this chapter.

3.2.3 Structural change in manufacturing production

For the OECD-8, the share of output generated in the five energy-intensive industries fell from 24% to 22% between 1973 and 1988 (Fig. 3.8). Particularly important was the decline in the share of ferrous metals – the

---

[6] By "autonomous" technological change we mean changes in production practices not caused directly by price-induced factor substitution; see Hogan and Jorgenson (1991).

Table 3.4. *Energy intensive industries' share of manufacturing output*
(% of value-added)

	1971	1979	1988
United States	23	23	21
Japan	28	27	23
Europe-6	22	23	22
West Germany	22	23	21
United Kingdom	22	23	24
France	21	22	23
Norway	23	26	30
Sweden	22	23	24
Denmark	15	16	17
OECD-8	24	24	22

most energy-intensive sector – from 5.6% to 3.3%. The one energy-intensive sector whose share grew (slightly) was chemicals – from 8.5% to 9.9%. However, the bulk of the growth was in relatively less energy-intensive finished chemical products rather than in basic industrial chemicals such as ammonia and chlorine.

There was little change in the share of energy-intensive industries in Europe-6, but significant decline in Japan and to a lesser extent in the US (Table 3.4). The share of energy-intensive industries is remarkably similar among the studied countries, with Norway and Denmark standing out somewhat. The case of Norway shows that low energy prices can favor the growth of energy-intensive industries. The hydro resources there provide electricity to many energy-intensive industries for less than \$0.01/kWh and have greatly encouraged the development of electricity-intensive industries such as aluminum smelting.

The observed shift away from energy-intensive industries is due in part to the long-term decline in the physical consumption of many basic materials relative to GDP (Williams et al., 1987). Traditional materials such as cement and steel have lost share to plastics and modern ceramics and composite materials, which often are lighter and less bulky than the materials they replace. There is also evidence that technological change has led to reductions in the proportion of materials wasted in the manufacturing process and in the quantity of materials embodied in finished products.

While the long-term decline in the role of steel and cement was a key force in manufacturing structural change, short-term trends also played a

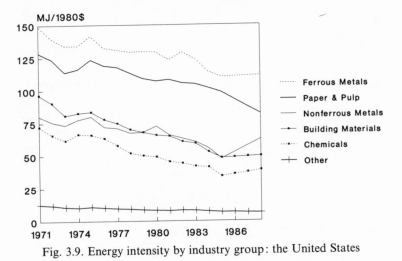

Fig. 3.9. Energy intensity by industry group: the United States

role. In the United States, for example, the value added of the ferrous metals sector fell by 37% just between 1981 and 1982, and was slow to recover from the recession. Since the period examined in this study witnessed significant recessionary periods, the structural change that occurred over the period may be a somewhat biased reflection of long-term trends.

3.2.4 Change in energy intensities

As shown in Fig. 3.9 through 3.11, significant reductions in energy intensity were achieved in all manufacturing subsectors in the United States, Japan, and Europe-6. For the OECD-8 average, the largest reduction was in chemicals (37%). The declines in other energy-intensive sectors were 27% in paper and pulp; 32% in stone, clay and glass; 27% in ferrous metals; and 26% in nonferrous metals. Interestingly, the decline in non-energy-intensive industries (37%) was comparable to that in the energy-intensive sectors. This development reflects the trend toward higher value per unit of physical output as light manufacturing shifted toward "high-tech" products.

We have not performed a formal analysis of the factors that contribute to the above intensity declines. Structural change *within* the sectors probably contributed to decline in energy intensities. That is, within these sectors there was some shift toward products that intrinsically require less energy to produce per unit of value-added. It appears that such a shift played a strong role in the chemicals industry; this could explain in part why the decline in chemicals was so much larger than in other sub-sectors.

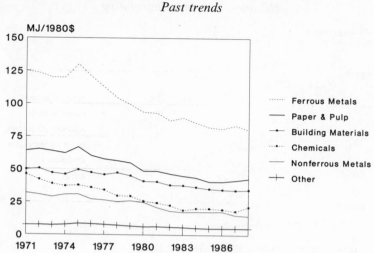

Fig. 3.10. Energy intensity by industry group: Japan

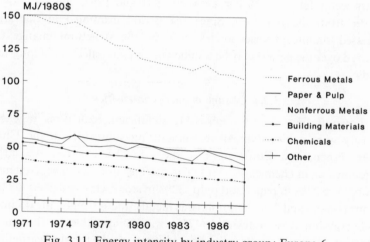

Fig. 3.11. Energy intensity by industry group: Europe-6

Structural change within sub-sectors is caused by change in the magnitude of production of different goods, as well as in the per-unit value of each type of good. For example, if the value of the average automobile produced increases, the share of the automotive industry in the value-added of the "metal products" subsector could increase even if the number of automobiles produced remained the same. In the advanced industrial countries, such increase in value per unit has played an important role in many industries.

While structural change within the sub-sectors played a role, it seems reasonable to conclude that most of the decline in intensities was due to reduction in the energy intensity of producing particular products. Three main factors contributed to such reduction: (1) improvements in operations and maintenance, and retrofits with low-cost equipment; (2) changes in process equipment or "add-on" energy technologies requiring significant investment; and (3) introduction of new production processes, often involving construction of a new facility. Shutting down older facilities, which generally are more energy-intensive than newer ones, was also a factor.

Decisions in each of the above areas are the result of many factors. The first area, which one might call "energy management," is motivated mainly by the desire to reduce energy costs, and is often a response to recent price changes (though incremental improvements are likely to be made over time once an energy management system has been institutionalized). "Add-on" energy technologies requiring significant investment are also motivated by the desire to reduce energy costs, though here expectations of future energy prices play an important role in the investment decision. Changes in process equipment and introduction of new production processes are primarily motivated by a desire to improve productivity generally; reduction in energy intensity is only one among several desired results. Introduction of new processes obviously depends on the cost-effectiveness of new techniques, which is related to past investment in R&D, capital availability and cost, and managers' assessment of the prospects for profitability in a particular industry. The rate of introduction is usually greater in industries that are growing than in those that are stagnant.

Clearly, manufacturing energy intensities are shaped by long-term technological trends that are only partially motivated by the desire to save energy, but analysis of this issue is difficult, and to our knowledge has not been adequately done. To shed some light on the question, we assembled data going back to 1960 or earlier for the United States and West Germany. The analysis shows that aggregate manufacturing energy intensity – holding sectoral structure fixed – fell at about the same average rate between 1960 and 1973 as between 1973 and 1988, despite there being almost no change in energy prices in the earlier period and major increase in prices in the latter. One might be tempted to conclude that the rise in prices had no effect at all, and that "autonomous" technological change alone was the cause of intensity decline. However, value-added grew substantially faster in the earlier period (5.4% in the United States and

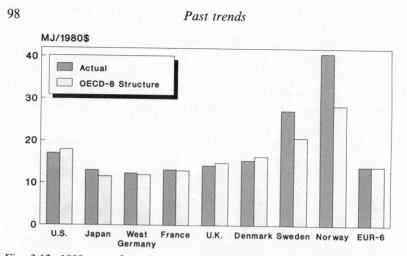

Fig. 3.12. 1988 manufacturing energy intensity: actual and adjusted to reflect OECD-8 industry structure

4.9% in West Germany) than the latter (2.6% in the United States and 1.0% in West Germany). Since a principal source of intensity reduction is investment in new capital facilities, we would expect there to be more intensity decline from this source in the earlier period. Other causes of intensity decline in the 1958–73 period include increase in the average scale of operations in some industries and improvement in efficiency related to switching from coal to oil and natural gas. Thus, the increase in energy prices obviously had an effect in the 1973–88 period, but it appears to have been smaller than the autonomous effect of technological change.

3.2.5 Energy intensity comparisons

To compare manufacturing energy intensity among countries, we calculated the aggregate energy intensity that would have resulted in each country in 1988 given its own subsectoral energy intensities, but aggregate OECD-8 sectoral structure. As shown in Fig. 3.12, this adjustment has only a small effect for most countries, but has a significant one for Norway and Sweden, which have relatively energy-intensive manufacturing structure. The structure-adjusted energy intensity of US manufacturing exceeds the Europe-6 and Japanese levels by 28% and 55% respectively. US energy intensities are equal to or higher than those in Europe-6 in all five of the energy-intensive industries, while Japanese intensities are lower than those of Europe-6 in all sectors. The low levels in Japan are due in part to the relative newness of Japanese factories and the accompanying incorporation of more advanced processes.

The above findings are suggestive of differences in technical energy efficiency, but structural differences among countries within the sub-sectors are not captured at this level of aggregation. Such differences are a reason for the relatively high energy intensity of Norway and Sweden. In the paper and pulp sector, Norwegian and Swedish energy intensities are far above the OECD-8 average because the share of raw pulp production, which is very energy-intensive, in total value-added of the sector is an order of magnitude higher than in the other countries. In the nonferrous metals sector, Norwegian and Swedish energy intensities are 250% and 100% above the OECD-8 average; the ratio of highly energy-intensive aluminum production to total sectoral value-added is over ten times the OECD-8 average in Norway and two times the average in Sweden. Thus, the high energy intensities observed in Norway and Sweden are not necessarily indicative of technical inefficiency in energy use.

The above examples underscore the danger of using energy per value-added to compare the technical energy efficiency of industries in different countries. An alternative approach is to look at energy use per unit of physical output (usually tonnes), but this is only possible in relatively homogeneous and narrowly defined industries. Even in such industries, however, problems arise in making comparisons, and careful analysis is required to understand why energy use per ton differs among countries. In the paper and pulp sector, it is problematic to add up products as different as pulp, newsprint, and corrogated boxes on a comparable physical basis. In iron and steel, summing tonnes of cast iron and tonnes of steel to obtain a single indicator of output is likewise questionable; there are also differences among steel products. A study of steel industries in 1980 found moderate international variation in energy intensity, as shown in Table 3.5, but it is not clear how much of this was due to differences in the mix of products. Steel industry energy intensity is also affected by the mix of inputs, since virgin ore requires more energy to process than scrap. For example, the Chinese industry is often cited as being very energy-intensive. One reason, however, is that there is a great deal of commerical pig iron flowing out and relatively little external scrap flowing in (Ross & Liu, 1991).

Comparison problems are less severe in certain industries. Vallance (1990) found a narrow range of physical energy intensities (4030–4244 MJ/tonne in 1987) in the cement sectors of France, West Germany, Switzerland, and Japan. Similarly, Lester (1987) found that the energy intensity of primary aluminum smelting in 1984 was roughly comparable

Table 3.5. *Steel production:*
1980 energy intensity

	GJ/tonne
Italy	17.6*
Spain	18.4
Japan	18.8
West Germany	21.8
United Kingdom	23.4
France	23.8
United States	25.9*

* Data for 1979.
Source: Meunier & de Bruyn Kops
(1984).

across world regions, ranging from 15.5 kWh/kg in East Asia to 17.3 kWh/kg in North America.

In some cases, one can compare changes in physical energy intensity with those in energy use per value-added, although the cautions about international comparison apply to some extent in trend analysis within a given country. A study of industrial energy trends in Japan between 1973 and the late 1980s found declines in physical energy intensity averaging 1.9%/yr in steel production, 2.7%/yr in cement, and 4.0%/yr in paper and pulp (Fujime, 1989). The corresponding declines in the energy/value-added ratios of the iron and steel; stone, clay, and glass; and paper and pulp sectors were 2.6, 2.3, and 3.1%/year respectively. The reasonably close correspondence between the two sets of figures suggest that the energy/value-added approach gives a good approximation to changes in underlying physical processes.

In the United States, physical energy intensity fell at a reported 3.0%/yr between 1972 and 1988 in the paper and pulp sector (American Paper Institute, 1989); the decline in the energy/value-added ratio was 2.6%/yr. In the steel industry, physical energy intensity fell at 3.0%/yr between 1972 and 1989 (Steiner, 1990), while the energy/value-added ratio dropped by 1.4%/yr.

3.2.6 Change in energy sources

Significant shifts in the mix of energy sources have occurred in manufacturing since the early 1970s. The key factor has been decline in the share of oil in response to higher prices. For the OECD-8, the share of oil in total final energy use fell from 31% in 1973 to 16% in 1988 (Fig. 3.13). The share

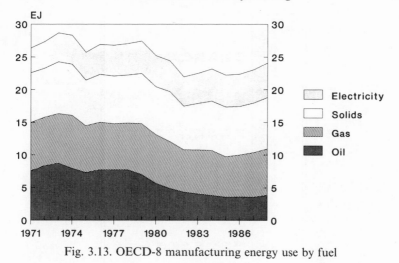

Fig. 3.13. OECD-8 manufacturing energy use by fuel

of solid fuels (mainly coal) grew from 26% to 33%, while the share of electricity increased substantially from 15% to 22%.

The decline in oil's share was most marked in Europe-6: from 44% to 21% (Table 3.6). The decline was less in Japan, in part because natural gas was not easily available to displace oil. In the United States, the share of oil was already relatively low in 1973 (15%), and fell further by 1988. The increase in the share of electricity was common across countries.

3.3 Former East Bloc

While the 1970s and 1980s were marked by considerable change in the manufacturing sector in the OECD countries, manufacturing in the Former East Bloc, relatively isolated from competition and administered by government policy, saw growth in production but relatively little technological change. Directed by national production targets, manufacturing continued to be dominated by the heavy industries that were established in the 1950s and 1960s. Relative to the level of GDP, production of steel, cement, ammonia, and nitrogenous fertilizers is high by international standards, while per capita production of paper and plastics is low.

In the Soviet Union, around 70% of manufacturing energy is used by industries that produce basic materials, a figure that remained relatively constant during the 1970s and 1980s.[7] Propelled by industrial policies that

---

[7] Information on the Soviet Union is based on the study described in Schipper and Cooper (1991).

Table 3.6. *OECD final energy use by type (%)*

	1973				1988			
	Oil	Gas	Solids	Electricity	Oil	Gas	Solids	Electricity
United States[a]	15	41	30	14	6	41	34	19
Japan	51	2	30	18	35	3	37	25
Europe-6	44	18	22	16	21	26	26	27
West Germany	40	24	17	17	16	32	23	27
United Kingdom	46	17	26	11	25	31	25	19
France	50	15	19	16	25	24	25	26
Norway[a]	29	1	19	52	14	1	24	61
Sweden[a]	46	0	30	23	18	0	42	37
Denmark	77	0	11	10	41	14	21	23
OECD-8	31	26	28	15	16	29	33	22

[a] Includes significant quantities of biomass in the paper and pulp sectors.

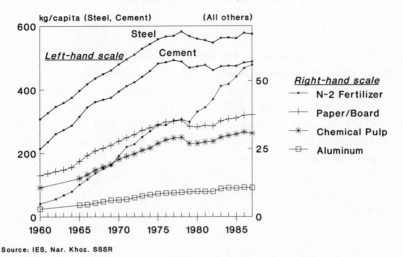

Source: IES, Nar. Khoz. SSSR

Fig. 3.14. Soviet Union: production of energy-intensive materials

favored the expansion of such industries, growth in manufacturing energy use averaged 4.3%/year between 1960 and 1973. Between 1973 and 1987, however, the growth rate fell to 2.0%/year as many industries became somewhat stagnant. Assessing the structural evolution of Soviet manufacturing is difficult since disaggregated data on output value are not available, but trends in per capita production suggest that there has been some decline in the role of the energy-intensive steel and cement industries since the mid-1970s (Fig. 3.14). Although per capita steel production has levelled off, steel production per unit GDP is still much higher than in the OECD countries; this is also the case for Eastern Europe (Dobozi, 1990). An incentive system that rewards quantity of production rather than quality is one reason why production is so high. Many steel elements are produced with extra bulk simply to assure the correct strength.

The lack of adequate indicators of sectoral activity precludes measurement of change in aggregate Soviet manufacturing energy intensity. For the steel, cement, pulp, and paper industries, however, data on energy use and physical production indicate that no significant reductions in intensity occurred between 1975 and 1985. Given the maintenance of low energy prices and lack of incentive for technological innovation, this is not surprising. There was some intensity decline in ammonia production due to addition of new facilities in the early 1970s.

3.4 Developing countries

The nature and evolution of the manufacturing sector has varied considerably among LDCs. Countries such as South Korea and Taiwan have developed industries that compete effectively in world markets. Others have relatively old industries that have seen relatively little capital investment in the past decade. Many small countries have remained primarily agrarian societies with modest manufacturing infrastructure.

Manufacturing output and energy use in the LDCs as a group is dominated by China, India, and Brazil. China alone accounts for nearly half of total LDC manufacturing energy use. Because of its importance, and because analysis of trends has been done, we discuss China separately.

3.4.1 China

The manufacturing sector in China, as in the Soviet Union, has been shaped by national policies that stressed development of heavy industries as the pathway to prosperity. Since 1970, however, the share of heavy industries in total industrial output has declined somewhat as industries devoted to export and consumer products have grown significantly. Total industrial output (gross output value) grew very rapidly between 1980 and 1988, averaging nearly 14% per year.

Data on industrial energy use are available only since 1980. Aggregate industrial energy intensity (including mining and energy industries) declined by a remarkable 45% (7.5% per year) between 1980 and 1988. Between 1980 and 1985, the years for which data allow analysis of the aggregate trend, intensity declined by 27%. In this period, decrease in the share of energy-intensive industries in total output (from 35% to 31%) contributed to the decline in intensity, but intensity reductions within particular industries accounted for around three-fourths of the total decrease (Liu et al., 1991). A key factor was a massive government program to improve energy efficiency. Energy conservation was made a high priority, and large amounts of capital were made available for energy-saving investments (Levine & Liu, 1990). As shown in Fig. 3.15, there was decline in all sub-sectors except food. Energy intensity actually declined more in non-energy-intensive industries (30%) than in energy-intensive ones (18%). This was likely due in part to the faster growth (and addition of newer factories) of non-energy-intensive industries, and perhaps to change in their product mix as well.

Despite the large decline in energy intensity, much of Chinese manu-

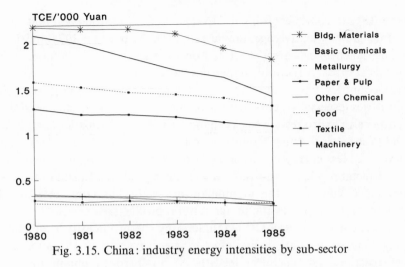

Fig. 3.15. China: industry energy intensities by sub-sector

facturing remains inefficient by international standards and still relies largely on outdated technology from the 1950s and 1960s. Steel production in China, for example, is 35% more energy-intensive than in the United States, and almost twice as energy-intensive as in Japan, once the data are adjusted for comparability (Ross and Liu 1991). Considerable differences also exist between Chinese and typical OECD energy intensity for other manufacturing products.

3.4.2 Other developing countries

For the LDCs as a group (not including China), manufacturing value-added averaged growth of 3.9% per year in the 1980s, well below the average of 7.4% registered in the 1970s.[8] Growth in manufacturing output has varied considerably among regions. In the 1980s, it grew at an average annual rate of 4.0% in Western Asia, 5.3% in North Africa, 6.2% in the Indian subcontinent, 7.5% in Southeast Asia, but only 1.6% in Latin America, and only 0.5% in Sub-Saharan Africa. Latin America and Southeast Asia accounted for 41% and 34% of total LDC value-added in 1987, respectively; the share of the Indian subcontinent was 12%.

The structure of manufacturing in the LDCs as a group has become somewhat more energy-intensive since 1975. The share of the five energy-intensive industries (see OECD discussion) in total value-added increased

[8] Data on value-added are from the United Nations Industrial Development Organization (UNIDO 1990). The data are expressed in real prices, converted to U.S. dollars using 1980 exchange rates. China is not included in the UNIDO statistics due to lack of comparable data.

from 18% to 20%, according to UNIDO data. While total manufacturing output grew at an average annual rate of 4.3% between 1975 and 1985, industrial chemicals grew at 5.1%, iron and steel at 5.6%, and nonferrous metals at 6.1%. Structural change has varied among countries. Many oil-rich countries have used their oil resources to build energy-intensive industries (especially petrochemicals), and other countries have used their inexpensive hydro resources to expand basic metals industries (e.g., Brazil and Venezuela). Other countries, especially in Southeast Asia, have developed less energy-intensive industries such as textiles, machinery and other metal products, and various consumer products. In still other, less-developed countries, the manufacturing sector has continued to rely heavily on basic processing of agricultural products.

Change in aggregate LDC manufacturing energy intensity is difficult to judge. Time-series data on manufacturing energy use (i.e., without mining, construction, and chemical feedstocks) are hard to obtain for most developing countries, as are data disaggregated into industry groups. On the whole, structural change probably contributed to increase in energy intensity, but there are indications that the energy intensities of particular industries declined in some countries due to technological change and, to a lesser extent, energy conservation activities. Such decline has probably been greater in those countries where industries face relatively unsubsidized energy prices and international competition, and where substantial investment in new facilities has occurred.

These features apply more in Southeast Asia than in other regions, and there is clear evidence that the rapid growth in manufacturing contributed to decline in energy intensity in that region. In Taiwan, output growth averaged 7.9%/year between 1978 and 1987, while aggregate energy intensity declined by 5.1%/year. Electricity intensity also declined (2.6%/year). The share of output in the production of machinery, electronic goods, transportation equipment, and other light goods rose, but the share of energy-intensive industries remained at about 35%. Thus, the net impact of structural change on manufacturing energy intensity was small (Li et al., 1990). The change in aggregate intensity was caused mainly by reduced intensities at the industry group level due to improved energy efficiency in older factories and introduction of new, more energy-efficient factories. Intensity decline was especially significant in basic metals production, where structural change – decline in aluminum production – played a role.

Rapid growth in output and substantial decline in aggregate energy intensity also occurred in South Korea. Between 1975 and 1987, sectoral output increased by over 11%/year, and aggregate energy intensity fell by 2.6%/year. While there was some decrease in the share of energy-intensive industries, an analysis found that structural change accounted for only 15% of the total reduction in aggregate energy intensity (Korea Energy Economics Institute, 1989). The decline in aggregate energy intensity, holding structure constant, averaged 2.1%/year. As in Taiwan, the pressure of competition spurred capital investment in modern industrial technologies. Indeed, Korea's steel industry is reportedly as energy-efficient as any in the world.

The trend has been quite different in Brazil, where aggregate energy intensity increased at an average rate of 0.9%/year between 1973 and 1988 (Geller & Zylbersztajn, 1991). The growth rate of sectoral output (3.3%/year) was much slower than in Taiwan and Korea, so there was less introduction of new factories and equipment. In addition, the availability of cheap hydroelectricity favored the development of energy-intensive industries in Brazil. Production of electricity-intensive products such as steel and aluminum (partly for export) increased by 245% and 450% respectively. While structural change contributed to increase in aggregate energy intensity, there was decline in energy intensity in particular industries. The energy intensity of the iron and steel, paper and pulp, nonferrous metals, and cement industries declined at annual rates of 0.5, 1.3, 2.1, and 2.3% respectively. In the chemicals sector, on the other hand, energy intensity increased by 35%, apparently because of intra-sectoral shifts towards more energy-intensive products.

3.5 Conclusion

Decrease in the relative importance of energy-intensive industries within the manufacturing sector contributed to a decline in energy use in the OECD countries. Much more important, however, were decreases in the energy intensities of the various industry groups that comprise the sector. For eight OECD countries that we analyzed in detail, the decline in average intensity in 1973–88 ranged from 27% in paper & pulp to 39% in chemicals. As a result of structural change and decline in intensities, total OECD-8 manufacturing energy use fell despite a 41% increase in output. These conclusions remain if we measure primary rather than final energy use.

In the Former East Bloc, there seems to have been relatively little structural change or change in intensities. In the Soviet Union, sluggish output growth, the continued dominance of energy-intensive industries, and lack of incentives to reduce energy intensity led to increased energy use. For the LDCs as a group (excluding China), there has been some increase in the share of energy-intensive industries, as would be expected. In China, on the other hand, less energy-intensive industries have grown in share of total manufacturing output. There is evidence that energy intensities have declined in a number of LDCs. In China, a major government program played an important role in reducing energy intensities.

Appendix: Quantifying impacts of changing activity, structure, and energy intensities

The approach we use to quantitatively describe the relative impacts of change in activity, structure, and energy intensity on aggregate sectoral energy use is rooted in the use of fixed-weight or Laspeyres indices.

Consider the evolution of the energy use of a country's manufacturing sector over time. Sectoral activity (A) is defined as total manufacturing output, measured in terms of real value-added. Structural change refers to changes in the share of value-added generated by each industry group $S_i = A_i/A$ where A_i is the output of the ith subsector. I_i is the energy intensity, or energy use per unit of output, of the ith subsector. Under these definitions, manufacturing energy use (E) may be written in the form:

$$E = A\sum_i S_i I_i.$$

It is clear from this identity that the impacts on E of changes in activity, structure, and intensity cannot uniquely be disaggregated in a linear fashion. We may, however, address the following question: If only activity, only structure, or only intensity had changed over time while the other two factors remained fixed, how would energy use have changed over time?

If the subscripts 0 and t refer to the values assumed by variables in the base and current periods, the following Laspeyres indices may be defined to answer these questions:

$$LA_t = \frac{A_0 \sum_i S_{i0} I_{i0}}{E_0}$$

$$LS_t = \frac{A_0 \sum_i S_{it} I_{i0}}{E_0}$$

$$LI_t = \frac{A_0 \sum_i S_{i0} I_{it}}{E_0}$$

LA measures the change in energy use that would have occurred given actual changes in output but fixed structure and intensity; *LS* measures the change that would have occurred due to structural change if output and intensity had remained constant; and *LI* captures the development of energy use given actual energy intensities with fixed output and structure. The changes in these indices generally do not sum to the actual change in energy use, because of the interactions between the various effects. These interaction terms are generally small.

We construct similar indices for each of the other sectors, except for the services sector, for which the data are insufficient. In other work, we used both the Laspeyres indices and Divisia indices to measure change and found similar results with each (Howarth et al., 1991).

References

American Paper Institute. 1989. *Annual Energy Report – U.S. Pulp, Paper, and Paperboard Industry*. Mimeo, New York, NY.

Boyd, G., McDonald, J.F., Ross, M. & Hanson, D.A. 1987. Separating the changing composition of U.S. manufacturing production from energy efficiency improvements: a Divisia index approach. *Energy Journal*, **8**, 77–96.

Dobozi, I. 1990. The centrally planned economies: extravagant consumers. In Tilton, J. (ed.), World Metal Demand: Trends and Prospects. Resources for the Future: Washington, DC.

Fujime, K. 1989. Energy conservation in the industrial sector in Japan – historical trends and outlook. *Energy in Japan*, **95**; 21–31.

Geller, H.S. & Zylbersztajn, D. 1991. Energy intensity trends in Brazil. *Annual Review of Energy*, Vol. **16**.

Hogan, W.W. & Jorgenson, D.W. 1991. Productivity trends and the cost of reducing CO_2 emissions. *Energy Journal*, **12**; 67–85.

Howarth, R.B. 1991. Energy use in U.S. manufacturing: The impacts of the energy shocks on sectoral output, industry structure, and energy intensity. *Journal of Energy and Development*, **14**; 175–91.

Howarth, R.B., Schipper, L., Duerr, P.A. & Strøm, S. 1991. Manufacturing energy use in eight OECD countries: Decomposing the impacts of changes in output, industry structure, and energy intensity. *Energy Economics*, **13**, 135–42.

Jochem, E. (ed.). 1990. *Systematische Analyse der Komponentenzur Energieintensitat und Effizienz in der Bundesrepublik Deutschland* 1970 *bis* 1987. Forschungsvorhaben im Auftrag des Bundesministeriums fur Forschung und Technologie. Karlsruhe, Germany: Fraunhofer Institute.

Korea Energy Economics Institute. 1989. *Sectoral Energy Demand in the Republic of Korea: Analysis and Outlook*. Seoul, South Korea.

Lester, M.D. 1987. Energy consumption in industrial processes: Aluminum. In *Energy Consumption in Industrial Processes: Aluminum, Cement, Glass, Pulp, Paper, Steel, Sugar*. London, UK: World Energy Conference.

Levine, M. & Liu, X. 1990. *Energy Conservation Programs in the People's Republic of China*. Berkeley, CA: Lawrence Berkeley Laboratory Report LBL-29211.

Li, J. W., Shrestha, R.M. & Foell, W.K. 1990. Structural change and energy use, *Energy Economics*, **12**; 109–15.

Liu, F., Davis, B. & Levine, M. 1991. *Energy in China*. Berkeley, CA: Lawrence Berkeley Laboratory draft report.

Meunier, M.Y. & de Bruyn Kops, O. 1984. *Energy Efficiency in the Steel Industry with Emphasis on Developing Countries*. Technical Paper Number 22. Washington, DC: World Bank.

Ross, M. & Liu, F. 1991. The energy efficiency of the steel industry in China. *Energy*, **16** (5); 843–8.

Schipper, L. & Cooper, C. 1991. *Energy Use and Conservation in the USSR: Patterns, Prospects, and Problems*, Berkeley, CA: Lawrence Berkeley Laboratory Report LBL-29831.

Steiner, B. 1990. American Iron and Steel Institute, personal communication.

UNIDO (United Nations Industrial Development Organization). 1990. *Industry and Development: Global Report* 1990/91. Vienna, Austria: United Nations.

Vallance, B. 1990. *Cross Time and Cross Country Comparisons of Energy Efficiency in Industrial Sectors using a Concept of Technical Efficiency – Case Studies of the Aluminum and Cement Industries*. Geneva, Switzerland: Centre universitaire des problèmes de l'energie, Université de Genève.

Williams, R., Larson, E. & Ross, M. 1987. Materials, affluence, and industrial energy use. *Annual Review of Energy*, **12**, 99–144.

4

Historic trends in transportation

Personal mobility and timely movement of goods have become increasingly important around the world, and energy use for transportation has grown rapidly as a consequence. Around 64% of total world transportation energy use in 1988 was accounted for by the OECD countries, which reflects the high level of automobile ownership and use. The shares of the Former East Bloc and the LDCs were only 14% and 22%, respectively. Since 1970, however, growth has been much faster in the LDCs (5.1% per year) than in the OECD countries (2.4%) and the Former East Bloc (2.0%) (Fig. 4.1).

Energy is used in transportation for two rather different activities: moving people, which we refer to as passenger travel, and moving freight. While freight transport is closely connected to economic activity, much of travel is conducted for personal reasons. In the OECD countries, travel accounts for around 70% of total transportation energy use. In contrast, freight transport accounts for the larger share in the Former East Bloc and the LDCs. Because of the different natures of travel and freight transport, we discuss and analyze them separately.

4.1 Analyzing transportation energy use

For eight OECD countries and the Soviet Union, we have divided energy use as it appears in most statistics (disaggregated into road, rail, air, water) into passenger travel and freight transport components. This was accomplished by careful bottom-up analyses of each country.[1] We primarily

---

[1] It is incorrect to assign all gasoline consumption in transportation to automobiles, or to assume that all diesel fuel is used by trucks. Instead, the consumption of each fuel by each mode can be estimated "from the bottom up" using data on vehicles by fuel type and assumptions about their average fuel use.

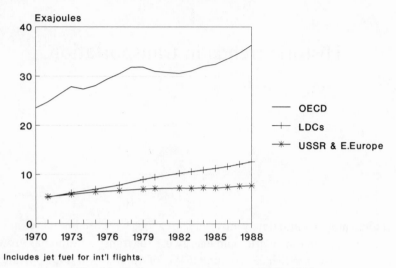

Includes jet fuel for int'l flights.

Fig. 4.1. Transportation energy use, 1970–1988

refer to *domestic* transportation (i.e., within national boundaries), but we briefly cover international air travel, which has grown considerably.

There is not a good measure of overall activity for transportation. While economic output in transport is a standard category in national accounts statistics, this quantity refers only to enterprises engaged in the transport business and does not correspond well to total transportation activity. The very different natures of passenger and freight transport make construction of a single transport indicator impossible.

A common indicator of travel activity is passenger–km (p–km), which measures the total distance that people travel. (A car trip in which two people travel 10 km counts for 20 p–km.) The usual indicator of freight transport activity is tonne–km (t–km), which measures the weight of freight and the distance it is moved. (A truck load of one tonne traveling 100 kilometers counts as 100 t–km). Tonne–kilometers is a problematic indicator because it does not consider the characteristics and value of the freight, merely the weight. As an economy evolves from economic output centered on agricultural and mining products to one in which manufactured goods predominate, the ratio of tonnage to GDP declines because manufactured goods have a higher value per tonne. However, the total distance traveled by freight vehicles may increase relative to GDP.

The statistics reported here refer to transport via motorized modes only. Walking and human- and animal-powered vehicles account for a significant amount of travel and freight transport in LDCs, and even a few

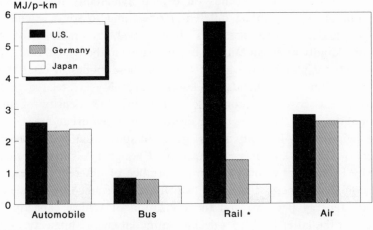

• Electricity counted as primary energy.

Fig. 4.2. Energy intensities by mode of travel in OECD countries, 1988

percent of total distance traveled in Europe. Indeed, the transition from these modes to mechanized ones is analogous to the transition in the residential sector from biomass fuels, which similarly are often uncounted or poorly counted, to fossil fuels.

The modal structure of travel and freight transport is very important because there are considerable differences in energy intensity among modes. Fig. 4.2 illustrates the average 1988 energy use per passenger–km of different travel modes in the United States, West Germany, and Japan. (Electricity use in rail travel is measured in primary energy equivalent.) With the exception of rail in the United States, bus and rail travel had much lower intensity than automobile and air travel.[2] What is perhaps surprising is that the intensity of air travel is only slightly higher than that of automobile travel. This reflects the much higher utilization of vehicle capacity in air travel, and the large share of automobile travel that takes place in urban traffic (automobile energy intensity in long-distance driving is much lower than the average over all types of driving).

Mode choice for trips is conditioned by a number of factors. The extent of the transport infrastructure (roads, rail, waterways, airports) is obviously a critical factor. Most of the transport infrastructure is built and maintained by the public sector. (Most waterways are a natural feature, though they may be modified for transport use.) The nature of a country's

---

[2] The high level of rail energy intensity in the United States is discussed in detail in a later section.

transport infrastructure depends on capital availability for transport, allocation of capital among different modes, and to some extent, the demand for particular modes. It is often argued, for example, that the lesser availability of urban rail and bus transport in the United States is a major factor explaining the lower share of these modes in total travel activity. On the other hand, the lesser availability is related to the relatively low cost of automobile usage, the low density of most metropolitan areas, and the greater distances between urban areas.

We define energy intensity at the aggregate and modal levels as energy use per passenger-km or per tonne-km. Change in aggregate energy intensity is shaped by shifts in the importance of different modes, and change in modal energy intensities, which are determined by vehicle fuel intensity (energy per km) and the utilization of vehicle capacity. An increase in the latter leads to a decline in modal energy intensity. A bus inherently has higher fuel use per kilometer than a car, but a fully loaded bus almost always uses less energy per p–km than a fully-loaded car. The difference in modal energy intensity between cars and buses thus depends in part on their occupancy ratio.

Average vehicle energy intensity is shaped by the characteristics of old and new vehicles, the rate at which new vehicles replace old ones, and by factors that affect in-use energy intensity. The latter include vehicle maintenance, driving habits, and the operating environment. Traffic congestion increases idling time and the amount of distance covered at very low speed, both of which increase fuel consumption per km. Congestion is also a factor for airplanes, as crowded conditions at airports increase time spent circling.

The quality of the transport infrastructure affects both mode choice and energy intensity. Quality factors that affect mode choice include the amount of time required for a trip and the level of amenity. Poor road quality, which is common in the LDCs, leads to higher energy intensity for the vehicles that use them. Where roads are in bad condition, truck operators often choose heavy vehicles that can stand the punishment to which they are subjected. If roads were better, lighter and more fuel-efficient trucks could perform the same tasks.

Fuel prices affect each of the three major factors that drive transport energy use: activity level, modal structure, and energy intensity. People may cut back on discretionary travel (such as long-distance vacations) if prices rise, but voluntary curtailment in activity is less likely for freight transport, the level of which is directly coupled to economic activity in the short run. Relative fuel prices affect mode choice to some degree, though other factors are usually more important. (The impact on mode choice depends on the importance of fuel costs in determining user prices for each mode.) Fuel prices have a large impact on modal energy intensity, as consumers and businesses tend to purchase more efficient vehicles or operate them more efficiently if prices are high or increase significantly.

Passenger travel

People travel for a variety of reasons including work commutes, on-the-job trips, shopping, social visits, and recreation. The number of trips people make in a given period is conditioned by their particular life situation and preferences, their income, the cost of travel, the amount of time available for travel, and the amount of time needed to accomplish various trips. The distance of various trips is affected by the spatial relationship between origin–destination pairs, such as home–work or home–shopping.

The choice of mode for a given trip is shaped by many factors, including purpose of trip, distance of trip, availability and cost of modes, speed of modes (or time available for the trip), quality of the travel experience, personal income, and preference. Some modes compete for certain types of travel, but not for others. Quality is somewhat intangible but important, particularly if income allows a wide choice among modes.

Recall that our focus is on domestic travel (i.e., within national boundaries). In Western Europe, air travel among countries is about 60% greater than total domestic air travel (Boeing, 1991). Domestic air travel in Europe has grown faster than intra-Europe international air travel, however, as travelers increasingly fly on routes for which rail or car were more common in the past.

4.2 OECD countries

Using a variety of national sources, we assembled a data base on travel energy use by mode for eight OECD countries. For details and data sources, see Schipper et al. (1991). Among these eight, the United States accounted for 70% of total travel energy use in 1988 (Table 4.1). (The US share in passenger travel is higher than in any other sector.) In our discussion, we focus on three entities: the United States, Japan, and a European six-country aggregate (Europe-6).

Table 4.1. *Travel energy use and activity in the OECD-8 in 1988*

	Energy use		Activity	
	(exajoules)	%	(bn p–km)	%
United States	14.50	71	5733	60
Japan	1.59	8	1081	11
Europe-6	4.44	22	2682	28
West Germany	1.36	7	659	7
United Kingdom	1.09	5	604	6
France	0.91	4	678	7
Italy	0.80	4	590	6
Sweden	0.19	1	103	1
Norway	0.09	0.4	48	0.5
OECD-8	20.53	100	9495	100

Table 4.2. *Growth in travel energy use, activity, and aggregate energy intensity, 1973–1988 (total % change)*

	Energy use	p–km	Aggregate intensity
United States	13	38	− 18
Japan	76	40	25
Europe-6	55	44	7
West Germany	56	33	17
France	50	46	3
United Kingdom	42	41	0
Italy	85	61	15
Sweden	37	39	− 2
Norway	80	56	15
OECD-8	23	40	− 13

4.2.1 Energy use and activity

Between 1973 and 1988, total energy use for travel grew by 13% in the United States, 55% in Europe-6, and 76% in Japan (Table 4.2). For all countries, growth in the amount of travel has been substantial (30-40%). In the United States, a significant decrease in aggregate energy intensity dampened growth in energy use for travel. In Japan, increased energy intensities contributed significantly to growth in energy use, while in Europe-6 intensity increased only slightly.

In the United States, all of the decline in intensity took place after 1976;

intensity had risen in the 1970-73 period. In Japan, there was an upward trend from 1970 through 1978, but thereafter it was more or less stable. In Europe-6, there was moderate fluctuation between 1972 and 1984, with a modest increase thereafter.

Growth in travel activity It is useful to consider the relationship between travel and GDP. In the United States, the ratio of p–km to GDP declined between the early 1970s and the early 1980s (reaching a low point in 1981), but has remained about the same since 1982 (Fig. 4.3). In Japan, travel has grown more slowly than GDP since the early 1970s, while in Europe the two have increased at about the same rate.

Travel increased in part because population grew. There was a decline in per capita travel after the 1973 oil embargo in the United States and Europe-6 (slightly), but not in Japan (Fig. 4.4). It fell again in the United States during and after the 1979-80 oil price rise, but not in Europe and Japan, where the increase in retail gasoline prices was proportionately smaller than in the United States. Between 1981 and 1988, however, per capita travel in the United States grew considerably.

Per capita travel is about twice as high in the United States as in Europe and Japan. While one might think that the average trip covers a longer distance in the United States, given its geography, it appears that this is not the case.[3] Rather, the annual number of trips per capita in the United States is higher than elsewhere.

Travel is affected by changes in disposable income, since people may take or curtail trips, or take longer or shorter trips, depending on their financial situation. The decline in per capita travel in the United States in the 1979-81 period was due to effects of the recession on disposable income and business travel as well as the influence of higher fuel prices. The change in highway vehicle-km per adult in the United States between 1960 and 1987 closely paralleled real disposable income per capita, corrected by a moderate fuel price elasticity of −0.1 (Ross 1989).

A major source of growth in travel has been leisure and vacation-related travel. Surveys show a major increase in driving for leisure activities in the United States, West Germany, Norway, and Sweden through the 1980s; and declining air fares for non-business travelers have contributed to a growth in air travel for leisure purposes.

[3] Surveys indicate that the average trip in West Germany in 1982 was 17 km vs. only 14 in the United States (both include walking and cycling). The average trip in the United Kingdom in 1985–86 was only 10 km. The average trip by car in the United States is shorter than in the other countries. While the average trip length increased in West Germany between 1976 and 1982, it decreased in the United States between 1977 and 1983.

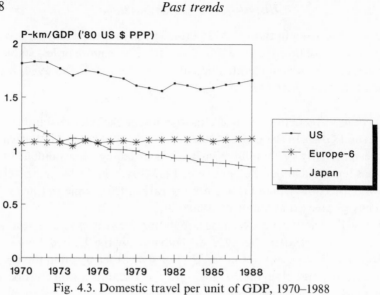

Fig. 4.3. Domestic travel per unit of GDP, 1970–1988

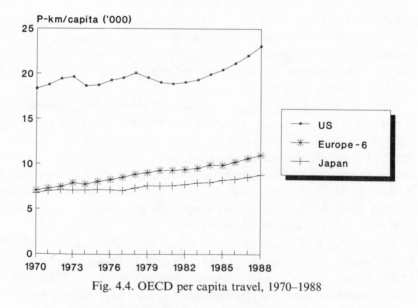

Fig. 4.4. OECD per capita travel, 1970–1988

4.2.2 Change in mode shares

Between 1973 and 1988, the fraction of travel p-km accounted for by automobiles declined from 91% to 86% in the United States, but increased from 43% to 53% in Japan, and from 79% to 82% in Europe-6 (Figs 4.5

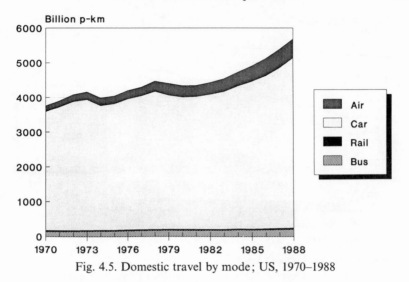

Fig. 4.5. Domestic travel by mode; US, 1970–1988

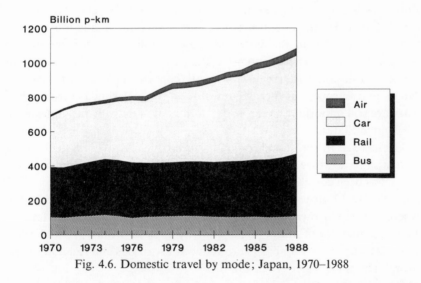

Fig. 4.6. Domestic travel by mode; Japan, 1970–1988

through 4.7).[4] In the United States, there was considerable growth in air travel. An increase in the share of air travel has also been important in Japan and, to a lesser extent, Europe. (Again, much intercity air travel in

---

[4] We use the term "automobiles" to include "personal" light trucks. Light trucks and similar vehicles are only included in the US data, since such vehicles are not commonly used as passenger vehicles in Europe and Japan. For the United States, we estimated the allocation of total light truck km between passenger and freight uses based on survey data.

Past trends

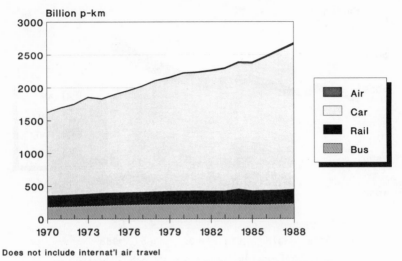

Does not include internat'l air travel

Fig. 4.7. Domestic travel by mode; Europe-6, 1970–1988

Europe is between countries and is thus not counted in the statistics we report here.) Air travel still accounts for a very small fraction of total travel, but growth in its share has major implications for energy use because it is more energy-intensive than other modes.[5] The volume of travel by rail and bus remained roughly constant in the United States, Japan, and Europe, but their shares of total travel declined considerably. They are still important modes in Japan, but are relatively insignificant in the United States.

The increase in automobile p–km has been due to growth in the number of cars and a decrease in persons per car trip. Growth in the number of cars in use has been considerable in Japan (from 15 to 31 million between 1973 and 1988) and Europe-6 (from 61 to 98 million). Even in the United States, where per capita ownership was already at the 1988 European level in 1970, the number of automobiles grew by 50% between 1973 and 1988. Increased car ownership has led to reduced use of bus and rail and higher overall travel (Webster et al., 1986).

Demographic and social factors have boosted car ownership. In the United States, the coming of age of the "baby boomers" caused a large growth in the driving age population. The percentage of eligible drivers with a driver's license also grew, in large part because of the movement of

[5] One reason why air travel accounts for a much higher fraction of total travel volume in the United States than in Japan and Europe is because destination points in the United States are so much farther apart (especially with major cities concentrated on the Atlantic and Pacific coasts).

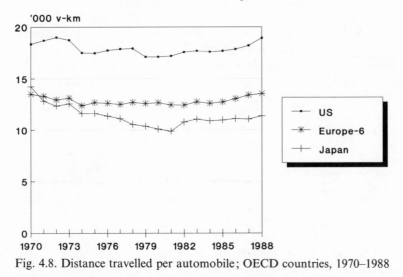

Fig. 4.8. Distance travelled per automobile; OECD countries, 1970–1988

women into the labor force. In 1969, 39% of adult women were employed, and 74% of them had licenses. By 1983, 50% were employed, and 91% of them had licenses (Ross, 1989).

Annual distance traveled per car has fluctuated in the United States, but was about the same in 1988 as in 1970 (Fig. 4.8). In Europe-6, there was a decline between 1970 and 1974, but there has been a slight increase since then (mainly due to growth in the United Kingdom).[6] Average distance declined in Japan through 1981 because private cars, which are driven less frequently than company cars and taxis, gradually accounted for a larger share of the total vehicle fleet; but it has increased since then as private cars came to be driven more frequently and farther. Usage per car might have increased more had there not been growth in household ownership of second or even third vehicles, which tend to be used less than the primary car. In the United States, kilometers per licensed driver increased more than did kilometers per automobile.

4.2.3 Energy intensity

The structural change described in the preceding section had only a modest net effect on aggregate travel energy intensity in the United States and Europe-6, but contributed significantly to an increase in intensity in Japan

---

[6] The increase in the United Kingdom can be largely explained by the rise in the use of company-provided cars and taxation policies that favor use of company cars (Ferguson & Holman, 1990).

Table 4.3. *Decomposition of the change in aggregate travel energy intensity, 1973–1988 (total % change)*

	Change in aggregate intensity	Decomposition		
		Structure	Intensity	Interaction[a]
United States	−18	3	−15	−6
Japan	25	20	5	0
Europe-6	7	4	4	−1
West Germany	17	5	12	0
Sweden	−2	4	−2	−4
Norway	15	14	7	−6
France	3	2	2	−1
United Kingdom	0	5	−3	−2
Italy	15	4	12	−1
OECD-8	−13	4	−14	3

[a] Because the structural and intensity variables interact in a nonlinear fashion, the two effects do not sum to the total change in aggregate intensity.

(Table 4.3).[7] In West Germany, which also had a big increase in aggregate energy intensity, most of the growth was due to an increase in modal energy intensities rather than the structural change away from rail and buses.

The effect of change in modal energy intensities described above reflects very different changes in the various modes, as we describe below.

Automobiles Automobile energy use per p–km declined by 18% in the United States; increased in Japan, West Germany, Italy, and Norway; and remained about the same in the rest of Europe between 1973 and 1988. A decline in the number of passengers per trip (partly due to a decrease in family size and increased numbers of cars per household) contributed to growth in energy intensity. In the United States, the average load declined from 2.2 persons per car in 1970 to 1.7 in 1983 and 1.5 in 1990 (US DOT, 1991). A decline also occurred in Japan (2.2 to 1.8), West Germany (1.7 to 1.5), Italy (2.0 to 1.7), and elsewhere in Europe.

Average fuel use per km fell more than did use per p–km. In the United States it declined by 29% between 1973 and 1988 (Fig. 4.9). The fuel

---

[7] The method used to decompose the change in aggregate energy intensity is the same as described in Chapter 3 for manufacturing. For further details in the case of transportation see Schipper et al. (1991).

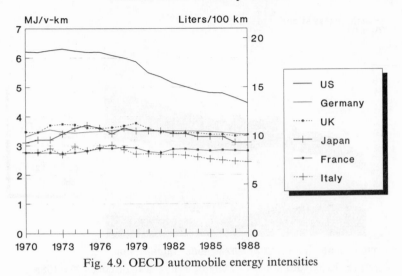

Fig. 4.9. OECD automobile energy intensities

intensity of cars fell by 33% (to 12 liters/100 km), but this was balanced somewhat by an increase in the use of light trucks as passenger vehicles (the share of light trucks in total automobile vehicle-km increased from 9% in 1973 to 18% in 1988). The fuel intensity of light trucks fell by 19% (to 18 liters/100 km), but remained well above that of cars.[8]

In Europe and Japan, there was little change in automobile fleet fuel intensity. While there were technical improvements in new cars that contributed to higher efficiency, this was counterbalanced by an increase in the size and power of automobiles and deterioration in operating conditions (more traffic congestion). In West Germany, for example, the fraction of all automobiles that had engine displacement of 1500 cm[3] and above increased from 40% in 1973 to 60% in 1987, and the average horsepower rose from 59 to 77 (DIW, 1991). By 1990, more than 80% of all cars sold in West Germany could reach 150 km/hr or greater, and 30% of them could surpass 180 km/hr. The average weight of an Audi 80 increased from 855 to 1050 kg between 1970 and 1991, while that of an Opel Kadett grew from 685 kg in 1963 to 865 kg in 1991. The average size of engines in the United Kingdom and France also rose. In the United Kingdom, growth in use of company cars has contributed to an increase in car size (Potter, 1991). As Fig. 4.10 illustrates for the United Kingdom, the trend in Europe toward larger cars had begun in the 1960s.

The turnover of the fleet had a different effect in the United States than

---

[8] Some of the improvement for light trucks reflects a shift within the light trucks category to smaller vans and pickup trucks.

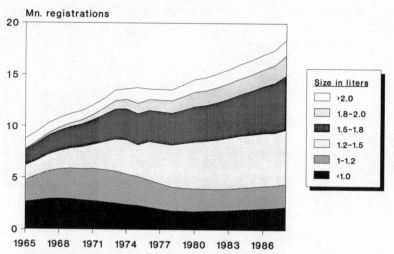

Fig. 4.10. Automobiles by engine size; Great Britain, 1965–1988

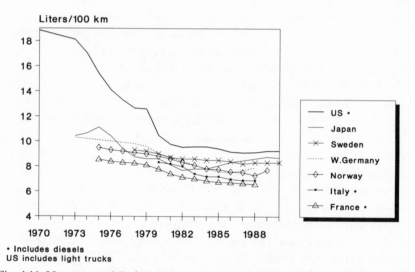

• Includes diesels
US includes light trucks

Fig. 4.11. New automobile fuel economy, test values; OECD countries, 1970–1990

in Europe and Japan. In the United States, the sales-weighted average fuel intensity of new automobiles (including all light trucks) declined by nearly 50% between 1973 and 1982 (Fig. 4.11), so turnover and growth of the stock strongly depressed fleet average fuel intensity. In Europe and Japan, the fuel intensity of new cars improved much less than in the United States, in part because it was already much lower in 1973, and in part because growth in vehicle size and power offset technical efficiency gains.[9] Test data show some decline in new car fuel intensity since 1975 in several countries, but intensity has increased since 1982 in Japan and since 1985 in West Germany as average size and power has risen. The continued decline in France and Italy is partly due to growing penetration of diesel-fueled cars, which have lower fuel intensity than comparable gasoline-fueled cars.[10]

In the United States, a shift to smaller cars contributed only slightly to the decline in new car fuel intensity after 1975. Average interior volume hardly changed between 1978 and 1988. (Since 1980, compacts have gained share at the expense of sub-compacts, but mid-size cars have also lost share.) Most of the change came from a decrease in fuel intensity within each size class. The average power of new cars fell by 25% between 1975 and 1980, contributing to a decline in intensity, but has increased since 1982, pushing intensity upward.

Fuel economy improvements have come from three main sources: propulsion-system engineering, other elements of vehicle design, and performance trade-offs. In the United States, engineering improvements are exemplified by the remarkable 36% increase in power per unit of engine size between 1978 and 1987. The ratio of vehicle weight to interior volume was reduced by 16% in this period, and reductions in air drag and rolling resistance (through introduction of radial tires) have also contributed to fuel economy improvement. Acceleration performance decreased in the 1980-82 period, which contributed to a decline in fuel intensity, but it has progressively improved since then.

The above discussion refers to passenger cars, but a significant factor in the United States has been an increase in the popularity of light trucks, vans, and jeeps for personal use. The share of such vehicles in total sales of light-duty vehicles rose from 19% in 1975 to 30% in 1988.[11] Since light trucks have higher average weight and power than cars, this shift has somewhat balanced the decline in fuel intensity of new cars.

Worsening traffic congestion has increased the in-use fuel intensity of the automobile fleet in most OECD countries. In the early 1980s, the US Environmental Protection Agency (EPA) determined that vehicles in use achieved 15% lower fuel economy than the nominal vehicle rating based

---

[9] Because of the difference in testing procedures, and differences in calculating averages, we caution against comparing new car fuel economy too exactly among countries. The trends over time within a country, however, give a valid picture for test performance, if not actual performance on the road.

[10] The diesel share of total automobiles (not only new ones) in France and Italy rose from 1% or less in 1970 to 12% and 14% respectively in 1988.

[11] The fraction of total light trucks that are used for personal use is estimated to be about 65–75.

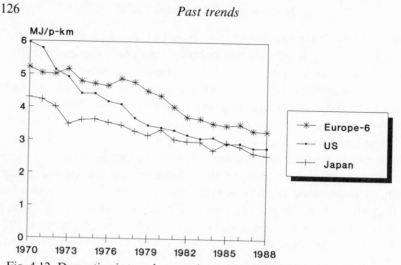

Fig. 4.12. Domestic air travel energy intensities; OECD countries, 1970–1988

on the driving cycle test (Ross, 1989; Westerbrook & Patterson, 1989). Some observers believe that the discrepancy has grown to as much as 25% as a result of increasing urban congestion, increasing share of urban driving (the EPA rating assumes 55% of vehicle–km are urban), higher speeds on open highways, and higher levels of acceleration in actual use than in the test.

In the 1970s, the reduction of highway speed limits in the United States reduced in-use fleet fuel intensity. In West Germany, conversely, lack of speed limits on expressways (and resulting high travel speeds) contributed to an increase in fuel intensity. While there are speed limits on motorways in the rest of Western Europe, relatively few drivers keep to these limits. Even when limits are observed, the fact that increasing numbers of cars are built to attain speeds in excess of 150 km/hr reduces the fuel economy of these cars at "ordinary" speeds (60-80 km/hr) (Dolan, 1991). Thus, it appears that both potential and actual high speed driving have reduced fuel economy from what tests of new cars indicate.

Air travel Energy use per p–km in domestic air travel declined considerably in the United States, Western Europe, and Japan between 1973 and 1988 (Fig. 4.12). The US decline – a remarkable 50% – exceeded that of Europe and Japan.[12] The drop in energy intensity was due to an increase in load

[12] Data supplied by several European airlines confirmed that this trend is also seen in international travel. Indeed, the long-range aircraft used by European and Japanese airlines for intercontinental travel, which account for more than half of their flown p–km, have significantly lower energy use per p–km than do smaller planes flown in domestic routes.

factor (passengers per available seats) and a decrease in energy use per seat-km. In the United States, load factor rose from 54% of available seats in 1973 to 63% in 1988. Load factors also increased in Western Europe and Japan. (For example, Air France reports that its system load factor rose from 53% in 1973 to 62% in the late 1980s.)

Energy use per seat-km declined for several reasons. First, new planes with significantly lower fuel intensity entered the fleets in large numbers. New planes were on average larger than those they replaced, and larger planes tend to use less energy per seat-mile than smaller planes with comparable technology.[13] There was also considerable decline in fuel intensity in planes of a given size (Gately, 1988). Technological changes included more fuel efficient engines, improvement in aircraft structural efficiency (lighter airframes), and improved lift/drag performance. Second, airlines retrofitted old planes with new engines (often for noise abatement reasons), and added seats. Third, airlines and airports instituted various operational improvements. As a result of these factors, energy use per seat-mile of US jet aircraft declined by one-third between 1973 and 1988.

Bus and rail travel Bus and rail travel combines urban transit and intercity service, with the former being much more energy-intensive than the latter. Bus energy use per p–km increased in Western Europe and the United States, and changed little in Japan. In the United States, energy use per vehicle–km increased by nearly one-third for transit buses between 1973 and 1988, reflecting operation in increasingly congested conditions.[14] Energy use per p–km increased even more due to a decline in average load factor. In Western Europe, congestion and declining ridership due to rising acquisition and use of cars increased the intensities of bus travel. In the United Kingdom, there has been a shift to smaller buses (which use more energy per seat–km) since deregulation in 1986.

The final energy intensity of rail travel declined slightly in Western Europe in the early 1970's, but has changed little since then. Rail energy intensity in Japan, which is much lower than elsewhere due to high levels of ridership, has remained about the same since 1970. The intensity declined somewhat in the United States between 1976 and 1981, but has risen since then. The increase is due to the growth in urban rail as a share of total rail p–km as new fixed rail systems have gone into service in several

[13] For US aircraft, available seats per aircraft increased from 111 in 1970 to 148 in 1980 and 161 in 1987 (Davis et al., 1991).
[14] School buses, which have low energy intensity due to their high load factor, constitute a significant portion of bus travel in the United States. The US energy intensity would be higher if they were excluded.

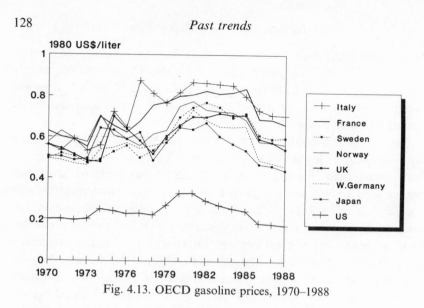

Fig. 4.13. OECD gasoline prices, 1970–1988

large metropolitan areas. Rail energy intensity is several times higher in the United States than in Europe and Japan (see Fig. 4.2) due in part to the relatively large share of urban and commuter rail in total rail travel. These types of rail systems are more energy-intensive than intercity rail because there are more trains running more frequently at a lower average speed, and especially during nonpeak hours, with lower load factors.

4.2.4 The effect of fuel prices on intensity trends

Although we have not performed a formal analysis of the impact of fuel price changes, some observations may be made. As shown in Fig. 4.13, the increase in real gasoline prices in the 1970s was fairly modest in most countries. Prices increased more in 1979-1981, but declined thereafter. In Western Europe and Japan, car buyers sought larger and more powerful cars, but the rise in prices and pressure from governments concerned about oil imports caused manufacturers to incorporate technical improvements that kept new car fuel intensity from rising, and in fact caused it to decline. The fall in real price after 1981-82 had an impact, however, especially in Japan, where new car fuel intensity began to rise. In the United States, the impact of rising prices is difficult to judge, since the government's fuel economy standards were an influential intervention in the market (Greene, 1990). The steady decline in real price since 1981 certainly contributed to lessened interest in fuel economy on the part of buyers. What is striking is that the real price of gasoline in 1988 in most countries was close to its 1970-1973 level.

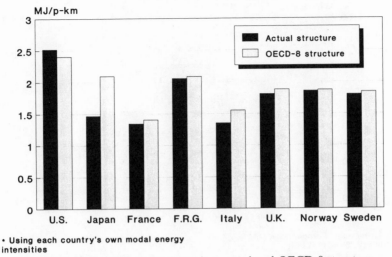

• Using each country's own modal energy
intensities

Fig. 4.14. 1988 travel energy intensity, actual and OECD-8 structure

4.2.5 Comparing travel energy intensity among countries

Although aggregate travel energy intensity in the United States declined considerably between 1973 and 1988, at 2.5 MJ/p–km it remained much higher than in the other OECD countries. The 1988 energy intensity of other countries can be divided into three tiers. "Low intensity" (1.3-1.4 MJ/p–km) countries include Japan, Italy, and France. "Medium intensity" (around 1.8) countries include Norway, Sweden, and the United Kingdom; West Germany was "medium high" at 2.1.

Differences in aggregate travel energy intensity are due to variation in the modal structure of travel as well as in modal energy intensities. Figure 4.14 shows the actual 1988 travel energy intensity in each country and what it would have been if each country had a modal structure equal to the OECD-8 average. In this hypothetical case, the US intensity declines, intensities in Western Europe increase slightly, while intensity in Japan increases considerably. The remaining difference among the countries is due to variation in the energy intensity of each mode, especially automobiles.

Per capita, if each country had a modal structure similar to that of Japan, but with their own modal energy intensities, total OECD-8 energy use for travel in 1988 would have been 11% lower than it actually was. If the US modal mix were imposed on all eight countries, total energy use for travel would have been 7% higher than it actually was. (Japan and the United States have the lowest and highest shares of energy-intensive

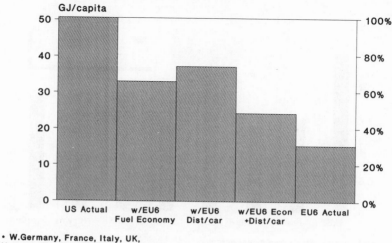

GJ/capita

• W.Germany, France, Italy, UK,
Norway, Sweden (EU6)

Fig. 4.15. Per capita energy use by automobiles; US and Europe in 1987

automobile and air travel of total travel.) The same experiments carried out in 1970 would have yielded 25% lower and 8% higher energy use respectively, indicating that the differences in modal mix have decreased over time.

A major reason for the higher energy intensity of travel in the United States is that per capita energy use in automobiles is around three times as high as the Europe-6 average. If US automobiles in 1988 had averaged the same fuel intensity as European ones, the US per capita value would have been about twice that of Europe-6 (Fig. 4.15). If US automobiles had been driven the same distance per year as European ones, US energy use would have been about two and one-half times that of Europe. If both of the above had been the case, the US value would have been only about 50% greater than that of Europe. The remaining difference is due to the higher level of automobile ownership per capita in the United States. While the United States remains above Western Europe in terms of automobile fuel intensity, average distance, and ownership, the differences have narrowed. Whereas in 1973 automobile energy use per km was twice as high in the United States as in Europe-6, by 1988 it was only about 50% higher (refer back to Fig. 4.9); and part of the difference is due to the popularity of light trucks in the US.

4.3 Former East Bloc

Despite the huge size of the Soviet Union, the level of per capita travel (about 6000 p–km in 1987) was half that of Japan, and one-third that of

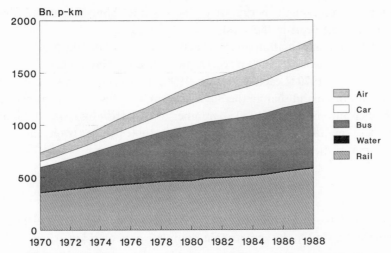

Fig. 4.16. Passenger travel in the Soviet Union, by mode, 1970–1988

Western Europe.[15] Growth in travel averaged 4.5%/year between 1973 and 1987, slower than the 7.5%/year between 1960 and 1973. Rail and bus dominate passenger travel (with 31% and 28% of total p–km in 1987, respectively), but automobile and air travel have grown twice as rapidly as total travel, and thus increased their shares to 21% and 12% by 1987 (Fig. 4.16).

The growing shares of automobile and air travel have contributed to an increase in aggregate travel energy intensity. Since the early 1970s, however, the intensity of automobile travel has declined as more small cars entered the fleet. From a technical standpoint, the energy intensity of the Soviet car fleet (estimated at around 9 liters/100 km) is not far above that of Western European cars (around 8 liters/100 km);[16] but Soviet cars have considerably less power than Western European ones. Furthermore, the actual on-the-road energy intensity of Soviet cars appears to be 11–12 liters/100 km due to the poor quality of fuel, vehicle maintenance, roads, and parts.

The energy intensity of air travel in the Soviet Union has declined by about 10% since the early 1970s due mainly to an increase in aircraft size. Load factors have remained constant at nearly 100%. For this reason, the energy use per p–km of Soviet air travel is low compared with OECD

---

[15] For details and data sources for the Soviet analysis, see Schipper & Cooper (1991).
[16] Large, inefficient official cars account for a considerable share of the Soviet fleet. If they were removed, the average fleet intensity would be lower.

countries, even though energy use per seat–km is about 50% higher than that of aircraft fleets in the West.

For Poland, information on travel activity is sparse, but the total level of travel appears close to 6000 p–km/capita, which is about where Western Europe was in 1970 (Leach & Nowak, 1990). Domestic air travel is far less important in Poland than in the Soviet Union, but this is compensated for by higher automobile travel. The number of cars grew by nearly 13% per year between 1970 and 1987. At 110 cars/1000 people, ownership was twice that of the Soviet Union.

4.4 Developing countries

Lack of reliable data limits our ability to analyze change in travel energy use in the LDCs. On the activity side, complete data on p-km are either not available or of questionable accuracy. On the energy side, time-series disaggregation of transport energy use between travel and freight transport is rare, and is more complicated than in the OECD countries because it is not uncommon for freight trucks to use gasoline. In addition, significant numbers of cars and trucks are used for both passenger travel and freight, often simultaneously.

While historical data on travel activity are generally not available, it is evident that per capita travel has increased considerably in the LDCs. Chinese data show growth in p–km per capita averaging nearly 11% per year between 1975 and 1988, but the 1988 level of 570 was still quite low even by LDC standards (less than half the 1985-86 level in India).[17] In much wealthier South Korea, the data show a three-fold increase from 970 to 3200 p–km/capita between 1970 and 1987. In Brazil, per capita road travel, which accounts for over 90% of estimated total travel, increased from 1600 p–km in 1973 to 3700 p–km in 1985 (Geller & Zylbersztajn, 1991).

Buses and, in a few countries, rail still account for a large majority of motorized travel in most LDCs. (Rail is particularly significant in China and India.) In South Korea, for example, the data show a decline in the shares of buses and rail between 1972 and 1987, but they still accounted for 60% and 24% of total travel in 1987, respectively, while cars (including taxis) accounted for only 14%. In China, rail still dominates travel, but its share of total p–km declined from 70% to 53% between 1970 and 1988, while the highway share increased from 23% to 41%. (The available data

[17] China's statistics on travel probably reflect the actual situation fairly well, since the number of private cars is very low.

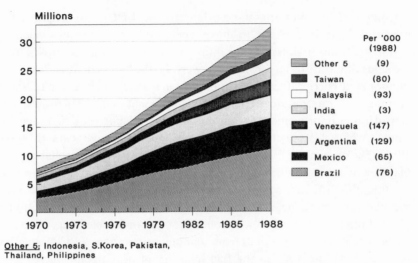

Fig. 4.17. Automobiles in selected LDCs, 1970–1988

do not distinguish between buses and automobiles, but it is clear that buses account for a large share of total highway p-km, as ownership of private vehicles is very low.) In India, the share of road modes in total travel (78%) is larger than in China, and has grown since the 1970s.

The structure of travel in most LDCs shows a highly skewed pattern, with most people relying on bus, rail, or nonmotorized modes, while the wealthier use cars (Sathaye & Meyers, 1987). Conventional and collective taxis (jitneys) are also important in most cities. Historically higher income and urbanization levels in Latin America and the Middle East have led to greater penetration of cars than in Asia and Africa. Business, government, and taxi operators own a considerable share of total automobiles in many LDCs.

Change in the modal structure of travel since the early 1970s in the LDCs is difficult to quantity, but several trends are evident. In much of Asia, there has been growing use of mopeds and motorcycles (the "poor man's car"). In most countries, there has also been considerable growth in the number of cars (Fig. 4.17). Lastly, domestic air travel has increased considerably in large countries in the past decade.

Change in the energy intensity of travel modes in LDCs is difficult to assess. Buses are generally old and inefficient – and thus have high fuel use per kilometer – but they are heavily loaded, so energy use per p–km may not be so high. Passenger rail systems have similar characteristics, though a growing number of LDC cities have relatively modern urban rail systems.

Compared to Western standards, cars in LDCs are smaller, less powerful, and less likely to have energy-intensive features such as automatic transmission and air-conditioning. In wealthier countries there are signs of a shift to larger, more powerful cars, however. It is likely that the nominal energy intensity of new cars has declined in most LDCs in keeping with international trends in vehicle technology. There has been considerable improvement in automobile energy efficiency in India and China, which historically produced very fuel-inefficient cars (Meyers, 1988). The opening of their automotive industries to foreign collaboration played a major role in this change. In Brazil, the transition from gasoline-to ethanol-fueled cars has reduced average fuel intensity, since the latter are inherently more fuel-efficient.[18] In addition, there has been reduction in the fuel intensity of both new gasoline- and ethanol-fueled cars (Geller & Zylbersztajn, 1991). Test data from manufacturers show a 10% reduction between 1983 and 1987 in the fuel intensity of new alcohol-fueled cars, which accounted for the majority of new cars sold in the mid-1980s. The average new car fuel economy of 9.65 liters of ethanol/100 km in 1987 is equivalent to about 6.8 liters of gasoline/100 km (35 mpg).

While data on average in-use automobile fuel intensity are lacking, it is probable that in many, if not most LDCs, the trend toward higher efficiency of new cars has been countered by a gradual worsening of urban traffic conditions. Increasing congestion has probably had a similar effect on the energy intensity of buses.

4.5 International air travel

International air travel throughout the world increased nearly six-fold between 1970 and 1990 – twice as much as the increase in domestic air travel.[19] The share of international travel in total air travel rose from 30% in 1970 to 44% in 1990. The largest absolute growth in international air travel has been in trans-Atlantic travel, but the relative increase has been much larger for trans-Pacific and Europe-Asia travel (Fig. 4.18). International travel within Asia has also grown significantly.

Most international travel markets are more competitive than domestic air travel markets, so air carriers tend to use the newest, most efficient equipment on these routes. The longer distances of most international routes also favor use of more modern, larger aircraft. As a result, these

---

[18] Alcohol fuel permits the use of engines with higher compression ratio due to the higher octane of ethanol relative to gasoline.

[19] Data in this section are from Boeing (1991).

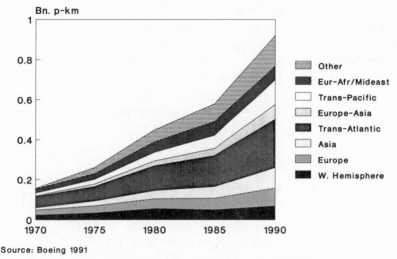

Source: Boeing 1991

Fig. 4.18. International air travel, 1970–1990

routes tend to be less energy-intensive than domestic ones. In addition, longer flights are less energy-intensive than short ones because the fuel used in take-off and ascent is a smaller share of overall fuel use.

Freight transport

Energy use for freight transport is strongly affected by the mode used, which is in turn affected by the type of goods that are moved. Mining and agricultural products, which have low value per tonne, are often transported via rail or ships (or barges), both of which have lower energy-intensity than trucks. As economies develop, intermediate and final goods take on a greater share of freight transport. Since trucks offer greater flexibility for such shipments, they tend to assume an increasing role in freight transport over time.

The data on freight reported in this section refer to domestic freight shipments. Fuel purchased by ships engaged in international commerce is counted as "marine bunkers" and is not included in national energy consumption.

The number of freight trips increases with agricultural and industrial output. Distance of trips is shaped by the physical geography of a country, as well as by the geographic patterns of economic production and consumption. Mode choice is strongly shaped by the type of freight to be moved, so the overall modal mix in a country depends heavily on the output of the economy. Other factors affecting mode choice include the nature of the transport network (roads, rail, waterways), cost factors, distance of trip, and time requirements for the trip.

Shipment of fuels comprises a significant share of total freight tonne–km in many countries. Change in the mix of primary energy sources thus can have a major impact on freight transport. In France, for example, the shift away from oil throughout the economy contributed to an absolute decline in total freight tonne-km between 1973 and 1987.

4.6 OECD countries

We assembled data on freight transport energy use and activity in the 1970-1988 period for eight OECD countries. Among these eight, the United States accounted for 62% of total freight energy use in 1988 (Table 4.4). Europe-6 accounted for 24%, while Japan accounted for 14%.

4.6.1 Energy use and activity

Energy use for domestic freight increased by 40%, 33%, and 48% between 1973 and 1988 in the United States, Japan, and Europe-6, respectively. Total freight transport activity (tonne–km) grew by 34% in the United States, 18% in Japan, and 32% in Europe-6 over the same time period. As shown in Table 4.5, increase in activity contributed more significantly to growth in energy use than did changes in aggregate intensity for most countries. The two exceptions are Sweden, where aggregate intensity was almost three times as important; and France, where there was a decline in activity and a large increase in intensity due to a substantial decline in the share of rail.

The ratio of freight activity to GDP has declined since the early 1970s in Japan, and since 1980 in the United States (Fig. 4.19). There has been essentially no change in Europe-6. Despite the decrease in the ratio since 1980 in the United States, it remains around three times as high as in Western Europe or Japan. The sheer size of the United States is partly responsible for this high level. There is also considerable shipment of bulk materials (including grain and coal for export) over long distances. Another reason – also related to size – is that various types of freight activity that are international (and therefore not counted in domestic activity) for Europe and Japan take place domestically within the United States.

While it is clear that because the mix of goods moved has changed over time, data on freight transport are usually not disaggregated by type of

Table 4.4. *Freight transport energy use and activity in the OECD-8 in 1988*

	Energy use		Activity	
	(exajoules)	%	(bn tonne–km)	%
United States	5.58	62	4100	74
Japan	1.21	14	483	9
Europe-6	2.14	24	947	17
West Germany	0.42	5	265	5
United Kingdom	0.51	6	209	4
France	0.53	6	171	3
Italy	0.54	6	229	4
Sweden	0.07	1	52	1
Norway	0.06	1	23	4
OECD-8	8.93	100	5530	100

Table 4.5. *Growth in freight transport energy use, activity, and aggregate intensity, 1973–88 (total % change)*

	Growth in energy use	Tonne–km	Aggregate intensity
United States	40	34	4
Japan	33	18	12
Europe-6	48	32	12
West Germany	17	24	−6
France	44	−2	46
United Kingdom	29	51	−14
Italy	134	77	32
Norway	43	34	7
Sweden	46	12	31
OECD-8	41	32	6

freight. In the OECD countries, it appears that freight tonnage is declining with respect to GDP, reflecting the growing role of lighter products. Data for West Germany indicate that growth in tonne-km has been dampened by a decline in the average number of tonnes per trip (DIW, 1991).

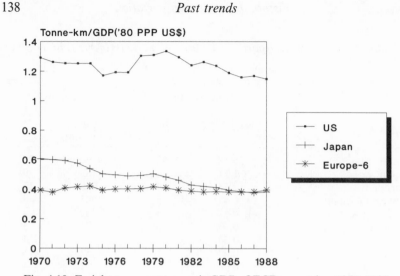

Fig. 4.19. Freight transport per unit GDP; OECD countries, 1970–1988

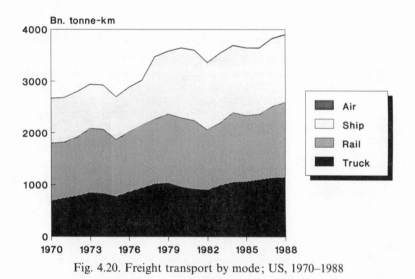

Fig. 4.20. Freight transport by mode; US, 1970–1988

4.6.2 Change in modal structure of freight transport

In the United States, there has been a slight shift of freight tonne-km from rail to trucks and ships (Fig. 4.20).[20] In Japan and Europe-6, there was a major shift from rail to trucks. The share of freight activity in trucks

[20] Time-series data on tonne-km shipped in pipelines are generally not available. In the United States, the amount of natural gas shipped in pipelines in 1985, as estimated by Ross (1989), was equal to around 7% of total freight tonne–km.

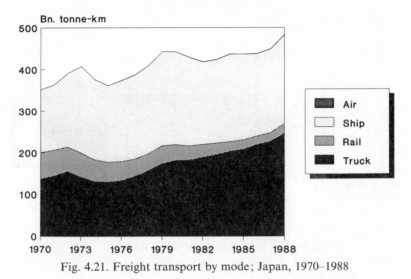

Fig. 4.21. Freight transport by mode; Japan, 1970–1988

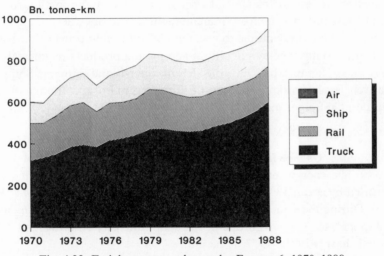

Fig. 4.22. Freight transport by mode; Europe-6, 1970–1988

increased from 35% to 51% in Japan (Fig. 4.21) and from 54% to 63% in Europe-6 (Fig. 4.22). The share of rail declined from 14% to 5% and 28% to 18%, respectively. A large drop in the share of rail in France, which was in part due to the decrease in consumption of oil products, contributed strongly to the Europe-6 trend.

The increase in the role of trucks reflects change in the composition of

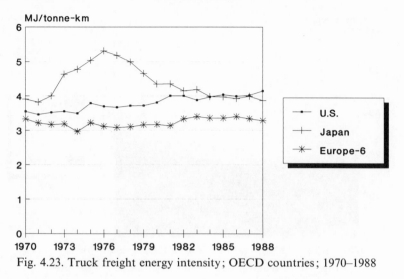

Fig. 4.23. Truck freight energy intensity; OECD countries; 1970–1988

freight toward products for which trucks have inherent advantages over competing modes (Grübler & Nakićenović, 1990). In addition, the growing use of "just-in-time" delivery in manufacturing favors trucks.

Ships and barges continue to be an important freight transport mode in the United States where agricultural and mining products rely heavily on them, and in Japan, which has considerable inter-island shipping. Ships are relatively less important for domestic shipments in Europe, though there is considerable freight moved by ship between countries.

4.6.3 Change in energy intensity of freight transport modes

Trucks The energy intensity of freight trucking (energy per tonne–km) increased by about 13% in the United States between 1973 and 1988 (Fig. 4.23). During the same period, there was a decline of 16% in Japan, and a slight increase (3%) in Europe-6. In Japan, intensity rose through 1976, but fell sharply in the 1977-80 period.

In the United States, the data show that average fuel use per km was the same in 1988 as in 1973 for both medium and heavy (tractor–trailer) trucks. It appears that improvement in technical efficiency was offset by several factors. One was the increase in operating speeds on major intercity highways. Another was increasing traffic congestion in urban areas. The overall increase in energy per tonne–km was apparently due to factors related to the operation of trucking fleets and the nature of freight carried. Despite deregulation of the trucking industry there is evidence that there

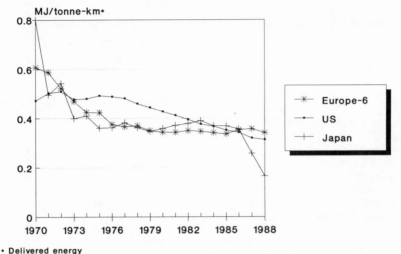

* Delivered energy

Fig. 4.24. Rail freight energy intensity; OECD countries, 1970–1988

was an increase in empty backhauls (Mintz, 1991), resulting in reduced tonnage per distance traveled.[21] In addition, it appears that the weight carried per volume of truck capacity declined. One reason for this is increased packaging for many goods (packaging is light-weight but takes up truck capacity).

Rail Between 1973 and 1988, the final energy intensity of rail freight transport declined by 34% in the United States, by 26% in Europe-6, and by 58% in Japan (Fig. 4.24). Electrification accounts for part of the decline. (The efficiency of a diesel locomotive is 20-25%, that of electric traction 90%; so replacement of diesel by electricity causes a significant decline in final energy intensity.) Other reasons for the use of stronger locomotives and the trend to cutting unprofitable lines, which presumably supported smaller trains with less than full loads. The large decline in Japan in the 1987-88 period is apparently due to the radical restructuring of the rail business that took place (Kibune, 1991).

4.6.4 Decomposing change in freight energy intensity

Between 1973 and 1988, aggregate freight transport energy intensity (energy per tonne-km) increased by 4% in the United States and by 12% in Japan and Europe-6. Structural change (modal shifts) contributed

---

[21] Increase in the volume of solid waste, and in the average distance of trips for disposal, also increased the distance of empty backhauls.

Table 4.6. *Decomposition of change in aggregate freight transport energy intensity, 1973–1988 (total % change)*

	Change in aggregate intensity	Decomposition		
		Structure	Intensity	Interaction
United States	4	3	1	0
Japan	12	30	−13	−5
Europe-6	12	13	−2	1
West Germany	−6	15	−20	−1
Sweden	31	4	27	0
Norway	7	15	−5	−3
France	46	22	18	6
United Kingdom	−14	1	−12	−3
Italy	32	15	14	3
OECD-8	6	8	−2	0

strongly to the increase in Japan, more than offsetting decline in energy intensities (Table 4-6). It accounted for all of the aggregate intensity increase in Europe-6.

Japan, West Germany, and the United Kingdom had a net decline in modal energy intensities. The United States showed little change, as the increase in intensity for trucks was offset by decreases in other modes. In Europe-6, the countries averaged out to almost no change. France, Italy, and Sweden had increases in the modal intensity of trucks, which offset decreases in the intensity of other modes.

4.7 Former East Bloc

Relative to economic activity, the level of freight transport in the Soviet Union (30,000 t–km/capita in 1987) was high because of the dominance of heavy raw materials and energy. Pipeline shipment of oil and gas (a substantial amount of which was destined for export) accounted for one-third of total freight tonne–km in 1987. (Even excluding pipeline transport for comparability with OECD countries, the Soviet level was still high.) In addition, the lack of real markets for intermediate goods means materials and goods are often shipped long distances because of the way the "buying" and "selling" ministries exchange goods.

Pipeline shipment of oil and gas accounted for most of the growth in total freight transport after 1974 (Fig. 4.25). The increase in other transport modes averaged only 2%/year between 1973 and 1987, compared to

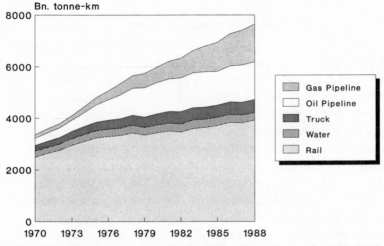

Fig. 4.25. Freight movements in the Soviet Union by mode, 1970–1988

6%/year between 1965 and 1973. This reflected the slowing of the Soviet economy. Excluding gas pipelines, rail accounted for around two-thirds of freight tonne-km. Trucks accounted for only 7% due in part to the relative unimportance of consumer goods and the lack of deliveries of these goods to consumer outlets.

Because of the dominance of rail, the overall energy intensity of Soviet freight transport is low relative to Western European levels. The energy intensity of each mode is close to levels in the West because of the importance of large shipments of bulk materials. The intensity for trucks has fallen, mainly because of an increase in the share of diesel trucks. Rail energy intensity (final energy) has also decreased, first through replacement of coal traction by oil, then through electrification.

Information on Poland indicates a pattern quite different from that of the Soviet Union. Total freight transport reached only 4250 t–km/capita in 1987, far below the level of the Soviet Union but close to that of West Germany (Leach & Nowak 1990). Given the much smaller per capita GDP in Poland than in West Germany, this comparison implies that Poland has a high level of freight transport relative to economic activity. Three-quarters of domestic freight was hauled by rail, a share that has remained stable since the late 1970s. The intensity of trucks and other freight modes is considerably higher than in the Soviet Union, Western Europe, or the United States.

4.8 Developing countries

Lack of reliable time-series data on total freight activity in developing countries makes it difficult to assess the historic relationship between freight transport and GDP. Even if reliable data on motorized freight transport were available, such data leave out animal-powered transport, which is significant in India and many other countries. Clearly though, change in the composition of economic output has had an effect on the ratio. In China, the ratio of freight tonne–km to GDP declined slightly between 1978 and 1988, perhaps reflecting some lightening of economic output. Data for South Korea show little change in the freight tonne–km/GDP ratio between 1970 and 1987. Brazilian data, on the other hand, show a significant increase between 1973 and 1985, which could be caused in part by greater transport of agricultural and mineral products from the Amazon region.

The modal structure of freight transport in LDCs varies across countries and over time depending on the composition of economic output and national geography. China and India have extensive (but outmoded and overburdened) rail networks. Both countries have considerable shipment of grains, coal, and other mineral products for domestic consumption. Rail has historically dominated freight transport in both, though the share of trucks has increased since the 1970s. Chinese data show the share of road transport in total tonne–km rising from 6% in 1978 to 13% in 1988. Waterways have also become more important in China, and accounted for the same share (42%) as rail in 1988.

Except for China and India, most LDCs have not built extensive rail networks, and much of the freight is moved by truck, or in some countries (especially those comprised of islands, such as Indonesia and the Philippines), via ships. Where the manufacturing sector has grown rapidly, the share of trucks in freight transport has risen also. In South Korea, for example, the share of trucks increased from only 11% in 1970 to 48% in 1987, while rail declined considerably in share.

Growth in the share of trucks leads to an increase in the aggregate energy intensity of freight transport. Data from South Korea show a 13% increase in this indicator even for the short period from 1983 to 1987. In Brazil, on the other hand, the data show a decline in the share of trucks from 62% in 1973 to 54% in 1985, and an increase in the share of ships. Since the latter have much lower energy intensity than trucks, this shift contributed to a decline in aggregate freight transport energy intensity.

Lack of data makes it difficult to assess how the energy intensity of particular freight transport modes has changed in LDCs. However, for trucks, two factors have contributed to a decline in energy intensity. One is a shift from gasoline to diesel-fueled trucks. The other is an increase in the share of heavy trucks, which use less energy per tonne–km than medium or light trucks. In Brazil, the fraction of diesel-fueled trucks increased from about 50% of the fleet in 1973 to over 85% by 1985, while the fraction of heavy and semi-heavy trucks rose from 15% to 28% (Geller & Zylbersztajn, 1991). In India, the transition from gasoline to diesel-fueled trucks is nearly complete, while in China most trucks still use gasoline. Trucks in India are larger than those in China, but are often overloaded, which leads to high energy use per km. In both countries, technological improvement affecting the fuel efficiency of gasoline and diesel trucks has been minimal.

The energy intensity of rail transport has declined in India due to increasing use of diesel and electric locomotives in place of inefficient steam locomotives using coal, but the efficiency of diesel and electric locomotives has not improved very much. In China, steam trains are still predominant, though a growing use of diesel locomotives has improved energy efficiency.

4.9 Conclusion

Between 1973 and 1988, per capita travel increased throughout the world. In all regions, there has been a shift in modal structure toward automobiles and airplanes, though the shares of the latter are much lower in the Former East Bloc and the LDCs than in the OECD countries. This shift has caused growth in the aggregate energy intensity of travel. The energy intensity of automobile travel declined considerably in the United States, but changed little in Western Europe and Japan, where growth in vehicle size and power counterbalanced technical efficiency gains. In the Former East Bloc and the LDCs, the extent of any decrease in automobile energy intensity is difficult to judge, but appears to have been modest. In contrast to automobiles, the energy intensity of air travel fell dramatically in the OECD countries.

Freight transport per unit GDP declined somewhat in the United States and Japan, but changed little in Western Europe and (probably) the Former East Bloc. In the LDCs, trends in the tonne-km/GDP ratio have differed among those countries for which we had historical data. As in travel, there has been a shift toward a more energy-intensive structure, in this case from rail to trucks, whose flexibility and convenience provide advantages for moving many manufactured products. In contrast to automobiles, there has not been much decline in truck energy intensity, at least in the OECD countries. In some LDCs, the shift from gasoline to diesel trucks has reduced energy intensity somewhat. There has been more decline in the energy intensity of rail freight transport, though some of this is also a result of fuel switching.

The combination of growth in activity, structural change toward more energy-intensive modes, and a relatively modest (in most cases) decline in modal energy intensities has led to a considerable growth in transportation energy use throughout the world. Since oil products account for almost all transport energy use, this growth has been the key factor pushing upward on world oil demand.

References

Boeing Commercial Airplane Group. 1991. *Current Market Outlook: World Market Demand and Airplane Supply Requirements.* Seattle, WA.

Davis, S. C. & Hu, P. S. 1991. *Transportation Energy Data Book: Edition 11,* Oak Ridge, TN: Oak Ridge National Laboratory, ORNL-6649.

DIW (Deutsches Institut für Wirtschaftsforschung). 1991. *Verkehr in Zahlen.* Bonn, Germany: Bundesminister für Verkehr.

Dolan, K. 1991. Tomorrow's clean and fuel-efficient automobile: Opportunities for East-West cooperation. A summary of the Berlin Conference, March 25-27, 1991, Paris, France: OECD Environment Directorate and Berkeley, CA: Lawrence Berkeley Laboratory.

Ferguson, M. & Holman, C. 1990. *Atmospheric Emissions from the Use of Transport in the UK* (Vol. 1). London, UK: Earth Resources Research.

Gately, D. 1988. Taking off: The U. S. demand for air travel and jet fuel, *Energy Journal*, **9** (4), 63-88.

Geller, H. S. & Zylbersztajn, D. 1991. Energy intensity trends in Brazil, *Annual Review of Energy*, **16**.

Greene, D. L. 1990. CAFE OR PRICE?: An analysis of the effects of federal fuel economy regulations and gasoline price on new car MPG, 1978-89, *Energy Journal*, **11** (3), 37-57.

Grübler, A. & Nakićenović, N. 1990. *Evolution of Transport Systems, Past and Future.* Laxenburg, Austria: International Institute for Applied Systems Analysis.

Kibune, H. 1991. Japan Institute of Energy Economics, Tokyo. Private communication.

Leach, G. & Nowak, Z. 1990. *Cutting Carbon Dioxide Emissions from Poland and the United Kingdom,* Draft Research Report, Stockholm, Sweden: The Stockholm Environment Institute.

Meyers, S. 1988. *Transportation in the LDCs: A Major Growth in World Oil Demand,* Berkeley, CA: Lawrence Berkeley Laboratory Report LBL-24198.

Mintz, M. A.M. 1991. Trends in demand for freight transportation, paper presented at the Conference on *Transportation and Global Climate Change: Long-Run Options*, Asilomar, CA, August 25-28, 1991.

Potter, S. 1991. *Company Car Report*, draft. Milton Keynes, UK: The Open University, Energy and Environment Research Unit.

Ross, M. 1989. Energy and transportation in the United States, *Annual Review of Energy*, **14**, 131-71.

Sathaye, L. & Meyers, S. 1987. Transport and home energy use in cities of the developing countries: a review. *Energy Journal*, **8**, Special LDC Issue.

Schipper, L. & Cooper, R. C. 1991. *Energy Use and Conservation in the USSR: Patterns, Prospects, and Problems*, Berkeley, CA: Lawrence Berkeley Laboratory Report LBL-29831.

Schipper, L., Steiner, R., Duerr, P., An, F. & Strøm, S. 1991. Energy Use in Passenger Transport in OECD Countries: Changes between 1970 and 1987. Berkeley, CA: Lawrence Berkeley Laboratory LBL-29830.

US DOT (US Department of Transportation). 1991. *National Personal Travel Survey*, draft report. Washington, DC.

Webster, F. V., Bly, P. H., Johnson, R. H., Pauley, N. & Dasgupta, M. 1986. Changing patterns of urban travel: Part 1. Urbanization, household travel, and car ownership, *Transport Reviews*, **6** (1), 49-86.

Westerbrook, F. & Patterson, R. 1989. *Changing Driving Patterns and Their Effect on Fuel Economy*. Paper presented at the 1989 SAE Government/Industry Meeting, May 2, Washington, DC.

5

Historic trends in the residential sector

The residential sector accounts for about 20% of final energy use in the OECD countries and about 15% in the Former East Bloc. For 13 major LDCs, its share is about 15% if one does not count biomass fuels, but around 40% if one does.[1] In poorer countries with large rural populations, the residential share, counting biofuels, is even higher.

The nature of residential energy use differs considerably among the three country groupings, and historic trends have also varied greatly. In the OECD countries, the ability of households to meet new services or increase amenity levels brought a rapid increase in use of oil, gas, and electricity in the 1960s. By the late 1980s (earlier in the United States), a large share of households had acquired desired services and improved amenity levels, which contributed to slower growth or decline in total residential energy consumption. Outside the OECD, however, there is very large unmet desire for new services and improved amenity levels. In some areas, rising incomes have allowed acquisition of home appliances, while in others stagnation or even decline in household income has frustrated people's desires.

5.1 Analyzing residential energy use

Energy use in the home depends on household demand for particular services that require energy, such as cooking, lighting, and space heating, as well as the manner in which energy is used to provide those services. The demand for services is shaped by a variety of factors, including climate, which is relatively fixed, and culture, which changes. A household's ability to satisfy its desires is constrained by its income.

---

[1]   The estimate is based on a rough disaggregation of combined residential/services energy use.

One can measure aggregate activity in the residential sector in terms of the number of people or households. (A household is comprised of the people that share a dwelling unit.) These measures of activity are different from those used in other sectors, for which the measures represent sectoral output (manufacturing and services) or physical levels of activity (transportation). In the residential sector, households or population are convenient for counting the number of units that demand household services, but are not measures of the service itself. Another measure, residential floor area, is indicative of the amount of housing, but also does not measure the services delivered in that housing. Use of households as a unit of account is problematic because the size of the average household changes over time. For this reason we use population as an aggregate measure of residential activity.

When considering structural change in the residential sector, we refer to three basic phenomena. One is change in home area per person, which affects the demand for space conditioning and lighting, and may allow for larger appliances. Another is change with respect to heating equipment. In most of Western Europe, for example, there has been a transition from use of room heaters to central heating, which improves the level of amenity and tends to increase energy consumption for space heating. The other key structural change is increase in ownership of major appliances.

Change in the energy sources used for particular end uses has affected energy intensity more in the residential sector than in other sectors. Sources differ with respect to their end-use conversion efficiency. Substitution of electricity for fossil fuels depresses energy intensity if it is expressed in terms of final energy, as does substitution of fossil fuels for biomass. To eliminate the effect of change in energy sources, we often use the concept "useful energy," which is equal to final energy use minus estimated conversion losses at the home.[2]

5.2 OECD countries

Over the past decade, our group has organized and analyzed data on residential energy use in nine OECD countries.[3] Among these countries, we discuss trends for the United States, Japan, "Europe-4" (an aggregate of

---

[2] In the OECD analysis, "useful energy" is calculated as 66% of final energy use for oil and gas, 55% for coal and wood, and 100% for electricity and district heat.

[3] For previous analyses of OECD residential energy use, and more extended discussion than is presented in this chapter, see Schipper, Ketoff & Kahane (1985); Schipper & Ketoff (1985); and Ketoff & Schipper (1990).

Table 5.1. *Residential energy use and population in nine OECD countries in 1988*

	Energy use		Population	
	exajoules	%	millions	%
United States	11.17	57.0	245.6	40.1
Japan	1.56	8.0	120.0	19.6
Europe-4	6.21	31.7	229.9	37.5
West Germany	1.82	9.3	61.1	10.0
United Kingdom	1.73	8.8	55.5	9.1
France	1.55	7.9	56.0	9.1
Italy	1.11	5.7	57.3	9.3
Scandinavia-3	0.66	3.4	17.7	2.9
Sweden	0.33	1.7	8.4	1.4
Norway	0.15	0.8	4.2	0.7
Denmark	0.18	0.9	5.1	0.8
OECD-9	19.60	100.0	613.2	100.0

West Germany, France, the United Kingdom, and Italy), and "Scandinavia-3" (an aggregate of Sweden, Norway, and Denmark).[4] The United States accounted for 57% of total final energy use in 1988 for the nine countries, though only 40% of the total population (Table 5.1).

5.2.1 Key factors

Before discussing the historic trends in residential energy use, it is useful to present some background on several key factors.

Decline in household size Between 1972/73 and 1988, the average household size fell from 3.1 to 2.7 persons in the United States and Europe-4, from 3.8 to 3.3 in Japan, and from 2.6 to 2.2 in Scandinavia-3. There are many reasons for this decline. Birth rates have fallen, and children tend to leave the family home to form their own households earlier in life than in the past. The period between leaving the family nest and beginning a new family has lengthened considerably, leading to more 1-person households. With increasing divorce, the number of single-parent families has grown. Lastly, the elderly live longer than previously, and it has become more

---

[4] In this chapter we separate the Scandinavian countries from the European aggregate because trends in these countries have differed somewhat from those in the "Big Four."

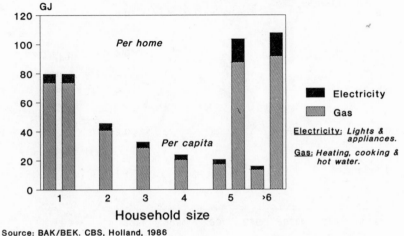

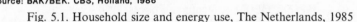

Source: BAK/BEK. CBS, Holland, 1986

Fig. 5.1. Household size and energy use, The Netherlands, 1985

common for them to live on their own rather than with their children. The longer life expectancy of women relative to men has contributed to growth in one-person households.

Change in household size is important because, all else being equal, per capita energy use rises as household size decreases. One reason for this is that a smaller household tends to have more floor area per capita than a larger one, which leads to higher space heating demand per capita. A small household has many of the same appliances as a large household, so therefore more appliances per capita. Energy consumption for some purposes varies with household size, but in other cases there is little variation. For example, the refrigerator in a small household may use nearly as much electricity as one in a large household. Larger households are also able to utilize economies of scale for certain end uses.

We illustrate the effect of declining household size with data from the Netherlands (Gasunie, 1986). Energy use per household increases very slowly with household size, but energy use per capita falls steeply (Fig. 5.1). We estimate that if the 1960 distribution of households by size had remained in 1985, total residential use would have been about 15% less for electricity, and 20% less for gas.

Decline in home occupancy Smaller families, greater numbers of households with two parents working, and increased numbers of single-person households where the occupant is gone during the day have all contributed to decline in home occupancy. With more people unmarried or married

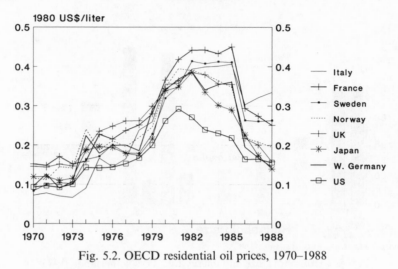

Fig. 5.2. OECD residential oil prices, 1970–1988

without children, there has also been increase in away-from-home activities such as eating out, going to movies and other entertainment, and leaving town on weekends. Occupancy has an important effect on energy use for space conditioning.

Growth in disposable income Income after taxes has allowed households to purchase larger homes, to improve levels of indoor comfort, and to acquire more and larger appliances. While there was little growth in real per capita income in the 1974–75 period, and stagnation or decline in 1980–82, over the 1973–88 period per capita income grew by nearly 50% in Western Europe and Japan and by about 20% in the United States. Historically higher income levels in the United States (partly due to lower taxation) help explain the higher levels of appliance ownership that existed in the United States in the early 1970s relative to Europe or Japan.

Change in energy prices This influenced fuel choice and energy intensities. Prices affect the latter through behavioral responses, decisions to improve existing homes and equipment, and decisions concerning new equipment and homes. The retail price of heating oil rose by about a factor of three in most OECD countries between 1973 and 1981 (Fig. 5.2). It declined thereafter in the United States and Japan, but remained high in Western Europe (due to taxation) until the crash of 1986. In the United States, the price of natural gas, the most important household fuel, rose less than the price of oil. In contrast to oil, electricity prices remained about the same or

Table 5.2. *Change in residential energy use, population, and aggregate intensity between 1972/73 and 1988* (*total % change*)

	Energy use[a]	Population	Energy per capita[a]
United States	+3	+16	−11
Japan	+78	+13	+59
Europe-4	+16	+3	+12
Scandinavia-3	+8	+4	+0.4
OECD-9	+10	+10	0

[a] Useful energy.

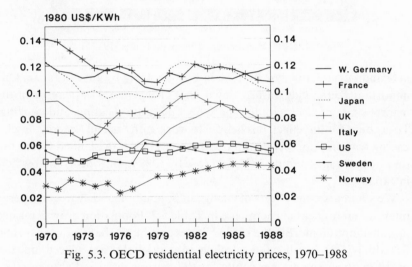

Fig. 5.3. OECD residential electricity prices, 1970–1988

declined in most countries in the 1970s (Fig. 5.3). There was a jump in some countries in 1980–82, but the general trend since then has been one of decline, with West Germany being an exception. Trends in Western Europe have differed among countries due in part to taxation policy.

5.2.2 Energy use and activity

Between 1972/73 and 1988, residential final energy use increased at an average annual rate of 0.9% in Europe-4 and 3.3% in Japan. Energy use was about the same in the United States in 1988 as in 1973, while it declined slightly in Scandinavia-3.

Final energy use per capita declined by 15% in the United States, by 8%

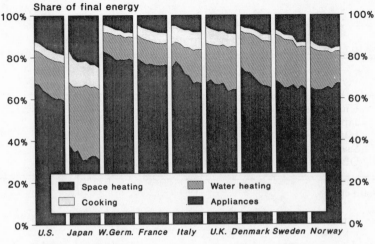

Fig. 5.4. OECD residential energy end-uses, 1972–1987

in Scandanivia-3, but rose by 10% in Europe-4 and by 46% in Japan. One important factor that pushed down final energy use was increase in the market share of electricity for space heating, water heating, and cooking. To account for this effect, it is helpful to work with "useful" energy. Useful energy consumption per capita declined by 11% in the United States but rose by 0.4% in Scandinavia-3, and rose by 12% in Europe-4, and by 59% in Japan.

We summarize the effects of change in population and aggregate energy intensity on residential energy use in Table 5.2. Population growth pushed up consumption more in the United States and Japan than in Europe-4 and Scandinavia-3. In all cases, decline in household size placed upward pressure on energy use per capita for the reasons discussed above.

To understand why aggregate energy intensity changed as it did, it is necessary to look at each of the main residential energy end uses. The importance of the different end uses in total energy use is shown in Fig. 5.4. Space heating accounts for 60% or more of total residential energy use in all of the OECD countries except Japan. Its share has declined in all countries, however, due both to growth in non-heating uses and reduction in space heating energy intensity. The largest growth has been for "electric-specific" uses (those for which electricity is the only or clearly favored choice over other fuels, such as lighting, refrigeration, and mechanical appliances).

5.2.3 Space heating

Structural change The average size of homes has increased moderately in the OECD countries despite the decline in household size. Growth in dwelling size had the most impact in Japan, and the least in the United States, where homes were already relatively large in 1972. Floor area per capita increased from 42 to 52 m² in the United States, from 21 to 28 in Japan, from 25 to 33 in Europe-4, and from 34 to 45 in Scandinavia-3.

A change that had a major impact on energy use for heating in Europe-4 was transition from room heating to central heating, which provides a greater level of amenity.[5] Between 1972 and 1988, the percent of homes with central heating grew from around 40–45% to 65–75% in Europe-4. This transition was not a factor in the United States and Scandinavia, where the share of central heating was already high in 1972. A comparable change in Japan was the considerable increase in the number of room heaters per home and in the size of heaters.

The mix of dwelling types in the housing stock has not changed very much in most of the OECD countries. All else being equal, it would seem that a housing stock with a high proportion of apartments would need less space heat per unit area than one with a low proportion, since apartments have less external surface area exposed to the elements. Data for Sweden and the United States show, however, that energy use per square meter is higher in apartments than in detached houses, particularly where centrally-heated apartments lack individual metering for space heating. The combination of higher heat distribution losses and higher use permitted by lack of metering more than offsets the reduction in heat losses due to less exposed surface area.

Energy intensity Space heating energy intensity – expressed as useful energy per m² of area – has declined in all of the OECD countries except Japan and Norway since 1972–73.[6] The decline was much greater in the United States (42% in 1973–88) than in Europe-4 (15%) and Scandinavia-3 (24%) (Fig. 5.5). In Japan, there was some decline between 1978 and 1980, but intensity rose thereafter for the reasons mentioned above. In the

---

[5] By "central heating," we refer to heating systems that are centrally controlled and are capable of heating the whole house.

[6] We include estimated energy use by secondary space heaters under space heating energy use. Increased use of secondary heaters was an important factor in the United States and Scandinavia. We also include estimated use of wood, much of which was used in secondary heaters. Energy use data for space heating have been normalized for climate within each country over time and among countries.

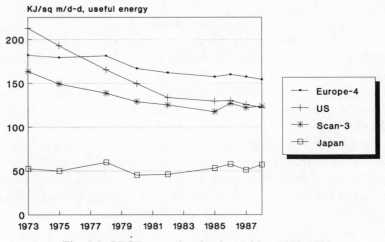

Fig. 5.5. OECD space heating intensities, 1973–1988

United States and Scandinavia, most of the reduction in energy intensity took place in the 1970s. There was much less decline in the 1980s, and intensity increased in Scandinavia after 1985.

The above intensity indicator includes the effect of growth in central heating. In Europe-4 especially, such growth obscures the improvement in technical efficiency and decline in indoor temperatures that took place. Growth in the number and size of heaters in Japan also pushed intensity upwards. If not for these changes, we estimate that space heating energy intensity would have declined by 28% in Europe-4 (instead of 15%), and by 35% in Japan (instead of a slight increase).

The decline in space heating energy intensity has been due to physical and behavioral changes in older homes, and introduction of new homes with improved thermal integrity and heating equipment. Estimates of the reduction in space heating energy intensity in pre-1975 homes between 1973–75 and 1985 are shown in Table 5.3 for four countries. The reductions were largest for oil-heated homes, as would be expected. These reductions were due to a combination of forces: lower indoor temperatures, improvements in thermal integrity and heating equipment, and greater use of secondary heating fuels.

The reduction in average indoor temperature has been an important factor in most countries, according to various surveys of reported practices. In the United States winter indoor temperatures declined considerably after the 1973–74 "Energy Crisis," and declined further after the oil price

Table 5.3. *Reduction in space heating energy intensity in pre-1975 homes between 1973–75 and 1985 (%)*

Heating fuel	United States	France	Sweden	Denmark
Oil	40	28	25	40
Gas	25	16	—	—
Electricity	25	17	10	25

Source: Authors' estimates based on surveys from each country.

rise in 1979–80. The percentage of homes keeping the daytime temperature at 70°F or above (when someone is at home) declined from 85% in the 1972–73 winter to 52% in 1974–75, and to 46% in 1981–82 (Meyers, 1987). Decline in nighttime temperature from 1972–73 to 1974–75 was even greater: from 51% of homes at 70°F or above to 29%. After 1981, temperatures increased, reflecting stabilizing of energy prices and lessened concern over energy. Thus, between 1972 and 1981, change in heating practices depressed heating energy intensity significantly, but thereafter it increased intensity.

Surveys in Denmark and West Germany show a similar pattern of sharp declines in indoor temperature in 1973–75 and 1979–81 followed by gradual rebound. The relative magnitude of the decline was probably not as great as in the United States, however, since the average temperature was higher in the United States in the early 1970s. In Japan as well, there were rapid declines in 1973–74 and 1979–80, but they were reversed within a few years.

The US data illustrate the importance of home occupancy. Since the average daytime temperature is much lower when no one is at home than when someone is, it follows that increase in the fraction of hours that a home is unoccupied decreases average temperature. In Europe and Japan, where apartments comprise a larger fraction of the housing stock, the impact of reduced occupancy has been smaller, since an unoccupied apartment gets heated somewhat by its neighbors.

The improvement of the thermal integrity and equipment of pre-1975 homes has been considerable in Western Europe and North America. Homeowners have increased insulation in ceilings and walls, added storm windows and doors, and reduced heat leaks, as evidenced by surveys in Denmark, Sweden, France, the United States, and the United Kingdom.

Retrofit activity was greater for owner-occupied homes (most detached dwellings) than for those occupied by renters (most apartments) for obvious reasons. Heating equipment tune-up and replacement of old equipment also played an important role. In the United States, new gas furnaces were about 15% more energy efficient in 1987 than in 1975 (US DOE, 1988). We estimate that the degree of improvement was somewhat less in Western Europe, in part because intensity was already lower than in the United States in 1975. In Japan, considerable growth in use of heat pumps pushed downward on heating energy intensity.

It is difficult to separate the impact of retrofit and equipment replacement from that of change in heating practices. In the United States, however, data from national surveys show that heating intensities declined considerably between 1981/82 and 1987 despite the rise in indoor temperature described above (US EIA, 1983, 1989). This indicates that retrofit and equipment replacement had an important effect in this period.

The introduction of new homes to the housing stock has decreased average space heating energy intensity. Dwellings built after 1974 have higher thermal integrity and more energy-efficient heating equipment than those built earlier. In France, for example, space heating energy consumption in 1985 in single-family houses was 16% lower in post-1974 oil-heated homes than in pre-1975 ones. The difference was 24% for gas- and electric-heated homes. In the United States in 1984–85, heating intensity in gas-heated single-family houses was 23% less in 1950–74 dwellings than in pre-1950 dwellings, and was 40% less in post-1974 dwellings (US EIA, 1987). The pre-1975 homes had been considerably improved through retrofit and equipment replacement by 1985. Thus, if we could compare the consumption in the post-1975 homes to what would have been the case in pre-1975 homes had retrofit not occurred, the difference would be greater than noted above.

The addition of new housing has been most significant in the United States because there has been more new construction than in Western Europe. Of all homes existing in 1987, about 25% were built after 1974 in the United States. In Western Europe, on the other hand, only 15–18% of the 1987 stock was built after 1974.

Decomposing change in space heating energy use Space heating useful energy per capita declined by 27% in the United States, by 4% in Scandinavia-3, and rose by 11% in Europe-4 and by 47% in Japan. The change consists of an increase caused by structural change (more area per capita and changes in heating equipment) and a decline in "adjusted"

Table 5.4. *Decomposition of the change in per capita space heating energy use, 1972/73–1988 (total % change)*

	Energy per capita	Structure	Adjusted intensity[a]	Interaction
United States	−27	+31	−45	−13
Japan	+47	+67	−35	+15
Europe-4	+11	+54	−26	−17
Scandinavia-3	−4	+33	−27	−10

[a] Useful energy per m²; values have been adjusted to remove the estimated impact of increased central heating and, in Japan, growth in number and size of heaters.

intensity (Table 5.4). Growth in floor area per capita pushed strongly upward on energy use in all countries. Growth in central heating pushed upward in Europe-4, as did increase in number and size of heating equipment in Japan. Structural change in these countries increased aggregate intensity more than the adjusted end-use intensity fell. Because the decline in adjusted intensity was the greatest in the United States, while the structural changes were the least, aggregate US heating intensity fell much more than in the other countries/regions.

5.2.4 Water heating

Hot water is used in the home for washing of people, dishes (by hand or machine), and clothes. Water heating accounts for 18–23% of total residential energy use in the United States, Scandinavia, Italy, and the United Kingdom; 11–12% in West Germany and France; but about 40% in Japan. The large fraction in Japan is not due to high water heating energy intensity, but is a result of the low share of space heating.

Structural change affecting water heating has several aspects. First, households must acquire running hot water; about 5% of homes in Europe-4 did not have this amenity in 1973. Another is change in equipment. Hot water may be provided by a central heater with storage (belonging to the house or to the building), small storage heaters, or by point-of-use heaters without storage. Clothes washers and dishwashers that heat their water internally, which are the rule in Western Europe but are rare in the United States, can be considered as point-of-use heaters. Since central systems have higher stand-by and distribution losses, a transition to them from point-of-use heaters tends to increase energy use. Another type of structural change is increase in the ownership of appliances that use hot water, particularly clothes washers and dishwashers.

The changes described above have pushed upward on energy intensity, but other factors have pushed downward. Some are behavioral, such as increase in clothes washing with lower-temperature water, use of energy-saving options on dishwashers, and reduction of thermostat settings on water heaters. Other measures concern equipment, such as installation of low-flow showerheads and better insulation of water heater tanks and distribution lines. There has also been improvement in the energy efficiency of new water heaters – about 15% between 1978 and 1984 for gas water heaters in the United States (US DOE, 1988). Improvements in clothes washers that have lowered water requirements per wash have also reduced the energy intensity of water heating.

Energy use for water heating is difficult to estimate without monthly or quarterly data on consumption because usually the same fuel (and sometimes the same equipment) is used for space and water heating. Our estimates, which are probably accurate to within 10–15%, show no net change in water heating useful energy per capita between 1972–73 and 1988 in the United States and Scandinavia-3, a 30% increase in Europe-4,[7] and 60% increase in Japan. The large increase in Europe-4 and Japan clearly reflects growth in ownership of clothes washers and dishwashers, as well as increased market share of storage heaters.

5.2.5 Cooking

Cooking accounts for around 5% or less of total residential energy use in most of the OECD countries, and around 10% in Italy and Japan. The low share of cooking in final energy is partly attributable to the high market penetration of electric cooking.

Cooking habits have changed considerably as a result of decline in household size, greater participation of women in the labor force, introduction of specialized cooking equipment, and change in eating habits. More meals are eaten away from home than in years past, and more ready-to-eat foods are brought into the home. There has also been some improvement in equipment efficiency. In West Germany, for example, there was a decline of 16% in the power consumption (kW) of new electric ranges between 1978 and 1985. Increase in ownership of microwave ovens

---

[7] We include estimated energy use for heating water by clothes washers and dishwashers that heat their water internally.

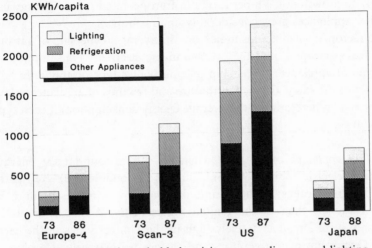

Fig. 5.6. OECD household electricity use, appliances and lighting

has also contributed to decline in cooking energy intensity. The proliferation of small devices such as food processors, coffee makers, and juicers has not had much impact on energy intensity, since these devices use minimal amounts of energy.

Because cooking involves various equipment that may use different fuels, it is difficult to accurately assess its energy intensity. Our estimates show a slight decline in useful energy per capita in the United States and Europe, and no net change in Japan.

5.2.6 *Electric-specific end uses*

The largest growth in residential energy use in the OECD countries has been for so-called "electric-specific" end uses. These uses include lighting, food preservation (refrigerators and freezers), clothes washing and drying,[8] dishwashing, entertainment (television, stereo), air-conditioning, and other miscellaneous uses.

Between 1972/73 and 1988, "appliance" electricity use per capita increased by 30% in the United States, 59% in Scandinavia-3, 125% in Europe-4, and 105% in Japan. (By "appliance" electricity use we refer to use for all electric-specific end uses, including lighting.) As shown in Fig. 5.6, this indicator was already far higher in the United States in 1973 than in other countries. Increase in energy use for refrigeration contributed

---

[8] While most dryers in OECD countries use electricity, around 20% of clothes dryers in the United States are gas-fired.

strongly to growth in kWh per capita in Europe-4 and Japan. Energy use for other appliances grew considerably in all countries.

Two factors determine this indicator: the average ownership of various appliances per capita, and the average in-use electricity consumption of each type of appliance. Ownership of a particular type of appliance tells only part of the story. Change in the size and features of appliances plays a major role in determining the average energy consumption of each type of appliance.

Growth in appliance ownership Ownership of the four largest energy-consuming electric appliances (not including air-conditioners) – refrigerators, freezers, clothes washers, and dishwashers – increased more in Western Europe than in the United States (Table 5.5). The levels of ownership were already relatively high in the United States in the early 1970s. Ownership of refrigerators and clothes washers was also high in Japan in the early 1970s, but freezers and dishwashers were and are still uncommon.

There has been substantial increase in ownership of air conditioners in the United States – from around 46% of homes in 1973 to 63% in 1988. The percent of homes with central air conditioners, which use more energy than room air conditioners, rose from around 16% to 34%. Japan has also seen considerable growth in penetration of room air conditioners – from only 12% of homes in 1973 to 63% in 1988. As in the United States, many of the Japanese units are heat pumps that are used for heating in the winter. In Europe, air-conditioning in homes remains uncommon, though units are becoming popular in Italy.

Lighting was present in nearly all homes in all countries in 1973, but there has been some growth in the number of lamps per household as a result of increase in home size.

Increase in the average number of appliances per person has been greater than that per household due to the decline in household size. Typical 1- or 2-person households have many of the same appliances as a larger household, so the distribution of the population into smaller households increases the number of appliances per person. Thus, decline in household size has pushed upward on appliance electricity use per capita.

Appliance energy intensity Change in unit energy consumption (UEC, which refers to energy use per year) for a given type of appliance depends on the characteristics of old and new appliances, the rate of new appliance purchase, and aspects of household usage. Estimates of how the stock

Table 5.5. *Appliance ownership in OECD countries (units per 100 households)*

	Year	Refrigerator	Freezer	Clothes washer	Dishwasher
United States	1973	99	34	70	25
	1988	113	35	73	43
Japan	1973	100	< 1	98	< 1
	1988	117	3–5	99	3–5
Europe-4	1973	83	13	69	5
	1988	114	45	89	21
Scandinavia-3	1973	93	50	56	7
	1987	106	72	77	27

Source: Tyler and Schipper (1990); Schipper and Hawk (1991).

Table 5.6. *Appliance unit energy consumption (kWh/year)*

	1973	1980–81	1986–87
Refrigerator[a]			
United States	1450	1380	1310
Japan	395	645	610
West Germany	770	670	600
Clothes dryer			
United States	1050	1050	990
Japan	~ 355	355	355
West Germany	475	425	270
Dishwasher			
United States	365	250	250
West Germany	800	625	310

[a] With separate freezer compartment.
Source: Tyler & Schipper (1990); Schipper & Hawk (1991).

average UEC has changed for several appliances, based on various national sources, are given in Table 5.6. The UEC for refrigerator-freezers declined by 10% in the United States and by about 20% in West Germany, but increased by 50% in Japan due to growth in size and features. For clothes dryers, the UEC dropped slightly in the United States, but considerably in West Germany, as a new generation of devices came to dominate the stock. A similar effect applied for dishwashers in West Germany and, to a lesser extent, in the United States.

Stock turnover had mixed effects on the average UEC of major appliances. Improvement in the energy efficiency of new appliances of a given size was considerable in most cases. The energy use per unit of size or service of the average new appliance (weighted by sales) declined considerably in several countries for which data are available (Table 5.7). The decline was largest for refrigerators, freezers, and dishwashers.

The efficiency effect was partially offset by increase in the size and/or features of the average new model purchased, however. Increase in average size was most significant for refrigerators. This effect was especially strong in Japan, where the market share of refrigerators of 170 liters and above increased from less than 5% in 1972 to 63% in 1988. The average electricity use per liter of one manufacturer's new refrigerators in the 150–350 liter class fell by 75% between 1972 and 1984,[9] but the average electricity use of all new refrigerators declined much less because of shift in the sales mix toward larger models. Similarly in the United States, electricity use per liter of the most popular model (top-mount, auto-defrost) declined by nearly 50% between 1972 and 1984, but increase in size and features has led to a smaller improvement in the average energy use per device of all new refrigerators.

Increase in size has also been a factor for TV sets. While the rated wattage of new TV sets of a given size fell by as much as 75% as vacuum-tube sets yielded to solid-state models, the average wattage of all new TV sets declined less because of increase in size and an almost complete transition from black and white models to more energy-intensive color ones.

Despite considerable exhortation to conserve energy, permanent change in household utilization has probably not been very significant for most electric-specific appliances. Behavior may have reduced lighting energy intensity somewhat, at least during periods of heightened energy aware-ness. There have also been short-term behavioral changes for air conditioners, since users tend to be more sensitive to the electricity costs of operation, and also have the opportunity to vary utilization. There is no sign that TV viewing has declined, however!

In the United States, the energy intensity of air-conditioning has been depressed by increase in the thermal integrity of older homes and incorporation of "new" homes into the stock. There has also been improvement in the energy efficiency of new room and central air

---

[9] One reason for the large decline is that the size of the refrigerator in 1984 (350 liters) was much larger than in 1972 (150 liters). Other things being equal, energy use per liter is inherently less in a larger refrigerator because it has a lower surface-to-volume ratio and thus lower heat losses per unit of volume.

Table 5.7. *Reduction in the energy intensity of new electric appliances (%)*

Country	Refrigerator	Freezer	Washer	Dishwasher	Oven	Air Cond.	Years
United States[a]	51	47	35	27	—	29/26[d]	1972–89
West Germany[b]	21	37	18	29	16	—	1978–85
Denmark[c]	25/47	32	30	39	13	—	1970–90

[a] Sales-weighted averages from Association of Home Appliance Manufacturers.
The refrigerator figure refers to a refrigerator-freezer.
[b] Sales-weighted averages tabulated by Zentral Verband der Elektroindustrie.
(Tabulations were discontinued after 1985.)
[c] Rough estimate of sales-weighted average improvement made by Moeller and Nielsen (1991).
The second figure for refrigerators refers to combination refrigerator-freezers.
[d] The figures refer to central (through 1987) and room air conditioners, respectively.

Table 5.8. *Decomposition of the change in appliance electricity use per capita between 1973 and 1987 (total % change)*[a]

	Change in kWh/capita	Structure[b] effect	Intensity[b] effect
United States	+23	+37	−13
Japan	+124	+74	+40
West Germany	+86	+82	−2
Norway	+44	+50	−4

[a] The appliances are refrigerator, freezer, refrigerator-freezer, clothes washer, dryer, dishwasher, and air conditioner for the United States and Japan. To estimate the impact of changing intensities (UECs), we multiplied the 1973 per capita ownership by UECs of each appliance. To estimate the impact of changing structure, we multiplied the actual per capita ownership levels of each appliance by the estimated 1973 UEC.
[b] The structure and intensity terms do not add up to the change in kWh/capita due to interaction between them.

conditioners. In Japan, the intensity of a typical window model fell by 43% between 1973 and 1988.

Summary For seven major appliances, we estimated the effects of change in ownership per capita and in UEC on total electricity use per capita for these appliances. Growth in ownership contributed strongly to increase in aggregate intensity in all four countries for which we were able to perform the calculations (Table 5.8). Change in UECs pushed downward on aggregate intensity in the United States, had little net effect in West Germany and Norway, but pushed strongly upward in Japan. Part of the reason for increase in UEC was growth in size and/or features of appliances, especially for refrigerators. In the United States it appears that much of the increase in aggregate intensity was due to increased ownership of air conditioners in general and central air conditioners in particular.

5.2.7 Comparing residential energy intensity among countries

Total residential useful energy per capita in 1988 was 34 GJ in the United States, 29 GJ in Scandinavia-3, 20 GJ in Europe-4, and only 10 GJ in Japan. To properly compare among countries, it is necessary to adjust heating energy use for climate differences. The long-run heating degree-day average (base 18°C) is about the same in the United States (2725) and Europe-4 (2743); it is considerably colder in Scandinavia-3 (3832), and

Table 5.9. *Household size and per capita home area in OECD countries*

		Household size	Home area per capita (m²)
United States	1973	3.1	41
	1988	2.7	52
Japan	1973	3.8	21
	1988	3.3	28
Europe-4	1973	3.1	25
	1988	2.7	33
Scandinavia-3	1973	2.6	34
	1988	2.3	45

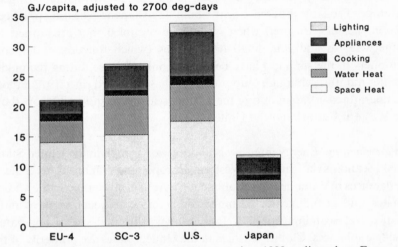

Fig. 5.7. OECD residential energy use per capita, 1988: adjusted to Europe-4 climate

much warmer in Japan (2040). With heating energy normalized to the average climate of Europe-4, per capita energy use declines somewhat in Scandinavia-3 to 27 GJ (Fig. 5.7).[10]

The differences in aggregate energy intensity are attributable to variation in the structure and energy intensity of each end use. Differences in household size and home area per capita also play a role (Table 5.9). The low level of household size in Scandinavia contributes to higher per capita energy use, and the opposite applies for Japan. High levels of home area

[10] The adjusted value for Japan does not increase very much despite the warmer climate; the reason is that space heating accounts for a relatively small share of total energy use.

per capita also contribute to higher energy use; larger home size not only means more area to be heated, cooled, and lit, but also allows for larger appliances. Home area per capita is especially high in the US and low in Japan, although Scandinavia is approaching the level of the US.

Space heating Energy intensity (normalized for dwelling area and climate) differs due to variation in heating practices as well as in the thermal integrity of buildings and the nature and efficiency of heating equipment. The intensity is about one-third greater in France and West Germany than in Sweden, Denmark, the United States, and Italy. Swedish homes have high thermal integrity and equipment efficiency, but also the highest average indoor temperatures. Still, Sweden has the lowest space heating intensity (except for Japan).

Heating energy intensity in Japan is only one-third of the European average. The large difference is attributable to Japanese heating practices. Room heaters are used when rooms are occupied, but are turned off otherwise. In addition, small heat pumps (which function as air conditioners in the summer) have become popular for use during the milder winter months. Although indoor comfort has increased since 1972, indoor temperatures in Japan during the heating season are still well below those in Western Europe and the United States.

Water heating Energy intensity is about twice as high in the United States and Scandinavia than in Western Europe and Japan. This reflects differences in water heating equipment (more central heating in the United States and Scandinavia), bathing habits (more frequent in the United States and Scandinavia), as well as in ownership and size of hot water-using appliances. Clothes washers in the United States are primarily of the vertical-axis type, which uses much more water than the horizontal-axis type that predominates in Europe.

Electric-specific end uses Despite the higher relative growth in appliance electricity use per capita in Europe-4 and Japan, intensity in these countries in 1988 (700–750 kWh) was still far below the US level (2350 kWh). Scandinavia-3 was intermediate at 1330 kWh. Removing air-conditioning would reduce the US value to around 2000. Part of the reason for higher energy intensity in the United States is greater levels of appliance ownership, including miscellaneous devices that are not trivial electricity users (e.g., swimming pool pumps, waterbed heaters). More important, however, is the larger size of most appliances in the United States, and the

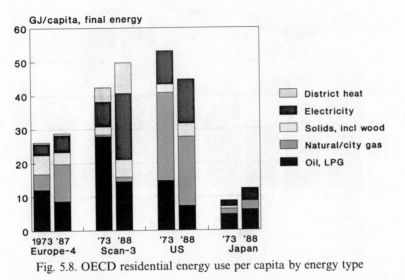

Fig. 5.8. OECD residential energy use per capita by energy type

larger size of homes, which leads to more lighting. Based on discussions with appliance manufacturers, it appears that new appliances of a given type and size have roughly the same energy efficiency in OECD countries.

5.2.8 Change in energy sources

Shifts among energy sources have been more significant in the residential sector than in other sectors. Most striking is the decline in the share of oil, which fell from 28% to 17% of final energy use in the United States, from 46% to 28% in Europe-4, from 66% to 29% in Scandinavia-3 (Fig. 5.8). It declined only slightly in Japan, however. The growth in share was greatest for electricity. In 1973, electricity accounted for only 13% of final energy consumption in Europe-4, 18% in Scandinavia, 19% in the United States, and 21% in Japan. By 1988, electricity had increased its market share to 17% in Europe-4, 40% in Scandinavia, 29% in the United States, and 29% in Japan. There was also considerable increase for natural gas in Europe-4.

The most important factor has been change in the sources used for heating. This shift was mainly due to fuel switching in older homes, though fuel choice in new homes also played a role (more so in the United States than in Europe). The share of homes using oil as primary heating fuel declined considerably in all of the countries except West Germany and Japan, where oil still predominated in 1988, and the United Kingdom, where it was never an important fuel (Fig. 5.9). The share of natural gas

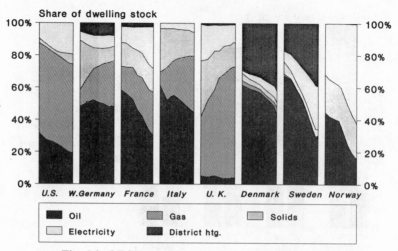

Fig. 5.9. OECD space heating fuel choice, 1972–1987

grew substantially in West Germany, France, and Italy, and it became the dominant fuel in the United Kingdom. The share of gas remained about the same in the United States as it gained share from oil in older homes but lost it to electricity in new homes. The penetration of electric heating increased in the United States, France (where it was strongly encouraged), and Scandinavia. In Western Europe, coal was practically eliminated as a heating fuel (this had occurred much earlier in the United States), but wood heating became increasingly popular in the United States, Sweden, and Norway, becoming the primary heating fuel in nearly 10% of homes.

For water heating, electricity gained in market share in the United States (to 35% of all households in 1988) and in Scandinavia-3, but not in Europe-4, where its share remained at about 40%. Gas remained the dominant fuel in the United States (55%), and gained share in Europe-4. As with space heating, oil lost share in most countries. For cooking, electric ranges gained market share in almost all countries, even where gas was cheap, as in the United Kingdom.

One reason for the increased share of electricity was the significant decline in the specific consumption of fossil fuels for space and water heating, which left electricity with a higher share of final consumption. But the nature of the electricity market also changed in each region. In Sweden and Norway, there were conversions in existing homes from other fuels to electricity for space and water heating and cooking. In other countries, increase in the stocks of electric-specific appliances was the main reason for increased electricity use, although electricity also captured a high share of space and water heating and cooking in new homes. In Japan, electricity holds a very small share of principal heating, cooking, or water heating, but provides secondary heating, which has grown recently as heat pumps became popular. Substitution of electricity for fossil fuels has depressed final energy use, since electricity does not have conversion losses at the building.

5.3 Former East Bloc

By OECD standards, living conditions in the Former East Bloc are spartan. Per capita home area in the Soviet Union and Poland in 1985 was less than 15m^2, compared to over 30 in Western Europe and over 50 in the United States. More than 10% of households lived in shared accommodations or simple barracks. Running hot water is found in only about 80% of Soviet homes.

Given the cold Soviet climate, it is not surprising that space heating dominates residential energy use.[11] In the Soviet Union in 1985, it accounted for about 75% of energy use (Fig. 5.10). About one-third of Soviet homes are heated by district heating systems fed by large boilers and combined heat and power systems. Another one-third are heated by smaller boilers, often serving a few neighboring buildings. Gas is the predominant fuel for both of these kinds of systems. The rest of Soviet homes are heated by stoves fueled principally by coal and other solid fuels. In Poland, coal provides most of the energy for space heating, although the use of gas and district heat has grown (Leach & Nowak, 1990).

The Soviet data for space heating are somewhat uncertain, but it appears that average heating intensity (adjusted for climate) is about 15% higher than in West Germany. This is misleading, however, since the Soviet housing stock has a higher share of apartments, and indoor temperatures are on average probably lower. Insulation levels in buildings are low, and windows are not well sealed. In other words, the energy requirements to heat a given area in the Soviet Union to a given temperature, corrected for climate, are far more than the difference in intensity would imply. Poland appears to have even higher space heating intensity, even though the penetration of true central heating is less and indoor comfort is lower than in the Soviet Union.

Energy use per capita in the Soviet Union for electric-specific appliances is low relative to Western Europe. Penetration of TV sets and refrigerators

---

[11] The Soviet winter averages about 4600 heating degree–days, more than the 4000 in Sweden, 3000 in Western Europe, and 2700 in the United States. Poland has approximately 3500 degree–days.

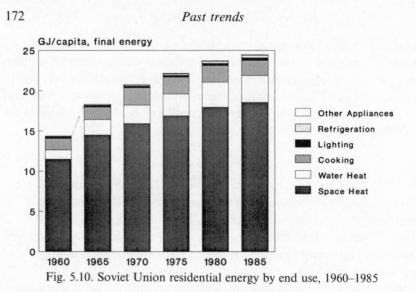

Fig. 5.10. Soviet Union residential energy by end use, 1960–1985

is over 90% of Soviet homes, but models are small by OECD standards. We estimate that only about 20% of the refrigerators are two-door style, and only about half of the TVs are color. Ownership of clothes washers is also high (80%), but most units are not fully automatic. The UEC of electric appliances appears high, relative to the services they deliver. In other words, residential electricity use in the Soviet Union is low yet inefficient. Polish homes have more appliances than do those in the Soviet Union and average consumption is also higher.

Lack of data makes it difficult to evaluate historic trends in energy intensity. In the Soviet Union, the intensity of space heating increased in the 1960s as central heating penetrated most urban buildings and comfort levels increased. During the 1970s and 1980s, there was a slow decline in intensity reflecting a gradual improvement in heating equipment and building practices, but little conscious effort to save energy. A similar trend is apparent in Poland between 1980 and 1985. In both countries, however, the efficiency of space heating and electric appliances is well below Western norms.

5.4 Developing countries

The structure of residential energy use differs considerably among and within LDCs depending on the level of per capita income and the degree of urbanization, which are strongly correlated. In lower-income countries (and among the poor in most countries), cooking accounts for the major share of residential energy use. One reason is that biomass fuels are still common and are used with low conversion efficiency. As income increases, water heating and electric-specific end uses come to play a more important role. Space heating is a minor end use in most of the developing world, with the important exception of China. Space heating energy intensity in China is relatively low; this is not because buildings have high thermal integrity, but because heating is severely restricted by decree. Most heating is done with relatively inefficient coal stoves.

5.4.1 Cooking

As urbanization and incomes have increased, there has been a major transition in the fuels used for cooking. The basic transition is from biomass to kerosene to LPG and, in some cases, electricity, though it is not uncommon for households to retain their old cooking devices and use them when desired. The transition away from biomass has occurred to a much greater extent in urban than in rural areas, both because incomes are higher and because fuel availability is better. Surveys in several countries show a gradual decrease in use of biomass as city size increases (Meyers & Leach, 1990). In rural areas of many poor countries, there is evidence that increasing scarcity of wood has led to longer collection times, more careful use, and greater reliance on less-desired biofuels such as agricultural residues.

The transition to LPG was already well-advanced in Latin American cities in the early 1970s. Since then, LPG has gained ground in cities and in rural areas as well. In Brazil, for example, the use of LPG or pipeline gas for cooking increased from 43% of households in 1970 to 78% in 1985 (Geller, 1991). In Asia, movement from biomass to kerosene, and especially from kerosene to LPG, has been occurring rapidly in the past decade (Sathaye & Meyers, 1985). In sub-Saharan Africa, there has been less change in fuels, and even movement back to biomass, due to decline in per capita income. Biomass in the form of charcoal is a common cooking fuel in African cities. The transition away from biomass decreases cooking energy intensity, measured in terms of final energy, because kerosene and LPG stoves are considerably more efficient than biomass cooking.

5.4.2 Electric-specific end uses

The electric end use that has seen the largest growth in LDCs is lighting. Rural electrification has advanced considerably in most countries, and lighting is the first end use that households acquire. After lighting, the largest growth in ownership of major appliances has occurred for TV sets and refrigerators. In Thailand, for example, the percentage of households owning a TV set increased from 11% in 1976 to 56% in 1986 (Meyers et al., 1990). For refrigerators, the increase was from 5% to 21%. In Brazil, ownership of TV sets grew from 39% to 66% between 1974 and 1988, while refrigerators increased from 36% to 63%. There has also been growth in ownership of clothes washers in upper-middle-income and upper-income countries. Air conditioners are as yet uncommon, except in Taiwan and the upper-income Middle East countries, but their ownership is growing.

Lack of data makes it difficult to estimate how UECs have changed for electric-specific end uses. Lighting energy use per household has undoubtedly risen along with the average number of lamps per home. For other appliances, there is evidence that efficiency has improved for new models of a given size and type. In Brazil, for example, test data show that the average energy use of new one-door refrigerators of 250–300 liter size declined from 490 kWh per year in 1986 to 435 kWh in 1989 (Geller, 1991). In South Korea between 1980 and 1987, manufacturers report a decline from 672 to 240 kWh per year for 200-liter refrigerators, and from 82 to 60 W for 14-inch TV sets (Meyers et al., 1990). In Brazil, the improvement occurred in part as a response to the government's testing program. In other countries, some improvement has occurred as manufacturers introduce new models. New models in most LDCs are still well behind the state-of-the-art in the manufacturers' home countries, however (Schipper & Meyers, 1991).

While there has probably been some improvement in the efficiency of new appliances of a given size, the average UEC of new appliances has likely increased in most countries due to the growing market share of larger and more feature-laden appliances. This change is especially important for refrigerators, for which the market share of two-door models (with separate freezer) is growing in upper-middle-income and upper-income countries. The average size of TV sets has also increased, as has the market share of color TV; both of these trends increase the average UEC of new TVs. The ownership of fully automatic clothes washers has also grown.

5.4.3 Fuel use trends

Time-series data on residential "commercial" energy use are lacking for most LDCs. (The difficulty is in separating residential and service sector use of non-electric fuels.) Our group has organized data for the combined residential/services sector for a number of countries, however, and in the case of kerosene, most of the consumption is in households. As shown in

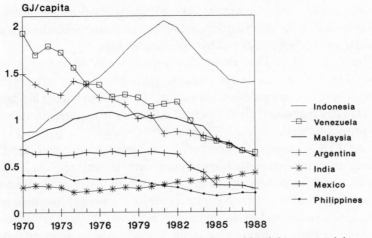

GJ/capita

— Indonesia
-□- Venezuela
— Malaysia
-+- Argentina
-*- India
-+- Mexico
-•- Philippines

Fig. 5.11. Per capita kerosene use in LDCs; residential/commercial sector

Fig. 5.11, the trend in kerosene use per capita has varied considerably among countries, and the actual levels are quite different also. In Latin America, use has declined considerably as LPG displaced kerosene for cooking and electricity displaced it for lighting. These phenomena have also taken place in Asia, but to a lesser degree, and there has been more movement in cooking from biofuels to kerosene, as illustrated in the case of India. In Indonesia, on the other hand, kerosene use grew rapidly in the 1970s because the price was heavily subsidized. When the price was increased in 1982, use declined sharply. While rural electrification played a role in the decline, there was also movement from kerosene to biofuels among poor households.[12]

5.5 Conclusion

In the OECD countries, income-driven growth in equipment ownership (heating, appliances) and home size drove household energy use up, but higher energy prices and conservation programs had a restraining effect. The results were mixed, with consumption per capita significantly lower in a few countries (United States, France), but higher in others. There was a significant decline in the intensity of space heating (30–40%) and a small decline in the intensity of electric appliances. Changes in the size and features of many appliances offset much of the improvement in technical energy efficiency. Not all of the decline in heating intensity was a result of

[12] The high level of kerosene use in Indonesia is somewhat misleading, because a significant amount of "residential/commercial" kerosene ends up being used as a transport fuel.

technical change; we estimate that about 25% was caused by change in heating behavior. In all, there were significant improvements in efficiency, but these were offset somewhat by structural change.

In the Former East Bloc, there is far less residential space and amenity than in OECD countries, and efficiency of space heating and water heating is low, in part due to lack of energy pricing. Electric appliances are simple and relatively inefficient. Unlike in the OECD countries, there is little sign of improved efficiency in the 1970s and 1980s.

In the LDCs, patterns of energy use, and changes in them, are very different in rural and urban areas, and vary among regions as well. Biomass is still the dominant fuel in rural areas. In urban areas, Western-like patterns of electricity (and even gas) use have emerged for appliances, cooking, and water heating among the affluent, and TV and refrigerators have become more common among the less-affluent. In many countries, especially in Southeast Asia, there has been very rapid growth in ownership of appliances. Most appliances are cheaply made and inefficient compared to similar appliances sold in the OECD countries, though there are signs of some improvement in the past decade.

References

Gasunie (Dutch Gas Authority). 1986. *Basisunderzoek Aardgas Kleinverbruik.* Groningen, Netherlands.

Geller, H. S. 1991. *Efficient Electricity Use: A Development Strategy for Brazil.* Washington, DC: American Council for an Energy-Efficient Economy.

Ketoff, A. & Schipper, L. 1990. Looking beyond the aggregate figures: what really happened to household energy conservation. In E. Vine, and D. Crawley (eds.) *State of the Art of Energy Efficiency: Future Directions.* Washington, DC: American Council for an Energy-Efficient Economy.

Leach G. & Nowak, Z. 1990. *Cutting Carbon Dioxide Emissions from Poland and the United Kingdom.* Stockholm, Sweden: The Stockholm Environment Institute.

Meyers, S. 1987. Energy consumption and structure of the US residential sector: Changes between 1970 and 1985, *Annual Review of Energy,* **12**, 81–97.

Meyers, S. & Leach, G. 1990. *Biomass Fuels in the Developing Countries: An Overview,* Berkeley, CA: Lawrence Berkeley Laboratory Report LBL-27222.

Meyers, S., Tyler, S., Geller, H. S., Sathaye, J. & Schipper, L. 1990. *Energy Efficiency and Household Electric Appliances in Developing and Newly Industrialized Countries,* Berkeley, CA: Lawrence Berkeley Laboratory Report LBL-29678.

Moeller, J. & Nielsen, B. 1991. *Ny Model for boligers elforbrug* (New Model for Household Electricity Use). Lynbgy, Denmark: Dansk Elvaerkers Forenings Udredningsinstitute (DEFU), Teknisk rapport 281.

Sathaye, J. & Meyers, S. 1985. Energy use in cities of the developing countries, *Annual Review of Energy,* **10**, 109–33.

Schipper, L. & Hawk, D. 1991. More efficient household electricity use: An international perspective. *Energy Policy*, **19**, 244–65.

Schipper, L. & Ketoff, A. 1985. Changes in household oil use in OECD countries: Permanent or reversible? *Science* **224** (6 December).

Schipper, L., Ketoff, A. & Kahane, A. 1985. Explaining residential energy use with international bottom-up comparisons. *Ann. Rev. Energy*, **10**, 341–405.

Schipper, L. & Meyers, A. 1991. Improving appliance efficiency in Indonesia. *Energy Policy*, **19** (6), 578–88.

Tyler, S. & Schipper, L. J. (1990). The dynamics of electricity use in Scandinavian households. *Energy*, **15** (10), 841–63.

US DOE (Department of Energy). 1988. *Technical Support Document: Energy Conservation Standards for Consumer Products: Refrigerators, Furnaces, and Television Sets, including Environmental Assessment (DOE/EA-03722) Regulatory Impact Analysis*, DOE/CE-02239. Washington, DC.

US EIA (Energy Information Administration). 1983. *Residential Energy Consumption Survey: Consumption and Expenditures, April* 1981 *through March* 1982, *Part* 1: *National Data*. DOE/EIA-0321/1(81). Washington, DC: US Department of Energy.

  1987. *Residential Energy Consumption Survey: Consumption and Expenditures April* 1984 *through March* 1985, *Part* 1: *National Data*. DOE/EIA-0321/1(87). Washington, DC: US Department of Energy.

  1989. *Residential Energy Consumption Survey: Household Energy Consumption and Expenditures* 1987, *Part* 1: *National Data*. DOE/EIA-0321/1(89). Washington, DC: US Department of Energy.

6

Historic trends in the service sector

The service sector consists of a wide range of activities whose common feature is the provision of services rather than the production of goods. It is often called the "commercial" sector, though some of the activities it includes are not really commercial in character (e.g., education, provision of social services). The types of activities included in this sector include wholesale and retail trade; finance, insurance, and real estate; business and personal services; education; and social and government services. Value-added in transportation (including warehousing), communications, and public utilities (which are separate sectors in national accounts) is usually included in the service sector, as is most of the energy used in the buildings that support such activities. Except for street lighting, water works, and a few other miscellaneous uses, all of the energy use in this sector takes place in buildings of one kind or another.

Across regions, the service sector accounts for a relatively small share of final energy use: about 10% in the OECD countries, 5% in the Former East Bloc, and even less in most LDCs. In all of the above groups, however, it accounts for a larger share of electricity than of energy use: about 25% in the OECD countries, and about 13% in the LDCs. In the United States and Japan, and in some of the wealthier LDCs, its electricity use is driving peak demand.

6.1 Analyzing service sector energy use

One measure of activity in the service sector is the sector's value-added. Time-series data on value-added are published in national accounts, but this measure is problematic because the sector typically includes activities such as transportation and public utilities, for which only a small part of the activity takes place in buildings. In addition, the activity that occurs in

178

the office buildings of industrial companies is counted as part of industrial rather than services GDP. Floor area is in our view a preferable measure, but time-series data on floor area are not available in many cases, and estimates of total floor area in the service sector are typically somewhat uncertain since there are rarely comprehensive surveys done. In some countries the data refer to total floor area, while in others they refer to conditioned or heated area. For these reasons, we rely on value-added as an aggregate measure of activity, but report data using floor area where available.

Structural change in the service sector may be viewed in a similar manner to manufacturing – that is, in terms of change in the role of the various subsectors in total sectoral value-added. As in manufacturing, there are differences in energy intensity among subsectors, though the differences among service subsectors are much smaller than among manufacturing subsectors. Unfortunately, we do not have matching data for energy use and value-added at the subsectoral level for any OECD country. For several countries, however, we do have matching data at the subsectoral level for energy use and floor area, so we are able to analyze the impact of change in subsectoral mix on aggregate energy intensity.[1]

The breakdown of energy use in the service sector by end uses is very uncertain. Clearly though, a substantial fraction of service-sector energy use goes for space heating and air-conditioning in OECD countries. Because service-sector buildings often have considerable internal heat gain from lighting, other equipment, and occupants, they have a much greater need for air-conditioning than do residential buildings in a similar climate. Indeed, internal heat gains may supply a high fraction of space heating requirements in many climates. Energy use for ventilation and lighting accounts for substantial shares of total energy use. Given these characteristics, and the common use of various types of electrical office equipment, electricity accounts for a much higher fraction of total energy use in the service sector than in any other sector. Thus, it is especially important to look at trends with respect to electric and non-electric energy forms separately (we refer to the latter as "fuels").

---

[1] Service subsectors are defined in a uniform manner among countries in national account statistics. The categories that are found in floor area estimates differ somewhat among countries, however, making it difficult to draw comparisons.

6.2 OECD countries

Using data from country sources, we organized a data base on the service sector for nine OECD countries.[2] The United States accounted for 61% of total OECD-9 energy use in 1988, Japan for 10%, Europe-4 (France, West Germany, Italy, and the United Kingdom) for 27% , and Scandinavia-3 (Denmark, Norway, and Sweden) for 3% (Table 6.1).

An important difference between the service sector and other sectors is that electricity accounts for a higher share of final energy consumption. Its share in 1988 ranged from 31% in Europe-4, where space heating is relatively more important and air-conditioning less important, to 56% in Scandinavia-3 where substantial substitution of electricity for heating fuel has occurred. The share in Japan is 46% and 43% in the United States. As in the residential sector, substitution of electricity for non-electric sources in space and water heating depresses final energy consumption, while primary energy consumption increases. While it would be preferable to express services energy use in terms of "useful energy," lack of end-use energy data precludes this. However, since space and water heating are relatively less important in the service sector than in households, and there has been less shift to electricity for heating, failure to use "useful energy" does not have a large effect.

6.2.1 Energy use and activity

Between 1973 and 1988, total service-sector energy use increased at an average annual rate of 3.9% in Japan, 0.8% in Europe-4, 0.6% in Scandinavia-3, and 0.8% in the United States.[3] As shown in Table 6.2, increase in activity pushed strongly upward on energy use in all countries. Annual growth in sectoral value-added averaged 3.7% in Europe-4, 3.1% in Scandinavia-3, 3.6% in the United States, and 4.2% in Japan.

Aggregate service-sector energy intensity, expressed in terms of final energy use per unit of value-added, declined by 27% in the United States, by 15% in Japan, by 28% in Europe-4, and by 26% in Scandinavia.

Energy use per square meter fell much less (16%) than did energy use per

[2] For a presentation of earlier results and more in-depth discussion of service-sector energy use, see Schipper, Meyers, and Ketoff (1986).

[3] We have not attempted to adjust service-sector energy use data for annual fluctuations in climate. Such adjustment would result in some differences in the trends described in this chapter, though annual fluctuation in weather has less impact on energy requirements in the service sector than in the residential sector because service-sector buildings are on average larger than residential buildings and so have less surface area exposed to the climate per indoor floor area than smaller "envelope-dominated" residential buildings.

Table 6.1. *Service sector energy use and value-added in the OECD-9 in 1988*

	Energy use		Value-added	
	exajoules	%	bn 1980 $	%
United States	6.63	61	2210	52
Japan	1.11	10	692	15
Europe-4	2.73	27	1279	30
West Germany	1.12	10	338	8
United Kingdom	0.78	7	312	7
France	0.69	6	331	8
Italy	0.35	3	229	7
Scandinavia-3	0.29	3	101	2
Sweden	0.15	1	47	1
Denmark	0.07	1	29	2
Norway	0.07	1	25	1
OECD-9	10.76	100	4282	100

Table 6.2. *Change in service sector energy use, value-added, and aggregate energy intensity between 1973 and 1988* (*total % change*)

	Energy use[a]	Value added	Aggregate intensity[a]
United States	+12	+54	−27
Japan	+58	+85	−15
Europe-4	+12	+55	−28
France	+18	+57	−25
West Germany	+1	+58	−37
Italy	+44	+54	−6
United Kingdom	+15	+51	−24
Scandinavia-3	+9	+46	−26
Denmark	−5	+38	−31
Norway	+43	+41	−16
Sweden	+4	+70	−26
OECD-9	+15	+59	−27

[a] Final energy.

services value-added in the United States. This is because services value-added grew faster than floor area; the ratio of services value-added to floor area grew by 15% between 1973 and 1988. In Japan and Sweden, value-added grew only slightly faster than floor area.

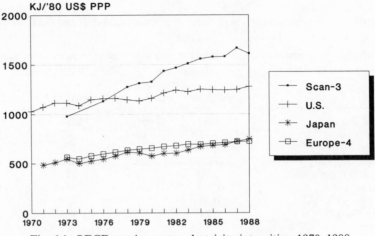

Fig. 6.1. OECD service sector electricity intensities, 1970–1988

The trends in total service-sector energy use mask very different trends for electricity and fossil fuels. Use of electricity grew considerably in all countries, while fuel use declined in the United States, Europe-4 and Scandinavia-3, and increased only slightly in Japan. Aggregate electricity intensity *increased* by 15% in the United States, by 36% in Japan, by 28% in Europe-4, and by 65% in Scandinavia-3 (Fig. 6.1). Fuel intensity, in contrast, *declined* by 43% in the United States, by 36% in Japan, by 40% in Europe-4, and by 56% in Scandinavia-3 (Fig. 6.2).

6.2.2 *Structural change in sector composition*

Service subsectors differ somewhat in their energy intensity. Health and food-related buildings tend to be more energy-intensive than other types, though the others are typically fairly close. Changes in the relative importance of subsectors can thus impact aggregate energy intensity somewhat. If the latter is measured in terms of floor area, and the area of more energy-intensive subsectors is growing faster than that of less energy-intensive ones, there is upward pressure on aggregate energy intensity. A calculation for Japan, the only country for which a lengthy historical data series on energy use and area by subsector is available, shows that structural change in the sector composition pushed slightly upward on aggregate energy intensity between 1973 and 1988. The survey data from the US confirm that there was little change in the relative size of subsectors. In general, we believe that structural change of this type has had relatively little impact.

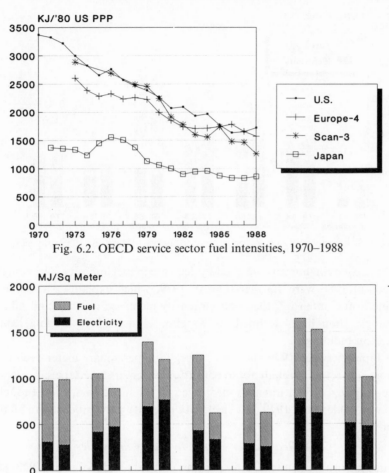

Fig. 6.2. OECD service sector fuel intensities, 1970–1988

Fig. 6.3. US service sector subsector energy intensities, 1979 and 1986

6.2.3 Energy intensity of service subsectors

Energy use per square meter declined in nearly all service subsectors in the United States between 1979 and 1986, according to data from national surveys (US EIA, 1983, 1988). As shown in Fig. 6.3, the exceptions were education and food sales and services, where there was slight increase. Energy intensity declined much more in warehouses (50%) and assembly buildings (34%) than in other subsectors. These two are subsectors in which energy management can and evidently did play a major role. There was moderate decline in trade and office buildings, which are the two largest subsectors in terms of floor area.

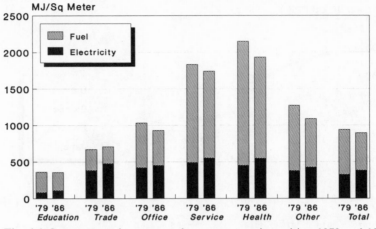

Fig. 6.4. Japanese service sector subsector energy intensities, 1979 and 1986

The declines in intensity were led by decrease in fuel intensity. The winter periods of 1986 were warmer than in 1979, which explains part of the decline in fuel intensity. Electricity intensity increased in trade and office buildings, though it declined in warehouse, assembly, lodging, and education buildings.

In Japan between 1979 and 1986, energy use per square meter declined in most subsectors, though not in education, where it stayed the same, and trade (Fig. 6.4). Fuel intensity decreased in all subsectors, even though 1986 was colder than 1979, but electricity intensity grew modestly in all of them.

6.2.4 Reasons for change in energy intensities

The decline in aggregate *fuel intensity* (per unit value-added) in the 1973–88 period ranged from a low of 30% in Italy to a high of 63% in Norway (Table 6.3). The decline for the OECD-9 aggregate was 42%. The main component of this change was a decrease in the energy intensity of fuel-based space heating, which accounts for an estimated 80–90% of total fuel use. In France, Sweden, and Norway, a decline in the fraction of floor area heated with fuels relative to electricity was also a major factor. In Sweden and Norway, there was considerable switching from heating with oil alone to oil supplemented with some other fuel; the single most common combination in Sweden was oil supplemented with electricity.

The decline in the energy intensity of fuel-based heating was due to addition of new buildings with lower heating requirements per square meter, retrofit improvements to heating equipment and building envelopes of older buildings, and improved energy management, including changes

Table 6.3. *Service sector energy intensity* (*MJ*/*1980* $)

	Fuel		Electricity		Total	
	1973	1988	1973	1988	1973	1988
United States	3.0	1.7	1.1	1.3	4.1	3.0
Japan	1.3	0.9	0.5	0.7	1.9	1.6
Europe-4	2.7	1.6	0.6	0.7	3.2	2.3
France	2.3	1.3	0.5	0.8	2.8	2.1
West Germany	4.4	2.4	0.8	0.9	5.2	3.3
Italy	1.0	0.7	0.3	0.4	1.2	1.2
United Kingdom	2.5	1.7	0.7	0.8	3.3	2.5
Scandinavia-3	2.9	1.3	1.0	1.6	3.9	3.9
Denmark	3.1	1.5	0.6	1.0	3.7	2.6
Norway	1.9	0.7	1.4	2.1	3.3	2.8
Sweden	3.2	1.4	1.0	1.7	4.2	3.1
OECD-9	2.6	1.5	0.9	1.0	3.5	2.6

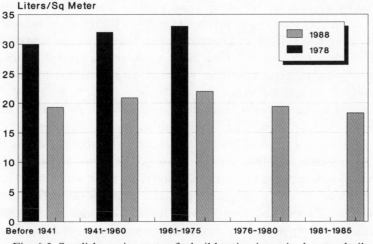

Fig. 6.5. Swedish service sector fuel oil heating intensity by year built

in indoor temperature. In the United States, the fraction of floor area in warmer climates increased significantly, and this contributed to decline in average heating energy intensity.

The effect of higher oil prices on energy intensity in older buildings, as well as the change in building practices brought on by higher prices and changed building codes, are illustrated in the case of Sweden in Fig. 6.5. Oil use per square meter in oil-heated buildings declined between 1978 and

1988 in each building cohort. Oil intensity was lower in 1988 in buildings built after 1975 than in those built before, though the difference was not so large. In part this was due to the conversion of many older, rather inefficient buildings to district heat, which brought down average intensity of pre-1976 buildings.

Changes in aggregate *electricity intensity* were more varied, ranging from increases of 7–16% in the United States, West Germany, Italy, and the United Kingdom to increases of 60–70% in France and Scandinavia. Japan was intermediate at 36%. The large increases in France and Scandinavia are evidence of the growing market share of electricity for space heating.

Several factors contributed to rising electricity intensity in the United States: growth in use of electricity for space heating (often supplementary), increase in air-conditioning requirements due to the growth in the fraction of buildings in warmer climates, increase in hours of operation in some subsectors, and growth in the saturation of electrical office equipment. The national surveys show that the fraction of total floor area that was 100% cooled increased from 27% to 32% between 1979 and 1986. In addition, 1988 was a very warm year, which means that the increase in intensity of 14% (relative to 1973) would be slightly less if electricity use were climate-normalized. Thus, the fact that the increase in the United States was relatively low suggests that there was considerable decline in electricity intensity at the end-use level.

Retrofit of conservation features in older buildings affected energy intensity the most in heating, ventilation, and air conditioning (HVAC) and lighting. In the United States (and implicitly in other countries), there was considerable retrofit activity (Table 6.4). The survey data suggest that there was more activity in the 1980s than before 1980, which could reflect the increase in utility programs and the activities of energy service companies. Two of the most popular features installed in the 1980–86 period were computerized energy management control systems (EMCS) and high-efficiency ballasts for lighting.

6.2.5 *Change in energy sources*

The share of oil in service-sector final energy use declined considerably between 1973 and 1988 in the United States, Japan, and Europe. Its share fell from 25% to 12% in the United States, from 61% to 38% in Japan, from 62% to 33% in Europe-4, and from 59% to 21% in Scandinavia-3 (Table 6.5). The share of natural gas declined somewhat in the United

Table 6.4. *Conservation features in US commercial buildings in 1986 (billion square feet with feature)*

Feature	Installed at construction	% of total area	Added before 1980	Added 1980–86	Total w/ feature	% of total area
HVAC:						
VAC	10.6	18	1.0	3.1	14.7	25
Maint program	26.7	46	5.6	8.6	40.9	70
Waste ht recovery	4.4	8	0.4	1.6	6.4	11
EMCS	4.1	7	0.8	6.2	11.1	19
Other HVAC	2.4	4	0.8	2.9	6.1	10
Lighting						
High-eff ballasts	10.3	18	3.3	10.9	24.5	42
Delamping	4.0	7	2.6	5.4	12.0	21
Daylt control	3.1	5	0.3	2.0	5.4	9
Other controls	7.4	13	1.6	3.6	12.6	22
Other features	0.7	1	0.5	1.0	2.2	4
Bldg shell						
Roof insulation	31.0	53	3.4	8.0	42.4	73
Wall insulation	42.1	41	1.8	3.4	29.3	50
Storm, multi glazing	14.8	25	2.8	4.1	21.7	37
Special glass	14.2	25	1.5	4.9	20.6	35
Ext, int shading	13.8	24	2.6	4.3	20.7	36
Wthr strip, caulking	28.5	49	3.5	9.5	41.5	71
Other shell	0.9	2	0.2	0.6	1.7	3

Source: US EIA (1988).

Table 6.5. *Service sector energy use by type (% of final energy)*

	Oil		Gas		Electricity		Other[a]	
	1973	1988	1973	1988	1973	1988	1973	1988
United States	23	12	47	44	27	43	3	2
Japan	61	41	9	13	29	46	1	0
Europe-4	62	33	12	30	18	31	8	6
France	72	39	9	22	16	36	2	2
West Germany	64	41	10	23	15	27	9	9
Italy	68	18	6	41	21	37	5	4
United Kingdom	45	22	17	40	22	31	15	6
Scandinavia-3	59	21	0	2	25	56	15	21
Denmark	56	20	1	7	17	40	26	33
Norway	54	23	0	0	42	75	4	2
Sweden	63	20	0	0	24	55	13	24
OECD-9[b]	37	21	33	36	25	40	5	3

[a] District heat and coal (mainly the former).
[b] The United States accounted for 90%/75% of OECD-9 gas and about 65% of OECD-9 electricity use.

States, grew slightly in Japan and Scandinavia, and grew considerably in Europe. The share of electricity grew substantially in all cases: from around 30% to 50% in the United States and Japan, from 18% to 31% in Europe-4, and from 25% to 56% in Scandinavia-3.

6.3 Former East Bloc

The service sector plays a small role in total energy use in the Soviet Union.[4] The main reason is the relative lack of personal services: area per capita for shopping, eating out, and lodging is low by Western standards. We estimate that the Soviet Union had approximately 4.2 m^2/capita of built space in 1985, up from less than 3 m^2 in 1970. Service sector value-added is not well defined in the Soviet Union, so this measure of activity is lacking.

The level of energy services in Soviet buildings is well below Western standards. While large buildings in most cities are well heated by district heating or large boilers, smaller buildings in cities and much of the countryside rely on stoves. Controls on temperature and ventilation are typically poor, and lighting levels are low. Electricity use for computing and other information-related technologies is also very low by Western standards.

The intensity of space heating in the Soviet Union is high. Although data are somewhat uncertain, it appears that energy use for heating (and some water heating) averaged about 225 kJ/m^2/degree–day in 1985, compared with 125–150 in Western Europe. Indications are that intensity has declined slightly since the 1970s. Part of the reason is the increased size of new buildings, which lowers heat losses per square meter of floor area. Electricity intensity is low, around 65 kWh/m^2 in 1985, but this is higher than the level of 50 kWh/m^2 in 1970.

6.4 Developing countries

Data on service-sector energy use in LDCs are sparse, but it is evident that the sector accounts for a small share of total final energy use in most LDCs. In part this is because there is little need for space heating (China being an important exception). Services account for a higher fraction of electricity use, however: 13% of total 1988 electricity use in 13 major LDCs (Meyers & Campbell, 1990). Further, electricity demand by this sector is growing rapidly in many countries, especially in Southeast Asia. For nine Asian

[4] The information in this section is based on Schipper & Cooper (1991).

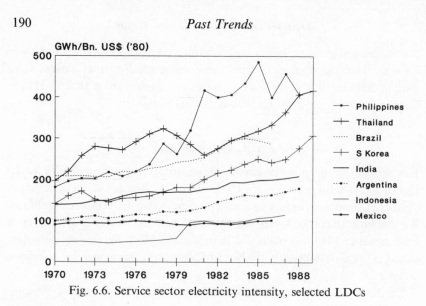

Fig. 6.6. Service sector electricity intensity, selected LDCs

LDCs (including China), growth averaged 10% per year between 1980 and 1988. It was much less rapid in Latin America: around 6% per year for the four major countries.

Growth in electricity use is partly due to rapid construction of new buildings, many of them more modern and energy-intensive than older buildings. In the warm, humid climates characteristic of many LDCs, air-conditioning is a major consumer of electricity. Many older buildings are cooled with window units, which are comparatively inefficient. As shown in Fig. 6.6, service-sector electricity intensity (kWh per US dollar of GDP in the "trade" and "other" categories) has risen considerably in a number of countries.[5] Increasing use of air-conditioning is evident in the Philippines and Thailand (as well as Malaysia and Taiwan, which are not depicted in the Figure), but not in Indonesia. In Latin America, the data show substantial growth in intensity in Brazil and Argentina, but no growth at all in Mexico. The higher intensity in very warm and humid countries like the Philippines and Thailand clearly reflects high levels of air-conditioning relative to countries like India and Mexico.

6.5 Conclusion

Service-sector energy use in nine OECD countries increased by 15% between 1973 and 1988, but the aggregate energy intensity declined by around 25%. The trends were very different for fuel intensity, which fell by

[5] Whether real GDP is accurately measured in the service sector of many LDCs is open to question, but data on floor area are lacking.

approximately 40%, and for electricity intensity, which rose slightly. The drop in fuel intensity was mainly due to considerable retrofitting of buildings and improved energy management practices, though entry of new, more energy-efficient buildings into the stock also played a role. There was also a trend away from fuel-based to electric heating in some countries. Electricity intensity was primarily shaped by two forces working in opposite directions. Addition of more office equipment pushed upward on intensity, while improvements in end-use efficiency, especially for lighting, pushed downward.

The service sector accounts for only a small share of total energy use in the Former East Bloc and the LDCs. In the USSR, there is some evidence of a slight decline in fuel intensity and an increase in electricity intensity, although the latter is quite low by Western standards. Services electricity intensity has risen considerably in most LDCs due largely to addition of modern, air-conditioned buildings.

References

Meyers, S. & Campbell, C. 1990. *Electricity in the Developing Countries: Trends in Supply and Use, 1970–1987*, Berkeley, CA: Lawrence Berkeley Laboratory Report LBL-26166 Rev.

Schipper, L. & Cooper, R. C. 1991. *Energy Use and Conservation in the U.S.S.R: Patterns, Prospects, and Problems*. Berkeley, CA: Lawrence Berkeley Laboratory Report LBL-29831.

Schipper, L., Meyers, S. & Ketoff, A. 1986. Energy use in the service sector: an international perspective. *Energy Policy*, June.

US EIA (Energy Information Administration). 1983. *Nonresidential Energy Consumption Survey: 1979 Consumption and Expenditures, Parts 1 and 2*, DOE/EIA-0318/1 and DOE/EIA-0318(79)2. Washington, DC: US Department of Energy.

1988. *Characteristics of Commercial Buildings 1986*, DOE/EIA-0246(86). Washington, DC: US Department of Energy.

7

Trends between 1973 and 1988: summary and key issues

The preceding chapters considered the evolution of activity, structure, energy intensity, and energy use between the early 1970s and 1988 for the manufacturing, transportation, residential, and service sectors. In this chapter, we summarize the changes that have taken place, and present results of a cross-sectoral analysis of the effect of change in sectoral activity, structure, and energy intensity on energy use in the three largest OECD economies: the United States, Japan, and West Germany. We also address a number of key issues that have important implications for future energy use in the OECD countries and elsewhere.

7.1 Summary of changes between 1973 and 1988

7.1.1 OECD countries

Between 1973 and 1988, final energy use in the five sectors we have studied (which account for around 90% of total energy use) grew by 3% in the United States, 15% in Japan, and 3% in West Germany (Table 7.1). Energy use in manufacturing declined by 10–20% in each country. Energy use for passenger travel grew by only 11% in the United States, but rose by 76% in Japan and by 56% in West Germany. There was moderate to strong growth in freight energy use in all three countries. There was very high growth in residential and services energy use in Japan, and moderate growth in these sectors in the United States and West Germany.

The effect of changes in activity levels, structure, and energy intensities on energy use varied among the sectors and countries. (See the Appendix of this chapter for a description of the method used to separate the effects of these factors.) Increases in sectoral *activity* placed strong upward pressure on manufacturing energy use in the United States and Japan, but less so in West Germany. Growth in activity played a major role in

192

Table 7.1. *Impacts of changing activity, structure, and energy intensities on sectoral final energy use, 1973–1988*

Indicator/sector	Definition/ description of factors	United States (%)	Japan (%)	West Germany (%)
FINAL ENERGY USE				
Manufacturing		−13	−10	−20
Passenger travel		11	76	56
Freight transport		40	33	17
Residential		−2	68	6
Services		12	32	6
Total		3	15	3
ACTIVITY				
Manufacturing	manufacturing value-added	52	64	18
Passenger travel	passenger–km	36	40	33
Freight transport	tonne–km	34	18	24
Residential	population	16	13	−1
Services	service sector value-added	54	71	58
Weighted average[a]		39	54	21
GDP		45	83	34
STRUCTURE				
Manufacturing	subsector value-added shares	−13	−16	−12
Passenger travel	modal mix	3	21	5
Freight transport	modal mix	3	30	15
Residential	heated area and appliance ownership per capita, central heating	42	64	61
Services	subsectoral mix	n.a.	n.a.	n.a.
Weighted average[a]		7	1	14
ENERGY INTENSITIES				
Manufacturing	subsectoral energy intensities	−32	−34	−23
Passenger travel	modal energy intensities	−19	5	12
Freight transport	modal energy intensities	1	−13	−20
Residential	useful space heat energy per heated area, electricity per appliance, useful energy per capita for cooking and hot water	−28	7	−32
Services	energy per unit of value-added	−27	−26	−33
Weighted average[a]		−24	−23	−23
Energy/GDP ratio		−29	−37	−23

[a] Weighted by end-use shares of 1973 energy use.
n.a. = not available (due to lack of data).

passenger travel and in the service sector in all three countries. Activity in freight transport had a somewhat weaker impact. The smallest impact of activity (as measured by population) was in the residential sector, especially in West Germany. Weighting the activity impact in each sector according to the sector's share of energy use in 1973, we find that the cumulative impact of growth in activity levels was to increase energy use by 39% in the United States, 54% in Japan, and 21% in West Germany. The weighted average of activity change was less than the growth in GDP, which was 45% in the United States, 83% in Japan, and 34% in West Germany.

Structural change led to significant reductions in energy use only in manufacturing, where decline in the importance of energy-intensive industries lowered energy use by 13% in the United States, 16% in Japan, and 12% in West Germany. Increases in the share of passenger travel in automobiles and airplanes raised energy use by 3% in the United States and 5% in West Germany, but by 21% in Japan. Changes in the modal mix of freight transport increased energy use by 30% in Japan and by 15% in West Germany, but by only 3% in the United States. Structural change had the greatest impact in the residential sector. Increases in per capita floor area, central heating or more heaters (Japan), and appliance ownership raised energy use by around 60% in West Germany and Japan and by 42% in the United States (where growth in central heating had occurred already prior to 1973). In the service sector, we were unable to quantitatively estimate the impacts of structural change (in terms of shares of value-added in different subsectors), but data from the United States and Japan indicate that there was little net impact. Overall, the net effect of structural change within the five sectors was to raise final energy use by 7% in the United States and 14% in West Germany. There was little net effect in Japan, as the impact of the shift toward less energy-intensive manufacturing was balanced by changes in other sectors.

Change in *energy intensities* had a significant impact in reducing energy use in most sectors. In manufacturing, intensity changes reduced energy use by 32% in the United States, 34% in Japan, and 23% in West Germany. In passenger travel, it reduced energy use by 19% in the United States, but increased it by 5% in Japan and 12% in West Germany. In freight transportation, intensity change reduced energy use by 13% in Japan and by 20% in West Germany, but had little effect in the United States. In the residential sector, change in intensities reduced energy use by 28% in the United States and 32% in West Germany, but increased use by 7% in Japan. We estimated the total impact of intensity changes by weighting each intensity by the share of 1973 energy use attributable to the

Table 7.2. *Percentage change in key subsectoral energy intensities,*
1973–88, United States, Japan, and West Germany

	Intensity measure	United States	Japan	West Germany
Manufacturing				
Paper and pulp	E/VA	−28	−33	−20
Chemicals	E/VA	−37	−55	−25
Building materials	E/VA	−39	−27	−30
Iron and steel	E/VA	−17	−33	−17
Nonferrous metals	E/VA	−15	−50	−20
Other sectors	E/VA	−43	−31	−26
Travel				
Automobiles	E/p-km	−17	+8	+13
	E/km	−29	−2	−3
Air	E/p-km	−43	−26	−38
Freight				
Trucks	E/t-km	+13	−16	−12
Residential				
Heating[a]	E/m^2	−43	−35	−38
Appliances[b]	UEC	−13	+44	−2
Services				
Fuels	E/VA	−43	−35	−45
Electricity	E/VA	+15	+33	+15

Note: Values refer to final energy, except as noted.
[a] Useful energy, climate-corrected (adjusted to remove effect of central heating in Germany and more heaters in Japan).
[b] Unit consumptions of 7 major appliances weighted by 1973 ownership; the end year is 1987.

associated end use. Cumulatively, changing energy intensities reduced final energy use by just under 25% in each country, but the manner in which each came to that decline differed.[1]

In the US and especially Japan, the decline in weighted-average intensity was less than the decrease in the energy/GDP ratio. Since these indicators do not measure the same thing, it is not surprising that they would differ. For one thing, macro-level structural change depressed the energy/GDP ratio somewhat, especially in Japan. In addition, our measures of activities are not exactly comparable with the measures used to construct GDP. Weighting the intensity effects in each sector to derive an overall effect also plays a role. Attempts to aggregate many different changes into a single

[1] The selection of 1973 energy use patterns for weighting affects the results somewhat, since the share of different end uses has changed over time.

measure are inherently problematic, but looking at energy intensities provides far greater insight than the energy/GDP ratio.

Energy intensities. The intensity changes discussed above reflect the net effect of trends in subsectoral energy intensities. Changes in these varied among sectors and countries (Table 7.2). In manufacturing, there was decline of 20% or more in nearly all subsectors in all three countries. In many cases, the percentage reduction was larger in the United States and Japan than in West Germany. Since the absolute level of most intensities (measured in energy use per US dollar of value-added) was lower in West Germany in 1973 than in the United States and Japan, such a result is expected.

Trends differed more in transportation. Automobile energy intensity (per km) declined by 29% in the United States, but remained about the same in Japan and West Germany. Air travel energy intensity declined in all three countries, though more in the United States and West Germany. For freight transport, truck energy intensity rose by 13% in the United States, but declined in Japan and West Germany.

In the residential sector, heating energy intensity (adjusted to account for increase in central heating in Germany and more heaters in Japan) fell by 35–45%. For appliances, the weighted-average intensity of seven major appliances decreased by 13% in the United States and by 2% in West Germany, but rose by 44% in Japan due to increase in average size and features. In the service sector, change in fuel intensity was similar among the three countries: a 35–45% decline. Electricity intensity rose by 15% in the United States and West Germany, and by 33% in Japan.

The impact of changing energy intensities is different if one looks at primary energy use, distributing power sector losses among end-use sectors according to their electricity use. The reason is that part of the decline in final energy intensities was due to the increasing share of electricity in final energy use. If electricity is measured in terms of primary energy, however, this phenomenon has the opposite effect. Using the same method as described earlier, the weighted-averaged decline in primary energy use from intensity changes was 21% in the United States, 16% in Japan, and 15% in West Germany. These values are considerably less than the decline in final energy for Japan and Germany, which was 23% in both cases. The reason is that electricity came to play a much larger role in both countries. The difference is smaller (21% vs. 24%) for the US.

The changes in energy intensities resulted in considerable energy savings. If 1973 intensities had been in effect in 1988, primary energy use would have been higher than it actually was by 25% in the United States, 17% in Japan, and 19% in West Germany. The net energy "savings" amounted to 17.7 EJ in the United States, 2.6 EJ in Japan, and 1.9 EJ in West Germany (Table 7.3).

The importance of different sectors to total savings varied among the three countries. Savings in the manufacturing sector amounted to 39% of total savings in the United States, 32% in West Germany, but nearly 100% in Japan. Savings in passenger travel account for 19% of total savings in the United States, but there were no savings in this sector in Japan and West Germany. Savings in the residential sector played a somewhat greater role in West Germany (42%) than in the United States (24%).

7.1.2 Former East Bloc

Total final energy use in the Soviet Union grew by around 60% between 1970 and 1987. Most of the growth was in heavy industry and freight transport. Activity increased in all sectors, though much less than in the OECD countries. Quantifying the magnitude of changes is difficult, however, due to lack of data. At the most aggregate level, it is hard to say how much national income increased, as official estimates are suspect. Use of official Net Material Product (NMP) figures results in a 15% decline in primary energy use per ruble between 1970 and 1985, while an unofficial Soviet estimate of NMP results in only a 4% reduction. The US Central Intelligence Agency estimated much slower growth in Soviet GNP, resulting in a 20% increase in the energy/GNP ratio over the same period.

Structural change had moderate effects on energy use in the Former East Bloc. In manufacturing, per capita steel and cement production in the Soviet Union rose through the mid-1970s, but was flat thereafter. In most of the Former East Bloc, ownership of automobiles and household equipment increased steadily. On balance, it appears that the impact on energy consumption of the decline in the growth in production of some energy-intensive commodities was offset by the increases in heating and travel from very low levels.

In the Soviet Union, a few end-use sectors exhibited a decline in their energy intensity. Among these were air travel and some heavy manufacturing. In other sectors there were slight reductions in energy intensities mainly as a result of upgrading of equipment and buildings. It appears that the intensity of space heating in homes and the service sector declined slightly. Trucks became less energy intensive as diesel trucks increased their penetration to almost 60% of the fleet.

Table 7.3. *Savings from energy intensity reductions between 1973 and 1988 (EJ of primary energy)*

Sector	United States	Japan	West Germany
Manufacturing			
1988 no EI change	25.3	10.6	4.2
1988 actual	18.4	7.6	3.6
Savings	6.9	3.0	0.6
Travel			
1988 no EI change	17.6	1.5	1.2
1988 actual	14.3	1.6	1.4
Savings	3.3	−0.1	−0.2
Freight Transport			
1988 no EI change	5.5	1.4	0.5
1988 actual	5.6	1.2	0.4
Savings	−0.1	0.2	0.1
Residential			
1988 no EI change	24.3	2.2	3.7
1988 actual	18.3	2.6	2.9
Savings	6.0	−0.4	0.8
Services			
1988 no EI change	14.6	2.1	2.3
1988 actual	13.0	2.1	1.8
Savings	1.6	—	0.5
TOTAL			
1988 no EI change	87.4	17.7	12.0
1988 actual	69.6	15.1	10.1
Savings	17.7	2.6	1.9

Note: "EI" = energy intensity.

For perspective, these changes can be compared with developments in the West. Decline in industrial energy intensities was much less than in the West. Change in space heating intensity was slight in comparison with the rapid changes that occurred in the West. The trend toward diesel trucks and buses currently underway in the East took place in the 1960s in the West. In all, the overall improvement in energy efficiency in the Former East Bloc was small compared to that which occurred in the West. Given the lack of incentive for efficiency improvement, this is hardly surprising.

7.1.3 Developing countries

Lack of reliable data makes it difficult to generalize about changes in the LDCs. Some trends are clear, but their extent is difficult to judge, and changes have varied among regions. In general, activity levels have grown faster in Asia, the Middle East, and North Africa than in Latin America and Sub-Saharan Africa, though per capita activity remains higher in Latin America. In most cases, increase in activity has caused considerable growth in energy use. Manufacturing output grew considerably, especially in Asia.

Structural change within sectors has contributed to growth in energy use. In manufacturing, there has been some shift toward energy-intensive industries, especially in countries with abundant energy resources. In passenger travel, there has been change toward greater use of automobiles and two-wheelers. In freight transport, trucks have grown in use relative to less energy-intensive modes. In the residential sector, there has been considerable growth in electric lighting and appliances per capita.

Change in energy intensities is difficult to judge. In manufacturing, there appears to have been decline in some countries resulting from adoption of more modern processes. In passenger transport, there may have been some decline in automobile and bus energy intensity in countries where newer vehicles have entered the fleet in large number, though worsening of urban traffic conditions has pushed intensity upward. In freight transport, growing use of diesel trucks has probably reduced energy intensity somewhat. In the residential sector, the energy intensity of cooking has declined due to the shift away from biofuels, and there are signs of some improvement in appliance efficiency.

7.2 Change in energy intensities: insights from the OECD experience

Our understanding of change in energy intensities and the reasons for change is much better for the OECD countries than for the Former East Bloc or the LDCs. In this section, we consider several issues related to the changes observed in the OECD countries.

7.2.1 Trend continuation or trend break?

For some end-uses, the trend in energy intensity in 1973–88 represented a break with the past, while in others there was more of a continuation of

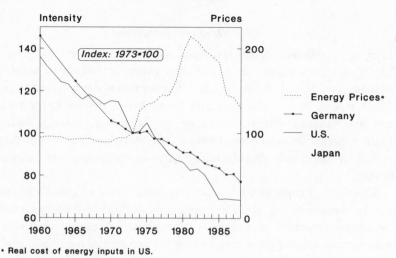

• Real cost of energy inputs in US.

Fig. 7.1. OECD manufacturing energy intensities, constant industry structure

prior trends. In the manufacturing sector, structure-adjusted energy intensity in the United States, Japan, and West Germany had been steadily declining before 1973. Indeed, it fell at about the same average rate between 1960 and 1973 as between 1973 and 1988, despite there being almost no change in energy prices in the earlier period and major increase in prices in the latter (Fig. 7.1). One might be tempted to conclude that the rise in prices had no effect at all, and that "autonomous" technological change alone was the cause of intensity decline. However, value-added grew substantially faster in the earlier period (5.4% in the United States and 4.9% in West Germany) than in the latter (2.6% in the United States and 1.0% in West Germany). Since a major source of intensity reduction is introduction of new facilities that incorporate new production techniques, we would expect that there was more intensity decline from this source in the earlier period. Other causes of intensity decline in the 1960–73 period include increase in the average scale of operations in some industries and improvement in efficiency related to switching from coal to oil and natural gas. The increase in energy prices obviously had an effect in the 1973–87 period, but it appears to be smaller than the autonomous effect of technological change.

Similar to manufacturing, data for US airlines show that energy use per seat-km had been declining in the decade prior to 1973 as a result of introduction of new aircraft (Gately, 1988). The declining intensities for manufacturing and jet airplanes are associated with long-term technological change. Changes were adopted in order to reduce production costs generally. Rise in energy prices may have accelerated changes that would in time have taken place anyway due to innovation and competitive pressure of the market.

By contrast, the energy intensities of automobiles, space heating, and household appliances had not been declining before 1973. In many cases, they were rising. Mostly, this was due to changes in the type of equipment in use rather than decline in technical energy efficiency. For automobiles, for example, weight and power increased steadily in Western Europe throughout the 1960s and early 1970s, raising fuel intensity. The rise in fuel prices, which was relatively less important in Europe due to high gasoline taxes, did not have much effect on the trend towards larger cars, although improvements in technical efficiency did restrain or in some cases reduce new-car fuel intensity. In Japan, the automobile stock in the mid-1960s was dominated by business cars and taxis, and the rapid increase in cars in the late 1960s was led by small vehicles. Size and power began to increase in the 1970s, however. The fuel intensity of new cars declined after 1975, then began to increase again after 1982. In the United States, the situation was quite different, since large and powerful cars had historically been the norm. There was a major break from historic trends in the first few years after 1973, as the mix of new cars shifted towards lighter and less powerful models. The weight and power of new cars declined further in 1979–1980. Since then, weight has increased slightly, but power has grown substantially. If light trucks are included, these trends are even stronger. Composite new car and light truck fuel economy increased considerably between 1973 and 1982, but has been about the same since then. The incorporation of new vehicles into the fleet has continued to depress fleet average energy intensity, however.

In households, a clear break can be seen after 1973 in almost every country studied. Space heating comfort (indoor temperatures, hours and fraction of the home heated) had been increasing, and in some countries, building practices appeared to have moved towards less energy-efficient construction. The oil price shocks reversed the comfort trends, at least temporarily, and led to great increases in the thermal integrity of both existing and new homes. The intensities of space heating in both existing and new homes declined considerably, despite continued growth in use of central heating in Western Europe. In Japan, indoor comfort had been very low by Western European standards before 1970, and was beginning to rise. There were interruptions caused by higher prices in 1974–75 and 1979–80, but the slow upward trend toward higher comfort continued otherwise. For new household appliances, there is some evidence that they were getting progressively less energy efficient in the decade prior to 1973, at least in the United States.[2] Increase in average size and features also contributed to rising energy intensity in many countries. In the years after 1973, the efficiency of most major appliances – and especially refrigerators – improved substantially.

In the service sector, the intensity of space heating was also rising before 1973, but thereafter leveled off or fell in almost every country. Electricity intensity, excluding space heating, was rising slowly before 1973. This increase was slowed but not stopped in the 1970s and 1980s in most countries.

Overall, trends in energy intensities in the household and service sectors exhibited important changes towards lower intensities after 1973. Trends in automobile intensity showed marked reversal only in the United States. The increases in automobile intensities in Europe and Japan were contained, but not reversed. Trends in the intensities of manufacturing and air travel, by contrast, were downward before 1973 and continued in this direction with some acceleration.

7.2.2 Reversible elements of intensity reductions

Most of the decline in energy intensities was a result of physical change in equipment and building stocks, but some was also due to behavioral practices, or changes in purchasing patterns, that in principle are reversible. While the equipment and building changes were caused by energy-price and non-price factors, the behavioral changes were prompted primarily by energy prices. Some of the behavioral changes continued in effect even after energy prices declined, while others were more short-lived.

In productive sectors (manufacturing, freight, services), firms reduced energy intensities through better management during periods when energy prices rose rapidly. Manufacturers instituted "energy housekeeping" practices, while airlines reduced energy intensity somewhat by flying slower or idling less on the ground. Higher energy prices were one of many factors stimulating airlines to increase load factors, but lower prices will certainly not bring back fuel-hungry 707s or DC8s.

For automobiles, carpooling and lower highway speed limits reduced

---

[2] In refrigerators, certain components were getting progressively less energy efficient as manufacturers turned to less-expensive wiring for coils and simpler compressors. It was not uncommon for combined refrigerator-freezers to have an internal heater to keep frost off the refrigerator compartment wall bordering on the freezer, rather than insulating the freezer better. Low electricity prices contributed to these changes.

energy intensity temporarily in the United States. The combination of higher fuel prices, fuel economy standards, and cost-competitiveness of smaller Japanese cars pushed car buyers toward smaller cars in the mid-1970s, which contributed somewhat to reduction in average fuel intensity. The market share of small cars declined in the late 1970s, but then rose again in 1980 and continued to be high through 1988. The trend in the 1980s is misleading, however, because light trucks accounted for an increasing share of total light-duty vehicle sales. There was also a trend in new cars toward lower average power in the 1979–82 period, but since then power has steadily increased (especially if one considers cars and light trucks together). The average horsepower of new cars and light trucks in 1990 was the same as in 1976; vehicle weight was nearly 15% less, and acceleration was much faster. Thus, some of the changes in purchasing preferences that lowered fuel intensity have reversed. Manufacturers have used technical improvements to increase performance more than to improve fuel economy, although many features that improved fuel economy continue to be incorporated into new cars.

The area of energy saving where behavior change played the largest role is space heating (Schipper & Ketoff, 1985). Of the 3.0 EJ of primary energy saved in space heating in the United States in 1987, we estimate that roughly one-third was a result of lower indoor temperatures (relative to 1973 practices). Of the 1.4 EJ saved in Europe and Japan, an estimated 25% came from behavioral change. In the service sector, we estimate that about 20% of the 1.5 EJ savings in space heating were due to lower indoor temperatures. In principle, some 2 EJ of saved demand could return if 1987 buildings were heated at 1973 levels. This is around 10% of the nearly 20 EJ consumed for space heating in 1987 in OECD buildings. In fact, indoor temperatures in homes did increase between 1982 and 1987. Households have gone partially but not completely back to 1973 practices.

7.2.3 Energy intensities and technical energy efficiency

The energy intensities that we have described are strongly affected by the "technical energy efficiency" of the equipment or buildings involved in each case, but other factors also play a role. In manufacturing, the energy intensity of each main subsector may be affected by change in its product mix. In the steel industry, for example, the product mix in several countries has shifted toward more refined, specialty products which require less energy use per unit of value than do simpler products. In the chemicals industry, there has been a shift away from industrial chemicals, which are more energy-intensive to produce, toward chemical products. A similar evolution has taken place in other industries.

Energy intensities in some industries are also affected by changes in the inputs to production. The energy intensity of producing steel and aluminum, for example, is affected by the extent to which scrap or recycled material is used. In the United States, and probably elsewhere as well, increase in use of scrap or recycled material rather than virgin ore contributed to decline in energy intensity in these industries, and in the paper industry as well.

In the transportation sector, changes in load factor affected modal energy intensities. For automobiles, decline in load factor contributed to increase in energy intensity, while in air travel the opposite occurred. Change in the operating environment has affected vehicle energy intensity (energy use per km). Increased traffic congestion contributed to higher energy intensity for automobiles in most OECD countries. Change in the type of vehicles also played an important role. Light trucks accounted for an increasing share of the automobile fleet in the United States, and larger cars did the same in Europe and Japan. For aircraft, there was a shift toward larger planes, which contributed to the decline in energy use per seat-km.[3]

For each size class of automobile, nominal energy use per km declined due to various technical improvements. But rise in the average power in each class contributed to increase in energy intensity. (It is an open question whether this should be considered a loss in "efficiency", since it could be argued that the more powerful car delivers more service to its user.) Engine efficiency, expressed as power per unit of engine size, has risen significantly in new US cars, but energy use per km has been relatively stagnant, since car manufacturers have used the gain in engine efficiency to boost performance rather than to improve fuel economy.

In the residential sector, space heating energy intensity was pushed upward in much of Western Europe and Japan by growth in central heating and more heaters. It was also affected by changes in household heating habits. Some of the behavior change was a reduction in wasteful habits, but there probably was also some reduction in comfort. Increase in average size and features affected the energy intensity of several important residential appliances, especially in Japan and Europe. As with cars, the

[3] If we could measure automobile energy in terms of energy use per seat-km, the increase in size might have had a somewhat similar effect. With aircraft, however, there was a strong economic incentive for operators to achieve high utilization of the larger capacity. Such an incentive does not work very well for car drivers.

technical energy efficiency (service per unit energy) improved considerably, but the average energy intensity of the stock increased because growth in size and features had more of an impact than improved efficiency.

In the service sector, the increase in penetration of equipment such as air-conditioning, office machines, and computers caused growth in electricity intensity (per square meter or per unit of GDP) in most subsectors. At the same time, there was improvement in the efficiency of electrical equipment and systems, and in the thermal properties of buildings.

7.3 The effect of energy prices in the OECD countries

Energy prices may affect the levels of activities and sectoral structure as well as energy intensities. It is important to distinguish between short-term response (mostly behavior or operations change), medium-term substitution away from energy, and long-term development and adoption of technologies that are less energy intensive. We have not attempted to formally model the relationship between energy prices and energy consumption. Nevertheless, a number of factors are clear from our time-series analyses and intercountry comparisons.

In general, large changes in prices, such as occurred for oil in 1973–74 and 1979–80, provoked a greater proportional response than did small, gradual changes. What often happens after a large change, however, is that prices gradually level off or move slowly back as the balance between energy demand and supply moves toward a new equilibrium. Because increase in energy prices affects all sectors of the economy, inflation may be stimulated so real energy prices may even decline. This is what happened to oil product prices in many OECD countries in the 1975–78 period. Thus, a response in the short-run may reverse in the medium-run.

7.3.1 Transportation

The record shows that automobile driving is mildly responsive to changes in fuel prices. In the short run, the price elasticity is low, probably less than -0.2 (Gately, 1990; Greene, 1991). In the United States, distance travelled per licensed driver fell by 5% in 1974 while the real price of gasoline rose by 25%. Driving returned to its 1973 level by 1977, however. The larger price increases of 1979–80 (about 50%) contributed to only a 7% decline in driving between 1978 and 1981. By 1987, driving was 15% above the 1981 level, and well above the peak level of the late 1970s. (Distance travelled per vehicle has increased much less than distance per driver due to greater ownership of second and third vehicles.) In Western Europe, there was also some decline in kilometers per car in 1974, but no decline in 1979–80. (The proportional price increase in this period was less in Europe than in the US since taxes make up a large part of the retail price of gasoline.)

The long-run price elasticity of gasoline demand has been estimated by numerous studies to be in the range of -0.8 to -1.0 (Dahl & Sterner, 1991). Over the long run, prices affect usage patterns as well as characteristics of cars that shape fuel intensity. Historically low gasoline prices in the United States have encouraged more driving per driver and vehicles with higher fuel intensity than in Western Europe and Japan. (Fuel prices may also have some effect on automobile ownership, although other factors are more important.) Americans have historically paid one-fourth to one-half the price that Europeans and Japanese have for motor fuels (due mainly to much lower taxation) and travel almost twice as far per capita by car. The higher marginal cost of car driving in Europe, along with better mass transit service, restrains automobile use and encourages use of alternative modes.

International comparison shows a nearly linear relationship between gasoline price and on-the-road fleet automobile fuel intensity (GJ/km) (Fig. 7.2). The correlation is less strong for gasoline price and total gasoline consumption per capita. There is relatively little difference among European countries and Japan in gasoline use per capita despite a wide spread in price. (Of course, this simple picture leaves out other explanatory variables such as per capita income and automobile price). The high level of the United States is partly due to higher income and lower automobile prices. Geographic factors, especially the low density of urban areas, also play a role. Surveys show that the average automobile trip in the United States in 1983 was shorter (12 km) than in West Germany (13 km) and the United Kingdom (17 km) (US DOT, 1986; Kloas & Kuhfeld, 1987; DOT, 1988). Americans apparently travel farther by car in a year by making more car trips, not because each trip is longer.

Fuel prices also affect the energy intensity of new cars. Historically lower prices in the United States are clearly a reason why cars in the early 1970s were so much larger and less fuel-efficient than in Western Europe and Japan. Prices rose proportionally more in the United States in 1973–74 than in Europe and Japan, and new car fuel intensity declined considerably more in the following years as well (before the Federal standards took effect).

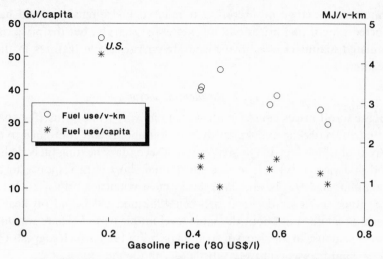

Fig. 7.2. Gasoline price vs. fuel use in OECD countries, 1987

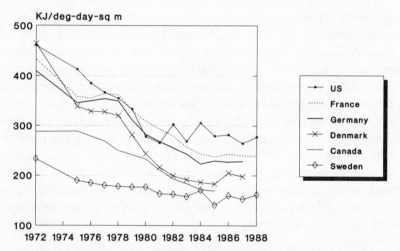

Fig. 7.3. Oil heating intensity in OECD countries, single-family homes with central heat

For air carriers, fuel represents a significant part of their operating costs. They responded to higher prices in the short run with operational changes to save fuel, (including cancelling some flights with low load factors), reduced speeds, and improved ground operations to reduce idling time. In the medium-term, they ordered more fuel-efficient planes. Some of the technological change in aircraft that reduced fuel intensity would have occurred even if fuel prices had not increased sharply, but the price rises stimulated manufacturers to incorporate energy-saving features in their new aircraft.

7.3.2 Residential sector

Increase in oil prices caused both switching away from oil use for space heating and a decline in intensity among households that continued to use oil (part of which was due to greater use of secondary heating fuels such as wood and electricity). There was substantial short-term response to the price jumps in 1973–74 and 1979–80 in most countries (Fig. 7.3). Prices were either stable or declined between 1982 and 1985, and oil heating intensity remained at about the same level in this period. There was little upward response to the sharp drop in oil prices in 1986, which suggests that indoor comfort was relatively satisfactory in the mid-1980s.

Changes in heating intensities among households using other fuels depended to a great degree on the price changes that occurred. Swedes and Norwegians using electricity faced almost constant prices, and made little adjustment; Danes saw large price increases (40%) and reduced electric heating intensity by 30% (in single-family houses) in the 1979–83 period.

Households who experienced price increases for electricity reduced consumption somewhat for uses such as lighting and air-conditioning. Moeller (1987) estimated that decline in utilization of lights and appliances in Denmark led to about a 5% reduction in electricity use after 1978, when prices began a steep increase, but this effect wore off by the late 1980s, when real prices had fallen. In general, however, increase in electricity prices did not have much effect on appliance efficiency (standards and agreements with manufacturers played a more important role). Over the long run, though, appliance electricity use per capita in countries with the highest prices (Denmark, West Germany) increased less than in countries with the lowest prices (Norway, Sweden).

7.3.3 Manufacturing

The impact of changing energy prices on manufacturing intensities is difficult to determine. Intensities in 1973–88 fell the most in the United States and Japan, which experienced increases in energy prices that were greater than those in Western Europe, but other factors complicate interpretation of these facts. Without a doubt, higher energy prices led firms to install energy-saving devices such as improved boilers, insulation, and heat exchangers that improved the energy efficiency of existing facilities. And they likely contributed to incorporation of higher levels of energy efficiency in new facilities, at least for energy-intensive industries. As we noted earlier, long-term reductions in energy intensities have occurred even during periods of stable energy prices as firms adopted new technologies designed to reduce requirements for all inputs, including energy.

7.3.4 Conclusion

The issue is not whether prices are important determinants of energy intensity and equipment efficiency, but rather how important, and over what time period. In the short term, for example, increased electricity prices stimulate homeowners to use air conditioners less, and in the medium term, to consider a more efficient model at replacement time. But a variety of factors – including higher electricity prices, imposition of standards, and international competition – led to the development of new air-conditioning systems which are both quieter and more energy-efficient. Not surprisingly, intensities for specific end uses are highest in countries with the lowest prices, which is an indication that energy prices have a strong impact in the long run.

The synergistic effects of higher prices and other factors can stimulate rapid improvements in energy efficiency. Rising energy prices also stimulate investment in the development of new technologies, and encourage their market penetration. Often, new and more efficient technologies remain standard even if energy prices fall. Higher energy prices contribute strongly to decline in end-use intensity, but they are not sufficient to ensure that most cost-effective steps to improve efficiency are actually taken within a desired period. For this result, government policy can play an important role.

7.4 The impact of energy conservation policies

Most governments in the OECD countries instituted energy conservation policies and programs in the 1970s and 1980s, though their nature and extent has varied considerably (IEA, 1989). In addition, many electric utilities in the United States and, to a much lesser extent, in Western Europe have implemented programs to encourage electricity conservation by their customers. Measuring the number of energy users directly affected is easy, but measuring the impact of programs on consumption is very difficult. Below we present an overview of the impact of two basic types of

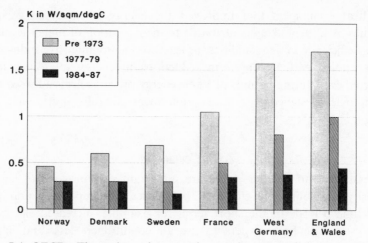

Fig. 7.4. OECD: Thermal requirements in new houses, wall heat transmission values

policies that have affected energy conservation: (1) regulations, standards, and incentives designed to require or encourage actions that lead to greater energy efficiency; and (2) energy taxation.

7.4.1 Regulations, standards, and incentives

The main target of government and utility programs designed to accelerate efficiency improvement has been the residential sector. The service sector has received attention in many countries also. In Western Europe and Japan, automobiles escaped the attention they received in the United States, although manufacturers and governments made agreements to improve the efficiency of cars. Few countries developed programs that had much affect on manufacturing, freight transport, or air travel.

Dissemination of information promoting energy conservation has been extensive, but assessing the impact is difficult. Energy labelling for new appliances was instituted in North America, but most evidence suggest that their impact on consumer behavior has been relatively small. Regulations on performance and maintenance of combustion equipment may have boosted efficiency somewhat. Regulations on indoor temperatures were tried in France, but were short lived.

Among all policies, efficiency standards probably had the strongest impact on energy use. Codes for new houses in Western Europe (Fig. 7.4) and some US states produced major improvements in construction practices. California, other US states, and, eventually, the Federal government adopted appliance efficiency standards that accelerated efficiency improvement considerably. The introduction of standards in California beginning in 1977 had a profound impact on appliances produced in the United States for other states, as manufacturers gradually withdrew products that did not meet California standards.

In Japan and West Germany, agreements between government and domestic manufacturers led to improvements in major household appliances (Wilson et al., 1989). The West German agreement forced manufacturers in other European countries to improve efficiencies in order to stay competitive in West Germany and abroad (Schipper & Hawk, 1991). The improvements generated by these agreements, however, were less than they might have been, as no Western European countries introduced standards to eliminate the low-efficiency models that were imported from Southern and Eastern Europe.

Subsidies (grants, tax credits) for retrofit of existing homes, and in some cases, for new buildings, also had an effect, although the degree to which the incentives increased energy-saving activities over what would have been adopted otherwise is uncertain. By stimulating retrofit activity by some households, however, the grants provided an important demonstration effect, showing that retrofits could indeed lower energy bills. One problem with the initial grants given out in Denmark, Sweden, and West Germany was that a large share of the funds went to retrofits of outside walls (essentially building rehabilitation) or exchange of useable single-frame windows for new double glazing. While each of these actions had clear benefits, the energy paybacks relative to the grants were small (Wilson et al., 1989). Based on a comparison of expected savings from retrofit programs with savings from all factors, we estimate that subsidy programs directly or indirectly accounted for no more than 20% of the reduction in space heating intensity in existing homes in Western Europe and North America.

In the United States, legislation enacted in 1975 required the sales-weighted average fuel economy of each manufacturer's new cars to meet a standard that began at 18 mpg (13 l/100 km) in 1978 and increased to 27.5 mpg (8.6 l/100 km) by 1985. (Standards for light trucks were also enacted.) The intent of the so-called "CAFE" standards was to stimulate technological improvements that would increase efficiency without substantially altering the size distribution of cars sold, but there has been considerable debate about the influence of the standards, as opposed to the gasoline price increases in 1973–74 and 1979–80, on new car fuel economy. A statistical analysis using detailed automobile manufacturer data for

1978–89 indicates that the standards were a significant factor for many manufacturers, and were at least twice as important an influence on them as gasoline prices (Greene, 1990). The study found no difference in the effect of the CAFE constraint during the 1978–82 period of rising prices and the 1983–89 period of falling prices. Although new car fuel economy fell slightly after 1988, it is still well above the values in the 1970s, even as gasoline prices fell to their lowest level in decades by early 1990. This suggests that one important effect of the standards was to induce permanent technological change.

7.4.2 Energy taxation

Energy taxation policies varied widely, and as a result, so did the prices users faced. All countries except the United States have had taxes that boosted the pump price of gasoline by a factor of two to three or more and diesel by a factor of two. Sweden, Denmark, and France taxed home heating oil by almost the same proportion. Denmark taxed household electricity by nearly 100%, other countries by lesser amounts. Taxes on gasoline were significant before 1973 in Western Europe and Japan; this reduced the relative change in the retail price caused by the increase in crude oil prices.

Aside from motor fuels, West Germany and Japan relied very little on energy taxes, although Germany taxes electricity to defray the cost of subsidizing domestic coal production. The United Kingdom and United States have placed very low taxes on energy. The United Kingdom, and, until recently, Sweden, exempted energy for households from value-added taxes, which made energy cheaper relative to efficiency improvement. The United States inhibited most energy prices from promptly reaching international levels in the 1970s, and the United Kingdom did the same with domestic natural gas.

Differences in oil taxation affected the response of oil-heating households in the 1981–85 period. French, Swedish, and Danish households faced large increases in taxes on heating oil, and reduced oil intensity and switched to other fuels (especially in Sweden). Japanese and West German households experienced low taxes (and the real price was level or declined); oil intensity remained constant in West Germany and increased in Japan, and there was not a great deal of switching away from oil.

7.4.3 Energy saved by conservation policies: a simple estimate

Although it is difficult to judge how much energy was saved by non-price energy conservation policies, we have made some estimates that we present below. These estimates were made by multiplying the change in energy intensity in an end use between 1973 and 1987 that we judge was "caused" by policy by the corresponding level of activity for that end use in 1987.

Automobiles Using Greene's findings as a rough guide, we credit the U.S. automobile standards with two-thirds of the reduction in fleet fuel intensity from 19 1/100 km (12 mpg) in 1973 to 12.5 1/100 km (19 mpg) in 1987. When this is multiplied by distance driven in 1987, policy savings amount to 2 EJ. In Europe, government pressure on automakers to improve efficiency probably had some impact, but is difficult to quantify.

Manufacturing Industrial energy-use management and reporting programs, as well as tax credits and other investment incentives, might be credited with a small part of the incremental reduction in energy intensities beyond the pre-1973 trend (0.2–0.3 EJ). The government's role in organization of energy management in Japan assisted efforts to improve energy efficiency, though it is not clear how much more savings were caused than Japanese industry, almost totally dependent upon foreign oil, would have done on its own in order to maintain its competitive position.

Space heating We estimate that retrofit of homes was responsible for about 50% of the overall drop in heating intensity, behavior and entry of new homes into the stock the rest. Retrofit subsidies for heat-saving measures raised participation and investment amounts in Western Europe and the United States. We estimate that half of the energy savings from retrofit subsidy programs occurred because of the existence of the programs.[4] The total amounts to 0.25 EJ in both Europe and the United States. Codes for new buildings raised levels of thermal integrity significantly in Western Europe. New homes in 1987 made up approximately 15% of the stock in Europe, and used roughly 40% less heat than homes built in the 1970–1973 period. Crediting the standards for 50% of

---

[4] While these programs did not play a significant direct role in the overall decline of heating intensity, it is likely that they did have an important demonstration effect, and probably induced many consumers to retrofit their homes without direct government aid. In Sweden, the number of single-family home dwellers undertaking retrofits with government subsidies or loans between 1977 and 1984 was roughly equal to the number carrying out similar measures with no direct aid.

the improvement (higher prices did the rest) yields a savings of about 0.2 EJ. Standards on heating equipment and commercial buildings may have contributed approximately 0.2 EJ in Europe as well.

Household appliances Efficiency standards on US appliances may be credited with at least one-fourth of the 0.4 EJ savings from more efficient appliances in the United States. West German and Japanese government/ industry efficiency agreements influenced appliances sold domestically and elsewhere; this saved approximately 0.3 EJ among the OECD-8. If we credit these agreements with one-third (because they were not binding), the impact is 0.1 EJ.

Adding up these estimated policy impacts, we obtain around 3.5 EJ of savings caused by energy conservation policies. The savings of primary energy are about 4 EJ. This amounts to approximately one-eighth of the total energy we estimate was saved by intensity reduction in the OECD countries in the 1972–85 period (Schipper, 1987). As rough as these estimates are, they provide some general insights about the impact of conservation policies. Policies that affected new capital stock (homes, appliances, cars) had a major impact on such equipment. Programs that subsidized retrofit of existing homes and commercial buildings were probably less important than changes in prices in stimulating retrofit. Policies aimed at manufacturing had a very small impact on energy intensity compared with the force of new technology, which continually reduced energy intensities.

7.5 OECD energy intensities in 1982–88: signs of plateau

Since 1982, the energy intensity of some end uses has ceased its previous declining trend or has even increased. It is not surprising that this should occur, since real prices of fuels have declined, and many energy conservation programs were discontinued or weakened (such as the US fuel economy standards). Much of the easy-to-cut "energy waste" was trimmed between 1973 and the early 1980s, and the easy technical improvements were made, so a slowed rate of decline is to be expected. Even though technology continually improves, the falling price of energy means that many investments that save energy are overlooked.

The plateau in energy intensities has been somewhat more apparent in "consumer" than in "producer" sectors. In the latter (especially manu-facturing), competitive pressures and technological change have contributed to continued progress in reducing energy intensities. The decline in fuel prices has lessened the interest of management in making capital investments for energy conservation, but addition of new capital stock has tended to reduce average energy intensities. Even so, the data show signs of a plateau in "producer" energy intensities in many cases. In manufacturing, the structure-adjusted energy intensity in the United States and Japan was unchanged in the 1985–88 period, although the trend in the United States is somewhat uncertain due to problems of data comparability between 1985 and 1988. For air travel, there was less decline in energy intensities in the 1984–88 period than previously despite the entry of new airplanes into the fleets. In freight transport, truck energy intensity shows a plateau in the United States and Western Europe since 1982, and has declined more slowly than before in Japan. In the service sector, the historical decline in fuel intensity has slowed since 1982 in the United States and Western Europe, and there has been no change in Japan.[5] Service-sector electricity intensity shows roughly the same trends after 1982 as before, but this result is difficult to interpret, since there has been increase in equipment per square meter and improvement in end-use efficiency at the same time. There is evidence that efficiency improvement is continuing due to retrofit (especially for lighting systems) and addition of new buildings with lower electricity use per square meter for cooling and lighting.

Consumer-dominated sectors show an evident plateau with respect to energy intensity. In the residential sector, space heating energy intensity has declined only slightly in the United States and Western Europe since 1982, and has increased in Japan. Addition of new homes has continued to decrease average stock intensity (very slowly in Europe), but retrofit of older homes has slowed, which is perhaps not surprising given the considerable activity between 1973 and 1982. Households have increased indoor temperatures somewhat, if the evidence from a few countries (the United States, Denmark) can be generalized. In the United States, there has also been a plateau with respect to the thermal integrity of new houses. The average ceiling insulation installed by builders nearly doubled between 1973 and 1982, but has been about the same since then.

For electric appliances, estimates of stock energy intensities are somewhat uncertain, but they show a plateau in recent years in many cases. In Western Europe and Japan, the plateau (or even rise in intensity) is

---

[5]   One would expect some decline in the United States due to the fact that more new buildings have been added in warm regions of the country than in cold ones, thus decreasing the average space heating requirements of the sector.

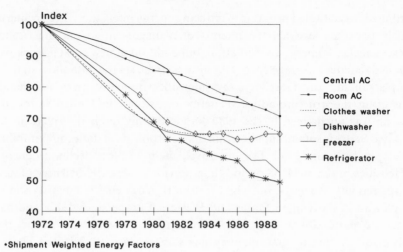

•Shipment Weighted Energy Factors

Fig. 7.5. US new electric appliances, specific energy intensity indices

partly due to increase in size and features. In the United States, the technical energy intensity (energy per unit of service or size) of new appliances has continued to decline in the 1980s for refrigerators, freezers, and room air conditioners, but not for clotheswashers and dishwashers (Fig. 7.5). This trend reflects the imposition of efficiency standards in California, New York, and other states, as well as anticipation of Federal standards.

The energy intensity of automobile travel followed the same trend after 1982 as before in Western Europe and Japan – essentially no change. In the US, however, the intensity has continued to decline as newer vehicles replace older ones. For new automobiles, however, there has been a plateau in average fuel economy in the United States and Sweden, and less decline than previously in Italy and France. In West Germany and Japan, new car fuel intensity has increased as consumers have moved to larger and more powerful vehicles. The plateau for new cars will have a larger effect on average fleet intensity in years to come than it did in the late 1980s.

How much did change in government energy conservation programs contribute to the plateau of energy intensity? Conversations with appliance manufacturers in West Germany and Japan suggest that expiration of agreements that set goals for efficiency of new appliances contributed to stagnation in improvement. For space heating, the expiration of many programs that subsidized retrofits probably had some effect. For auto- mobiles, the weakening of standards in the United States has contributed to the plateau in new car fuel economy. Since programs overall only caused a small part of the overall gain in efficiency, however, their expiration could not be a major cause for the plateau.

Appendix: disaggregating changes in sectoral energy use

Here we describe the methodology used for deriving the values given in Table 7.1. If we take E_{it} as the energy use of sector i in year t, A_{it} as its activity level, and S_{it} and I_{it} as vectors of structural and intensity parameters, we may write sectoral energy use symbolically as:

$$E_{it} = E_i(A_{it}, S_{it}, I_{it}).$$

The details of this calculation for each sector have already been discussed. The relative change in energy use driven by changes in activity between periods 0 and t, given constant structure and intensity, is:

$$\%\Delta E_{Ai} = \frac{E_i(A_{it}, S_{i0}, I_{i0}) - E_i(A_{i0}, S_{i0}, I_{i0})}{E_i(A_{i0}, S_{i0}, I_{i0})}$$

The corresponding structure and intensity indicators are:

$$\%\Delta E_{Si} = \frac{E_i(A_{i0}, S_{it}, I_{i0}) - E_i(A_{i0}, S_{i0}, I_{i0})}{E_i(A_{i0}, S_{i0}, I_{i0})}$$

and:

$$\%\Delta E_{Ii} = \frac{E_i(A_{i0}, S_{i0}, I_{it}) - E_i(A_{i0}, S_{i0}, I_{i0})}{E_i(A_{i0}, S_{i0}, I_{i0})}$$

To derive indicators of the impacts of changes in sectoral activity levels, structure, and energy intensities on aggregate energy use, we construct weighted sums of the sector-specific indicators so defined. The aggregate change in energy use induced by changes in sectoral activity levels, given constant structure and intensity, for example, is:

$$\sum_i [E_i(A_{it}, S_{i0}, I_{i0}) - E_i(A_{i0}, S_{i0}, I_{i0})].$$

Dividing this expression by aggregate base-year energy use, the relative change in energy use due to changing activity levels is:

$$\%\Delta E_A = \sum_i w_i \%\Delta E_{Ai}$$

where the weight w_i is sector i's share of base-year energy use. The same formula may be used to construct aggregate structural and intensity indicators.

From these equations we may also derive an expression for measuring energy saved in a sector or activity. We define such a quantity as the extra energy that would have been required, relative to actual use, in year t in sector i if the intensity of that sector had remained at its 1973 ($t=0$) value but the levels of activity and the mix of activities had evolved to their actual values in year t. Savings are calculated using the following formula:

$$S_i = E_i\,(A_{it},\,S_{it},\,I_{it})\left(\frac{E_i(A_{i0},\,S_{i0},\,I_{i0})}{E_i(A_{i0},\,S_{i0},\,I_{it})}-1\right).$$

References

Dahl, C. & Sterner, T. 1991. Analysing gasoline demand elasticities: a survey. *Energy Economics* (July).

DOT (Department of Transport). 1988. *National Travel Survey*: 1985/86 *Report*. London.

Gately, D. 1990. The U.S. demand for highway travel and motor fuel, *Energy Journal*, **11** (3), 59–73.

1988. Taking off: The U.S. demand for air travel and jet fuel, *Energy Journal*, **9** (4), 63–88.

Greene, D.L. 1991. Vehicle use and fuel economy: How big is the rebound effect? *Energy Journal*, **13** (1), 117–143.

1990. CAFE OR PRICE?: An analysis of the effects of federal fuel economy regulations and gasoline price on new car MPG, 1978–89, *Energy Journal*, **11** (3), 37–57.

IEA (International Energy Agency). 1989. *Energy Policies and Programmes of IEA Countries*, Paris, France.

Kloas, J. & Kuhfeld, H. 1987. *Verkehrsverhalten im Vergleich*. Berlin, Germany: Deutsches Institut für Wirtschaftsforschung.

Moeller, J. 1987. Elbesparelser i boligsektoren (in Danish: Electricity Conservation in the Household Sector). Lynbgy, Denmark: Dansk Elvaerkers Forenings Udredningsinstitute (DEFU), Teknisk rapport 258.

Schipper, L. 1987. Energy saving policies of OECD countries: did they make a difference? *Energy Policy*, **15** (6), 538–42.

Schipper, L. & Hawk, D. 1991. More efficient household electricity use: An international perspective. *Energy Policy*, **19**, 244–65.

Schipper, L. & Ketoff, A. 1985. Changes in household oil use in OECD countries: permanent or reversible? *Science*, **224** (6 December).

US DOT (US Department in Transportation). 1985. *National Transportation Statistics: Annual Report*. DOT-TSC-RSPA-85. Washington, DC.

1986. *Personal Travel in the US, Volumes I and II*. Washington, DC.

Wilson, D., Schipper, L., Tyler, S. & Bartlett, S. 1989. *Policies and Programs for Promoting Energy Conservation in the Residential Sector: Lessons from Five OECD Countries*. Berkeley, CA: Lawrence Berkeley Laboratory Report LBL-27289.

Part II
Future prospects

8

Outlook for activity and structural change

As population grows and economies expand, the level of energy-using activities will increase in all regions of the world. The relative importance of different types of activities will change as technology, lifestyles, and consumption habits co-evolve. Certain activities will decline in importance in some regions, and increase in others. Because the pace of change seems to quicken year by year as economic, political, and cultural integration rises, predictions of future activity become increasingly uncertain the further ahead one attempts to look. What is clear, however, is that the rates at which different activities grow will have a profound impact on future energy use.

In this chapter, we present an overview of the prospects for activity growth and structural change over the next 20 years. As in the historical analysis, our understanding of future prospects is better for the OECD countries than for other parts of the world. For the Former East Bloc, and especially the former Soviet Union, there is considerable uncertainty regarding the future, but one can describe patterns that are likely to emerge if the transition to a market economy is reasonably successful. In the LDCs, assessing the future outlook with precision is difficult due to lack of information about recent trends and the rapid pace of change that is occurring in many countries. Here too, however, some reasonable observations are possible.

Our discussion of structural change mainly focuses on change *within* sectors, but shifts in production *among* economic sectors will also have a major impact on energy use. In the OECD countries, the service sector is expected to continue growing in importance, but manufacturing and services are becoming more interdependent. Many of the growth industries in manufacturing support service-sector activities, and the "service content" of manufactured goods is increasing as well. In the Former East

Bloc, the dominant role of manufacturing is an artifact of the old centrally-planned economies; the service sector is likely to expand significantly from its current low level. The service sector is also growing in the LDCs, but the continuing shift from agriculture to manufacturing will increase the overall energy-intensiveness of many economies.

8.1 Population

Growth in population will contribute to increasing activity in all sectors – more demand for goods and services, transport of goods, personal travel, and residential services. Projections of population increase are among the most certain elements of the future, since past trends exert a somewhat predictable force on the future. But birth and death rates depend on a complex of cultural, economic, and technological factors, as well as government policies. Decline in birth rates in the LDCs is linked to increase in economic security, opportunities for women, as well as access to birth control techniques. The pace of urbanization, which depends in part on the extent of economic development in rural areas, will also affect birth rates. The size of the population in any one country is also shaped by migration, which could increase as global integration grows.

The impact of population growth on world energy consumption very much depends on where the additional people live. The average child born today in the wealthy countries will be associated with far more activity by the time it is 25 than will a child born in a poor country. Of course, the evolution of activities in any particular geographic area is quite dynamic, so the 25-year old in the poor country will be associated with much more activity in the year 2010 than its parents are today. It is also quite possible that the differences in per capita activity among various parts of the world will be smaller in the future than they are today. (Whether the degree of difference *within* countries will be more or less than today is also an important issue.)

Recent projections from the World Bank show average annual growth for the 1988–2000 period of 0.6% in the OECD and Former East Bloc countries, and 1.9% in the LDCs (Table 8.1). Among the LDCs, projected growth is much slower in China (1.3%/year), which has historically had strict population control policies, than in the middle and low-income countries (1.9% and 2.2%, respectively). Among all three groups, the rate of growth is expected to gradually slow over time, but the projections still show an increase of two-thirds in world population – to around 8.5 billion

Table 8.1. *Population growth: historic and projected (millions)*

Country group	1965	1988	2000	2025
OECD countries	681	828	888	979
AAGR (%)	—	0.8	0.6	0.4
Developing countries	2295	3843	4850	7004
AAGR (%)	—	2.2	1.9	1.5
Low-income	1016	1796	2345	3634
AAGR (%)	—	2.5	2.2	1.8
China	725	1088	1275	1566
AAGR (%)	—	1.8	1.3	0.8
Middle-income	540	959	1230	1804
AAGR (%)	—	2.3	1.9	1.4
Former East Bloc	350	423	447	494[a]
AAGR (%)	—	0.9	0.6	0.4
World total	3314	5095	6186	8478
AAGR (%)	—	1.9	1.6	1.3

[a] Authors' estimate.
AAGR = average annual growth rate.
Source: Derived from data in *World Development Report 1990* (World Bank).

– between 1988 and 2025. By then, the share of the world's population that is in today's LDCs will have grown from 78% in 1988 to 84%.

8.1.1 *The changing age structure of the population*

In the OECD countries, the aging of the population due to continuing decline in birth rates and rising life expectancy will affect energy use in all sectors. It will shape the type of goods that are produced and the services that are provided. It will affect where people live: having grown accustomed to changing residences, the future elderly may be more likely to retire in warmer locations where the need for heating is less. The elderly tend to spend more time at home, and maintain higher levels of indoor comfort than younger people. The higher life expectancy of women relative to men may lead to further increase in the share of single-person households, which would increase energy use per capita.

The effect of the aging of the population on passenger travel is uncertain. Elderly singles currently drive considerably less than other singles (although the cars they drive are usually older, larger, and less fuel-efficient). The present may not be a good guide to the future, however, as a new generation of perhaps more active and mobile elderly is forming. The future elderly will be quite accustomed to driving. The "baby boom" generation is the first in which a large majority of all members have had a driver's license for their entire adult lives. By comparison, in West Germany or France 25% of the people over 60 have no driver's license. When baby-boomers retire, they may continue to drive frequently. In addition, more people may work into their late-60s and continue commuting, and also have more income for leisure travel.

The biggest unknown concerning future energy use of the elderly is how they will use their free time. If pensions, savings, and social security permit them to enjoy economic security, and better health care allows more years of active retirement, their energy use could grow from increased travel and participation in away-from-home activities. National policies towards the social security system and care for the elderly will have an effect. In Scandinavia, generous pensions permit retirees to travel and lead an otherwise active life on their own, but in Japan concerns about retirement security force families to save more and live together longer. If pension systems permit the elderly to live independently in their own homes, household size will be low. Limited pension resources could lead to more elderly living with their children, or increased collective living in communities or rest homes. This would decrease energy use per capita if shopping, travel, and other functions occur collectively.

In contrast to the OECD countries, in the developing countries a large share of the population is entering into the energy-using activities that characterize young adulthood. The new generations of adults will have been more affected by modern influences than their forebears. The resulting desire for personal mobility and other aspects of Western lifestyles could greatly expand the demand for many energy-intensive activities.

8.2 Economic prospects

Change in the levels of energy-using activities and in the way they are carried out is shaped by economic growth. The growth and development of economies is a complex phenomenon. Understanding prospects in particular national economies has become even more complicated than in the past due to the increasing interdependence of the global economy. A growing network of links in trade, finance, investment, and technology is binding nations to one another as never before. Global interdependence will increase further as a result of the collapse of the Communist system in the Former East Bloc and the adoption of more outward-oriented policies by many countries in Latin America and Sub-Saharan Africa.

Because of rising interdependence, the long-term growth prospects of industrial, developing, and former East Bloc economies depend heavily on the world trading environment. The forces for trade liberalization have weakened since the mid-1970s, when many industrial countries began to establish new (mainly nontariff) barriers to trade. The protectionist trend in LDCs (including a variety of regulations that effectively operate as barriers) has been similar. If the current round of multilateral trade negotiations does not have a successful outcome, the trend toward regional trading blocs in Europe, North America, and East Asia could accelerate. If such regional integration leads to discriminatory trading practices, long-term global economic prospects could be diminished, and a retrogression into trade war cannot be ruled out.

OECD countries The OECD countries account for roughly two-thirds of world GDP and provide the market for much of the developing world's exports. Since the labor force will grow more slowly in these countries in the future, labor-productivity growth must accelerate if they are to maintain historic rates of economic growth. Most economic forecasters believe that some acceleration is likely, but expect GDP to grow somewhat more slowly (on average) in the 1990s than in the 1980s. If however, macroeconomic policies are unfavorable, high interest rates and trade imbalances persist, and protectionism increases, international trade and investment would be hurt, and economic growth would falter.

Developing countries The economies of the developing countries are strongly linked to the industrial economies. Their prospects will be shaped by the growth of industrial country markets, the international cost of capital, the availability of external capital for investment, and the terms of trade. For severely indebted countries (which include Argentina, Brazil, Mexico, Egypt, the Philippines, and much of Sub-Saharan Africa), the level of debt relief will be an important factor. Access to export markets in the OECD countries will also play a key role. If the emerging regional blocs erect trade barriers, many developing countries outside these blocs could be hurt severely.

While external circumstances will play a critical role in shaping the economic prospects of the LDCs, national economic policies are also important. An analysis of historic performance by economists at the World Bank found that domestic policies and institutions have been a much greater factor in long-term growth than international conditions (World Bank 1991). In many LDCs, the faith in the ability of the state to direct economic development has given way to a greater reliance on markets, and there is increasing recognition of the importance of international trade and finance for economic growth. A growing number of countries in Latin America and Sub-Saharan Africa are implementing major policy reforms toward greater openness and private sector development. Even with domestic reforms and favorable international conditions, however, many LDCs face formidable obstacles to economic growth. The prospects are considerably brighter for the dynamic economies of East Asia than for much of Latin America and Sub-Saharan Africa.

Former East Bloc Economic prospects in the Former East Bloc will depend on the success of the *laissez-faire* approach that governments are adopting, the ability of people and institutions to adapt to the ways of a market economy, and the degree of foreign investment in existing and new industries. The ability of governments to maintain a course of reform in the face of rising unemployment and declining real income is also a key factor. As with the LDCs, establishment of a reasonably stable environment for investment and fair access to Western markets will also be important.

Expectations for the global economy Economic growth in the 1990s and beyond will depend on conditions in the areas of finance, trade, and macroeconomic policy. Both domestic and international policies will be important. Oil prices will also play a role. In its 1991 *World Development Report*, the World Bank presented four scenarios for the global economy in the 1990s that reflect different conditions in these areas. The "baseline" scenario envisions average annual GDP growth of 2.9% in the high-income OECD countries and 4.9% in the LDCs (including most of Eastern Europe). In the "downside" scenario, the comparable figures are 2.2% and 4.1%[1] Within each of these large groupings, the prospects differ. In the "baseline" scenario, GDP grows at 2.5% in the US, 3.7% in Japan, and 3% in Western Europe. Among the LDCs, the average growth is 3.6% in Sub-Saharan Africa and Europe, Middle East, and North Africa; 3.8% in Latin America; 4.7% in South Asia; and 6.7% in East Asia.[2]

Missing from the above scenarios is the former Soviet Union, for which the economic outlook is highly uncertain. On the one hand, the

---

[1] The "baseline" scenario was subjectively judged as being somewhat more likely than the "downside" scenario. The Bank also developed "low" and "high" scenarios that reflect conditions judged as possible but relatively unlikely.

[2] Europe includes Greece, Portugal, Turkey, and Eastern Europe. East Asia includes China.

disintegration of the USSR into independent countries could speed the pace of reform. On the other hand, a lack of economic integration in the new Commonwealth, along with hesitation in carrying through policies that may be unpopular, could lead to a slow recovery of the economies from the current state of near-collapse. Internal ethnic unrest or conflict among the newly independent nations could also undermine the stability necessary to attract foreign investment.

Beyond the year 2000, the outlook for most regions is for somewhat slower economic growth (in part due to lower population growth). The exception to this trend is the Former East Bloc, which will almost certainly grow faster after the transition to a market economy is consolidated and new trade links are well-established. While the evaluation of economic prospects presented above is not destiny, it suggests that GDP in the year 2010 could be 1.5–1.8 times its 1990 level in the OECD countries, and 2–2.5 times its 1990 level in the LDCs. The levels of aggregate activity in each energy-use sector may rise faster or slower than GDP, and the structure of activity may tend toward higher or lower energy intensity. The prospects vary in different parts of the world, as we discuss in the sections below.

8.3 Manufacturing

The evolution of the manufacturing sector in any given country is becoming increasingly dependent on the international developments discussed above. In general, growth in a nation's manufacturing output and the composition of output is shaped by the role of locally produced goods in domestic consumption and the ability of manufacturers to export their products. Both of these depend on the comparative advantages of domestic manufacturers, government policies, and the openness of national (or regional) borders to imported goods. Since the share of total world manufacturing output that is traded is growing, manufacturing is becoming more dependent on trade policy. It is generally agreed that a world in which borders are relatively open to trade will result in higher levels of overall manufacturing activity than a more protectionist world. The development of regional trading blocs which are somewhat closed to the outside but are open within might produce a mixed outcome. What is clear is that it is impossible to consider the size and composition of the manufacturing sector in any one country in isolation from global trends.

8.3.1 OECD countries

Growth in manufacturing value-added is expected to roughly parallel or be somewhat less than growth in GDP in most OECD countries. In the United States, however, the share of GDP from production of goods for investment and export is expected to rise, with the result that manufacturing would grow faster than GDP. With domestic markets saturating for many products, non-OECD demand will influence prospects for OECD production.

Growth in output will be increasingly concentrated in products with relatively low energy requirements per unit of value, such as computers, medical instruments, pharmaceuticals, and other "high-tech" goods. The role of energy-intensive industries such as steel, nonferrous metals, and cement will continue to decline. And in cases where the basic metals sector is expected to rebound somewhat, as in the United States, the growth will be primarily in the production of higher-value-added products. Among other energy-intensive sectors, chemicals is expected to grow most rapidly, but less of the sector's value-added will come from energy-intensive industrial chemicals. Production of paper products is also expected to continue to grow in response to the demands of the service sector.

The decline in the role of energy-intensive industries is partly the result of a long-term trend toward lower consumption of certain basic materials relative to GDP. Consumption of steel and cement per unit GDP has been declining since 1960 in the United States (since 1950 for steel) and since 1970 in West Germany (Fig. 8.1).[3] Aluminum use per GDP has been roughly the same in the United States since 1965 (compared to a rising trend in earlier years), but has been rising moderately in West Germany. Use per GDP of the basic chemicals ammonia, ethylene, and chlorine has been declining (in the United States) or stagnant (in West Germany) since the mid-1970s after having risen steadily since 1950. Paper use per GDP has been roughly steady in both countries. Trends in France and the United Kingdom have been fairly similar to those in the United States and West Germany.

The trend away from steel, which is especially important because its production is so energy-intensive, reflects several developments (Williams et al., 1987). One is the maturation of transport and buildings infra-

---

[3] The data were assembled by Eric Larson of Princeton University from a variety of sources. They reflect "apparent consumption" of materials, but do not include materials "embodied" in goods that are imported, or exclude materials "embodied" in exported goods. For discussion, see Williams et al. (1987).

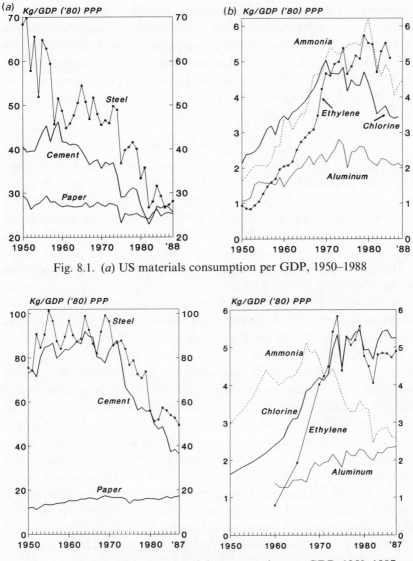

Fig. 8.1. (*a*) US materials consumption per GDP, 1950–1988

Fig. 8.1. (*b*) West German materials consumption per GDP, 1950–1987

structure. A second factor is technological improvements that reduce waste in manufacturing processes. Qualitative improvements in the strength and durability of materials have permitted manufacturers to do more with less; many consumer goods contain smaller quantities of material today than they did in the past. Yet another factor is the shift away from steel towards modern composite materials.

Among non-energy-intensive industries, growth is likely to be faster in sectors such as office, medical, and communications equipment than in traditional sectors such as transport equipment. Food and textiles are likely to experience a relative decline, especially if barriers to imports from the LDCs are lowered. These changes will have some impact on energy use, since the faster-growing sectors are less energy-intensive than the others.

8.3.2 Former East Bloc

The manufacturing sectors in the Former East Bloc are likely to undergo a radical transformation over the next 20 years, but the pace of change is difficult to assess. Much of the old manufacturing infrastructure is obsolete in terms of competing in the world market, or is simply not needed in the new economic order being established. The extent of investment from the OECD countries will strongly influence the development of the manufacturing sector. In the near-term, it seems likely that manufacturing production will decline considerably before recovering, especially in the former Soviet Union. With time, however, manufacturers will target the tremendous latent domestic demand for consumer products, and industries have the capacity to be a low-cost supplier to OECD markets.

The composition of manufacturing output will shift strongly away from the energy-intensive industries that dominated the old centrally-planned economies. The production of energy-intensive materials will decline as a share of total output, or even in absolute terms, as the subsidies for steel, cement, and other commodities are reduced or eliminated. The shifts from bureaucratic production targets to market signals will remove the artificial incentives to produce many basic materials. Russia and a few other countries have relatively low-cost energy and other natural resources that may be utilized in manufacturing, but in general light manufacturing will grow much more rapidly. Demand for products for commercial buildings such as office equipment and computers will increase in the near-term, and domestic demand for consumer products could skyrocket as economic reforms succeed.

8.3.3 Developing countries

Much of the developing world is in a stage of industrialization in which manufacturing is growing faster than GDP. Prospects differ considerably among regions, however. Manufacturing is growing very rapidly in Southeast Asia. In China, growth has slowed recently from the rapid pace of the mid-1980s, but is expected to be strong over the long run. In both regions, links with Japan will enhance the prospects for local manufacturing. In South Asia, prospects are less buoyant than in East Asia, but the recent changes toward more liberal economic policies in India may attract the investment needed for industries to modernize and compete internationally. In Latin America, growth in manufacturing will likely be higher in the 1990s than in the stagnant 1980s due to the more liberal economic policies being instituted. Prospects are less bright in Sub-Saharan Africa, although increasing links with South Africa as it democratizes could act as a positive force.

Overall, it seems likely that the importance of energy-intensive industries will increase somewhat in the LDCs, but change in the composition of manufacturing will vary. Basic materials such as steel and cement will be needed to build infrastructure, but the intensity of raw material use in LDCs could decline as new technologies are disseminated from the industrial countries. In many countries, the low cost of labor will encourage growth of export-oriented assembly and other light industry. More industrialized LDCs without natural resources will move increasingly toward manufacturing activities with higher value added. A recent projection for South Korea envisions a trend toward less energy-intensive industrial structure due to faster growth in the machinery and equipment industries and slower growth in heavy industries (Korea Energy Economics Institute, 1989). On the other hand, large countries such as China, India, Brazil, and Mexico will probably continue to produce most of their basic materials. In addition, countries richly endowed with relatively low-cost energy resources are likely to use them to support expansion of energy-intensive industries such as steel, aluminum, and petrochemicals.

8.4 Passenger travel

Travel is expected to grow all over the world due to several factors: increasing leisure time in the wealthy countries, opening on the inside and to the outside in the Former East Bloc, rising income (and migration in search of income) in developing countries, and increasing international travel resulting from economic integration and tourism. Travel is being encouraged by technological change that is allowing people to travel greater distances in less time and at lower real cost than in the past.

Future levels of per capita travel will depend on the number of trips per person and the average distance per trip. These elements are shaped by the amount of time available for travel and the speed of travel, among other factors. Car ownership, which is rising in most of the world, encourages trips because of the convenience and relatively low variable cost of travel once a car is acquired, and it may increase average distance for similar reasons. Growth in car ownership both facilitates and results from urban sprawl, and ownership tends to lead to use for most trips unless another mode is significantly more convenient.

8.4.1 OECD countries

It is likely that travel will grow more rapidly than GDP in most OECD countries. The combination of more leisure time, likely decrease in real travel prices, and growth in both personal and business long-distance links will probably increase long-distance travel substantially. International travel will grow especially rapidly. Urban travel may grow less rapidly due to congestion and increase in work at home. If urban areas continue to expand, however, there will be upward pressure on passenger-km since the distances of potential trips will be greater.

For travel to and from work, some of the historic growth has come from women commuting to work in greater numbers. This will be less of a source in the future, since the percentage of women who are in the labor force is already high in most countries. In addition, increase in the percentage of people who work at home will reduce commute trips, as will movement toward a four-day work week. On the other hand, increase in the number of workers past the traditional retirement age of 65 will add to the number of commuters, as may growth in part-time work.

The average distance of many types of trips will depend heavily on the spatial evolution of urban areas, as well as the proximity of destinations to home. Expansion of metropolitan areas, driven partly by the desire for affordable yet low-density housing, is increasing the distances that people travel for commuting, shopping, and entertainment. Changes in land-use planning to encourage higher density and multiple uses in proximity to one another could have an important impact on local travel, as we discuss further later in this chapter.

The type of trip with the greatest potential for growth is leisure and vacation. If leisure-related destinations are close to home, trip distance is reduced, but the number of trips may increase. Movement toward four-day work weeks and more vacation will give people time to travel longer distances. The average distance of vacation trips seems likely to increase as foreign travel becomes more familiar and comfortable.

International integration of economic activity is likely to increase the frequency and average distance of work-related trips. Intra-Europe business trips in particular will increase greatly throughout the 1990s. On the other hand, advances in telecommunications technology promise to make some business-related travel unnecessary, especially for large corporations with offices spread over a wide area or around the world.

Travel structure The structure of travel in OECD countries is shifting towards greater energy intensity as auto use continues to grow in urban travel and air travel increases for longer distances. In the United States, growth in automobile travel has been greater than expected, but there are reasons to expect some slowing. Nearly all adults are licensed drivers, and there are already 0.9 vehicles per driver. The average annual distance travelled per driver rose by around 20% between 1980 and 1988, but time constraints and traffic congestion may limit growth in the future. In Western Europe and Japan, on the other hand, there is still considerable room for increase in the ratio of drivers to adults. Automobile ownership per person has been rising steadily, but is still well below the level of the United States (Fig. 8.2). While it will probably not reach the US level because of higher urban density and better mass transit service, substantial growth is likely during the 1990s. Annual distance travelled per automobile is about 25% less in Western Europe than in the United States, and is lower still in Japan, suggesting room for growth in this area also.

Air travel will increase in all regions. In Western Europe, decontrol of air fares and economic integration are likely to bring strong growth. As high-speed rail expands, however, it will compete with air (and automobiles) for medium-distance trips in Western Europe and Japan. For distances up to a few hundred km, automobile and high-speed rail compete. For distances of a few hundred to 1000 km, high speed rail and air compete. For many business travellers, the time required in getting to and from airports will favor rail lines that connect city centers. For distances beyond 1000 km, air is most attractive.

Since differences in the energy intensities of high-speed rail, automobiles (on the open road), and air travel have diminished, small shifts among modes in the range of distances where they compete will have a relatively modest impact on energy use. The energy intensity of the French TGV (very high speed rail) is twice that of conventional rapid rail over the same routes (Benard, 1989), but this intensity (roughly 1 MJ/p-km of primary energy) is still well below that of air travel (2.5–3.0 MJ/p–km).

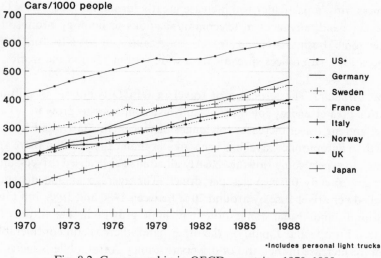

Fig. 8.2. Car ownership in OECD countries, 1970–1988

8.4.2 Former East Bloc

Since per capita travel in the Soviet Union in the late 1980s was only 40% of the Western European average despite the country's great size, there is clearly considerable room for growth. People are only beginning to travel within the Union with almost no bureaucratic hassles, but disintegration of the Union will slow this trend until suitable border arrangements are worked out. In Eastern Europe, the rules that inhibited international travel have fallen or are falling, and travel to Western Europe for shopping and business will increase greatly.

In the former Soviet Union, rail and bus will continue to carry most passengers, but the automobile and air shares of travel are likely to rise. Reforms may result in higher prices for rail and bus, which could encourage more people to buy and use cars. There is a large latent demand for cars likely to be served by joint ventures with Western car manufacturers. Car ownership in the former Soviet Union could easily grow from the 1985 level of around 50 per 1000 persons to 100–150 (less than the level of East Germany before reunification) over the next 20 years, depending on the speed and success of economic reform. Similar forces are apparent in Hungary, Czechoslovakia, and Poland.

Growth in air travel following the breakup of the Soviet Union will depend on the evolution of economic relations among the new nations. As the economies recover, travel within the former Republics may increase. There may be less growth in travel between the new nations, but there will be much more traffic between each new nation and the world outside the borders of the former USSR. New airlines may spring up to meet the demand. Aeroflot, once the world's largest airline, has entered into an agreement with at least one Western carrier to start a joint venture airline. In Eastern Europe, growing business and tourism links with the West are also likely to encourage substantial increase in air travel.

8.4.3 Developing countries

A number of factors are causing considerable growth in per capita domestic and international travel in the LDCs. Acquisition of automobiles and motorcycles permits greater travel for people who formerly relied on walking, bicycles, or mass transit. Change in the structure of the economies plays an important role. Workers employed in manufacturing are more likely to need to travel to work than are those employed in agriculture (especially traditional agriculture). The urbanization associated with economic change generates more travel between cities and the countryside – at least until the urban fraction of the population is relatively high. Travel in search of work is likely to increase within and among countries. In addition, rising participation of women in the labor force will increase travel. Traffic congestion and inadequate transport infrastructure could constrain travel to some extent. Improvements in phone systems will reduce trips somewhat; in many countries people are forced to make certain trips because basic telecommunications are inadequate.

As in other parts of the world, automobiles and air seem destined to account for a rising share of travel. The rate at which automobile ownership rises will depend on many factors, including income growth and distribution. Asian countries currently have low levels of automobile ownership relative to Latin America (Fig. 8.3). This suggests considerable room for growth, though congested conditions in cities may favor two-wheelers over cars for some people. Indeed, motorcycle ownership is growing rapidly in much of Asia.

Government investment in mass transit and policies affecting automobile purchase (import duties, taxes, registration fees) will play an important role in shaping travel structure. South Korea has very low car ownership relative to its GDP per capita due in large measure to restrictive government policies, as well as investment in mass transit. China, with only about one automobile per 1000 persons, is an extreme example of limiting automobile ownership. The Chinese government intends to continue to discourage private automobile ownership, but some growth is likely.

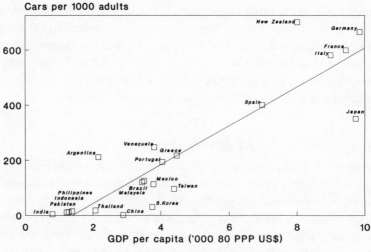

Fig. 8.3. Car ownership vs. GDP per capita, 1987

Throughout the developing world, land-use planning – or the lack of it – will have a major impact on the overall amount of travel and its modal structure.

For inter-city travel, the poor state of highways and rail systems in many countries could encourage air travel. Recent years have seen considerable growth in the number of flights between cities in the larger LDCs. High-speed rail could be an attractive option for busy corridors, but its development is constrained by lack of capital.

8.5 Freight transportation

The evolution of energy use in freight transportation is strongly affected by the mix of different transport modes, which in turn is conditioned by the type of goods that need to be moved, the size of shipments, the requirements of businesses receiving them, and by considerations of cost and quality of service. Use of rail and ships depends on the geographic coverage of railroads and navigable waterways. Trucks are much more flexible, a feature that gives them an important advantage for shipment of many types of goods.

The mix of goods moved in freight transport is primarily a function of production and consumption of industrial and agricultural products. The composition of freight depends not only on domestic patterns, but also on the extent of international trade of various goods. Economic development from agriculture and minerals production to manufacturing leads to a reduction in the ratio of freight tonne–km to GDP, since the products of manufacturing have higher value per unit weight. Growth of the service sector has an even stronger effect in the same direction, since it produces very little that needs to be shipped.

8.5.1 OECD countries

Freight transport activity, measured in tonne-km, is likely to grow slower than GDP as heavy, low-value goods (grain, coal, stone, mineral ores) come to play a smaller role in the OECD economies. The extent of this phenomenon will vary among countries. The United States will continue to be a major exporter of grains, and is expanding coal production. In general, however, smaller goods with higher value per unit weight will comprise a growing share of freight. Shipments will be smaller and more time-sensitive as the use of "just in time" manufacturing, in which both inputs and products are dispatched and received as available and needed, increases. Origin and destination points seem likely to increase in number and become more dispersed. The combination of improved communications systems, computer-aided design and manufacturing, flexible manufacturing, and increases in labor productivity is changing economies of scale on the producer side, and the growth of suburbs and small cities is increasing the number of destination points for goods (Sobey, 1987). In the United States, most new warehouses and small manufacturing plants are located in industrial parks, very few of which have direct rail access. In addition, intra-urban movement of goods associated with business and personal services (including home delivery) is on the rise. All of these changes point to an increasing role for trucks and vans.

Changes in technology may help truckers to divert inter-city goods movement from rail. Use of rigs consisting of a semi-trailer and a full trailer (called "double-bottoms" in the United States) is allowing truck lines to reduce rates. On the other hand, development of hybrid truck/rail equipment could make it possible for rail to compete more effectively. Hybrid rail and highway trailers, such as the "Roadrailer" used by General Motors for some inter-plant shipments, can combine the ease of pickup and delivery on highways with the performance advantages of rail, and provide single-vehicle shipment from door to door.

A sometimes overlooked freight mode that is likely to become more important is pipelines for moving gas. This is especially true in Europe, where the role of natural gas in the energy mix will increase substantially. Pipelines may also become more common for moving coal in the western

United States. On the other hand, shipment of energy by wire (i.e., generating electricity near the coal rather than transporting the coal) could increase, especially in the long run if superconducting technology becomes viable for power transmission.

8.5.2 Former East Bloc

The ratio of freight tonne–km to GDP, which is very high in the countries of the Former East Bloc, is certain to fall as the role of bulk raw materials, including fuels, declines. The disintegration of the Soviet Union could lead many fomer Republics to build new facilities to produce locally what was previously produced in other Republics. Such decentralization would place more factories closer to markets, reducing the distances that finished goods travel. On the other hand, the growth of trade with the West will bring significant increase in freight shipments destined for export. In this regard, the openness of Western countries to Eastern agricultural commodities will be a factor. The overall tonnage of goods shipped may fall, but the distances may increase, so total tonne-km could even rise (Grübler & Nakićenović, 1990).

While trucks have played a much smaller role in freight transport than in the West, especially in the former Soviet Union, the share of truck freight has been rising. This trend is likely to accelerate as consumer and other finished goods come to play a larger role in the economy, and as production becomes more decentralized.

8.5.3 Developing countries

Freight activity will increase substantially with economic growth, but tonne-km may grow slower than GDP in many countries. While movement of heavy agricultural and mining products (including fuels) will continue to be important in the LDCs, over time manufactured goods with higher value per unit weight will come to play a greater role. The shift toward lighter products, along with expansion of internal distribution networks and other factors, will favor use of trucks. Except for China and India, most LDCs lack extensive rail infrastructure, and are unlikely to develop it substantially due to capital constraints. Rail will continue to be important in China due in part to the reliance of the economy on domestic coal, but trucks will increase their share of total tonne–km. In India, the importance of rail is declining. Water-borne freight transport will be significant in some countries.

8.6 The residential sector

Future structural change in the residential sector will depend on income growth and distribution. The cost of housing and home appliances will also be a factor. The real cost of most appliances has declined over time, which means that households outside the OECD countries can acquire them at lower income levels than was the case for OECD households in the past.

The residential sector is different from others in that a saturation of energy-using equipment begins to appear at higher income levels. In this situation, marginal household expenditures tend to go toward other things (or toward improvements in quality that have only minor impact on energy use).

8.6.1 OECD countries

As described earlier, the population is growing relatively slowly in the OECD countries. The projected average rate of growth between 1988 and 2000 is higher in the United States (0.8%/year) than in Japan (0.4%) and Western Europe (0.2%). In the case of Western Europe, greater immigration from Eastern Europe and Africa could lead to somewhat higher growth, depending on immigration policies.

The number of households is increasing faster than population due to decline in household size. A recent forecast for the United States projects a fall from 2.69 persons per household in 1988 to 2.52 in 2000 and 2.36 in 2010 (Terlecky & Coleman, 1989). Household size in continental Western Europe is at about the same level as the United States today, but will likely decline faster unless immigration increases very significantly. An average of around two persons per household may be a practical lower limit; Sweden is already at 2.16, the lowest among OECD countries. Decline in household size will have the largest impact in Japan, which has much bigger households today than the other wealthy countries.

The structural changes that pushed upward on residential energy use in the 1970s and 1980s will have much less impact in the future. The average size of homes is increasing very little in the United States, slowly in Europe, and somewhat faster in Japan. Area per capita will grow faster due to decline in household size. It is likely to increase and perhaps level off by 2010 at 70–75 m²/capita in the United States, 55–60 m² in Western Europe, and 40–45 m² in Japan. (The 1988 levels were 52, 34, and 28 m², respectively.)

Further growth in central heating penetration in Western Europe will be slow, mostly as homes with modern systems replace older homes without central heating. Among the countries we studied, only Japan and the United Kingdom show significant potential for much increase in heating equipment. Running hot water is by now nearly universal, but decline in household size may increase per capita consumption somewhat.

Growth in per capita ownership of the major appliances, which pushed household electricity use up considerably in the 1970s, will have a much smaller impact in the future. Ownership of refrigerator-freezers, freezers, and clothes washers is approaching saturation. Color TV sets are nearly universal. Of the other major appliances being added, only clothes dryers represent a significant use of electricity. Dishwashers are also growing in saturation, but the mechanical energy supplied by dishwashers is relatively small, and the thermal energy used for heating water is probably only slightly more than that which is already used for dishwashing by hand. Air conditioning, prominent in the United States and Japan, has begun to appear in Italy and southern France, but its penetration is expected to be limited in these regions.

Ownership of home electronics is growing, but these devices use relatively little electricity. The one important development in electronics that could be a major consumer of electricity is high-definition TV. This new technology requires up to ten times more power than conventional TV of the same size, though this figure will doubtlessly fall as the technology improves.

8.6.2 Former East Bloc

Population is growing very slowly in the Former East Bloc, but household size (currently between 3 and 3.5) will fall as more housing is built by the private sector. House area may grow, particularly if private initiatives lead to increased construction of low-rise and detached (or semi-detached) housing. In addition, growth in second homes is likely in the former Soviet Union (which will also increase travel). Central heating penetration will increase, particularly outside of large cities with district heat. Ownership of various electric appliances will also grow. All of these changes will push upward on energy use.

8.6.3 Developing countries

The combination of rising population, urbanization, increase in per capita income, and further spread of household electrification in rural areas will lead to tremendous growth in demand for residential energy services in much of the developing world. Household size will fall with urbanization and decline in fertility rates, as has already occurred in the Asian NICs, for example. As in other parts of the world, this will increase *per capita* energy use for many purposes.

Change in equipment and fuels will have a major impact on cooking, currently the most important residential end use in the LDCs. The transition from biomass fuels to kerosene and LPG (or electricity in some cases) has already occurred to a large extent in Latin American urban households, and is proceeding rapidly among Asian urban households (Sathaye & Tyler, 1991). In much of Sub-Saharan Africa, on the other hand, the transition has been slowed due to decline in incomes and fuel distribution problems. The transition will be much slower among rural households, who have greater access to biomass resources and also lower income. Growing shortage of biomass fuels could accelerate the transition, but many poor households may lack the income to purchase modern stoves and fuels. Government policies toward subsidization of kerosene and LPG (or of equipment to use these fuels) will play an important role in determining the speed of the transition away from biomass. In addition to changes with respect to the primary cooking device, there will be increasing proliferation of small kitchen appliance such as rice cookers, electric frypans, and the like.

Electrification may have somewhat less impact than in the past in much of the developing world, since efforts over the last two decades have brought electricity to a relatively high percentage of rural settlements (particularly in Latin America and China). But even in villages that have electricity, there are many households that are not yet connected, and growth in rural population will add to that number. Among already-electrified households (and those that will be electrified), there is enormous room for growth in appliance ownership, especially in Asia and Africa. Market penetration of TV sets is already relatively high, but penetration of refrigerators is still low (20–25%) even in middle-income countries like Thailand and the Philippines, and is lower still in populous countries such as India, China, and Indonesia (Meyers et al., 1990). The relationship between refrigerator ownership and per capita GDP suggests that penetration will grow rapidly once countries move into upper-middle-income status (Fig. 8.4). TV ownership penetrates rapidly at a lower income level. The other major appliance whose penetration is likely to grow considerably is the automatic clothes washer.

A key uncertainty is the extent to which use of air conditioning will grow. Its market penetration is currently very low: 2% of homes in the Philippines, 1% in Thailand, and 4% in Brazil. But the experience of

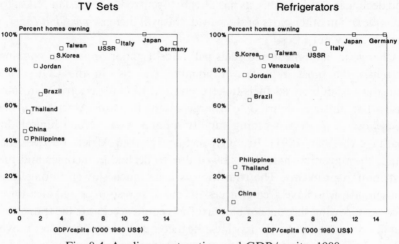

Fig. 8.4. Appliance saturation and GDP/capita, 1988

Taiwan, where saturation rose from 12% to 29% of households in the 1980s, or northern Mexico, where it has also grown substantially, suggests that use of air conditioning could rise in warm climates as households reach upper-middle income levels.

Space heating is a relatively minor demand, except in China, where some increase in central heating is likely. (Space heating is already well developed in Korea and Argentina, two other developing countries with heating needs.) Water heating equipment will become more common, though penetration of storage heaters will be relatively low.

8.7 The service sector

8.7.1 OECD countries

Between 1973 and 1987, service sector value-added grew 1.3 times as fast as total GDP in the United States, 1.4 times as fast in Japan, and 1.8 times as fast in West Germany. It is expected that it will continue to grow more rapidly than GDP in most OECD countries. Floor area may grow more slowly, however. The US Department of Energy's 1991 forecast shows commercial-sector floor space growing somewhat slower than GDP to 2010 (partly due to overbuilding in the 1980s) (US EIA, 1991). In Western Europe and Japan, floor area per employee is 20–30% less than in the United States. While this suggests room for growth, the high cost of real estate could constrain that.

Expansion of service-sector floor space is likely to be especially strong in two subsectors: health care (due to the aging of populations), and leisure-related buildings (due to increase in leisure time among workers and growth in numbers of retired persons). Both of these sectors are relatively energy intensive.

8.7.2 *Former East Bloc and developing countries*

The Former East Bloc countries currently have less than 5 m^2/capita of service sector area, about half the level of Italy. Office and retail space will increase considerably to meet the demands of the emerging private sector. Growth will also occur in lodging and restaurants.

Even larger increase in service-sector floor area will occur in the LDCs, particularly in the private sector. The many financial services that have fed growth of the service sector in the OECD countries are just beginning to expand in many LDCs. Hotels and other facilities for tourists will also grow substantially. In the public sector, growth in population will require substantial increase in education and health-care buildings.

8.8 The role of lifestyle

"Lifestyle" issues will have a major impact on future energy use. In the OECD countries, a key issue is whether people conduct activities at home or travel away from home to accomplish them. Even if travel uses modes with low energy intensity, the resulting energy use is almost always much higher than if the individual had stayed at home. For example, the energy used in the process of preparing a meal at home (including the energy used for space conditioning) is usually far less than the energy used in driving to eat out or pick up already-prepared food. Growth in away-from-home activities such as eating out, going to movies and other entertainment, and leaving town on weekends increases transportation energy use and also causes a transfer of energy use from the residential sector to the service sector. The shift of children from the home to day-care centers, a consequence of both parents working, has a similar effect.

The home is increasingly becoming a place of work. Almost 10% of the US population already does some work at home; by the year 2000, this share is predicted to grow to 15% (Ambry, 1988). This trend will be fostered by the introduction of new information technology, the continuing decrease in the cost of office equipment, and the rising time cost of commuting. It will increase energy use in the home for office equipment (not a particularly energy-intensive use) and for heating and cooling (that would not be required if the home were unoccupied), but will reduce commuting (Kitamura & Mokhtarian, 1991). Technological change will also enhance the ability of people to shop (via mail catalogues or cable television) from their home. Increase in working or shopping from home alters both travel and home energy use patterns, while reducing space needs (or occupancy) in places of employment and services. While providing a comfortable work environment for 100 people in an office requires much less energy than if those same 100 people each worked at home, the relatively high energy intensity of commuting (especially in automobiles in congested traffic) usually results in much greater overall energy use.[4]

A key issue is how people will use their leisure time, which most observers expect will increase in the wealthy countries. The aging of the population and innovation in home electronics such as high-resolution TV and increasingly user-friendly personal computers could lead to a higher share of leisure time at home. But out-of-home leisure activities (especially for "three-day weekend" trips or vacations) will probably grow in absolute terms, which would increase energy demand for transportation (Schipper et al., 1989).

Decisions about staying at home or going out will be affected by the time cost of travel and the quality of the experience. Under adverse conditions, the stress associated with travel may encourage people to conduct activities from the comfort of home. In the long run, increase in the ability of people to make better-quality electronic interaction with distant people and places could lessen the need to travel to them, although it would be unwise to underestimate people's desire to have direct experience.

The radical changes in the Former East Bloc will profoundly affect people's lifestyles. *Glasnost* has permitted most people their first real look at how people live in the West. Citizens can now travel abroad much more easily. While the economic outlook for the 1990s is not bright, income growth in the longer term should allow consumers to own more cars and appliances and live in larger, more comfortable homes. The situation in the former East Germany suggests that while differences in lifestyles between East and West have been enormous, the Easterners take on Western-like habits quickly, as incomes permit. By Eastern standards, these lifestyles are energy-intensive.

4 Whether conditioning extra space in commercial buildings for workers would require more energy than using existing space in homes depends on whether the homes would have been heated or cooled otherwise.

In the LDCs, lifestyles will be affected by urbanization, income distribution, and the influence of mass media. These factors will shape overall consumption patterns as well as demand for energy-using goods such as appliances and motor vehicles. Urbanization increases exposure to modern tastes, and may also provide income-earning opportunities to cater to them. Income distribution leading to growth of a well-established middle class, as has occurred in the Asian NICs, can result in rapid increase in ownership of devices such as refrigerators and automobiles. If on the other hand, income distribution remains highly skewed, ownership of high-priced goods will be more limited to the wealthy few.

8.9 Policies that affect activity and structure

The levels of different energy-using activities are affected by a variety of government policies. Energy pricing policy is important: subsidies for particular fuels encourage activities that use them. Often, however, the energy impacts are more subtle. For example, policies that subsidize the cost of home ownership, such as tax deductions for mortgage interest, encourage larger homes, which use more energy. Policies that encourage end-user recycling of materials reduce the demand for virgin materials from mining and forestry, and also provide scrap input to manufacturing processes.

Public investment in the transportation infrastructure has a major impact on mode choice in travel and freight transport. Public policies face difficulty in pushing against market forces or consumer preferences, however. In the industrialized countries, for example, the potential for encouraging use of rail for freight transport is somewhat limited by the market forces that favor trucks. Some modal shifts are possible, but will be successful only when there is an overall gain in transport productivity. These factors also apply in developing countries, but since much of the future transport infrastructure is still to be built, there is more room for government to shape the freight system. Building or expanding rail systems may have long-term advantages, but roads require less capital and can be constructed incrementally, and are therefore easier to build for most LDCs. Careful coordination of new infrastructure for rail or ship transport with siting of new industrial plants that will utilize those modes is needed to ensure effective use.

8.9.1 *The case of passenger travel*

Policies can have considerable impact on the level of travel activity and its modal structure. In the long run, land-use planning can play a major role in reducing the distance of many types of trips. Policies that favor high-density housing and location of services closer to residences, transit centers, and work places result in shorter distances to many of the common destinations to which people travel in their daily lives, and facilitate "trip chaining". Mixed-used developments that include services like restaurants and banking at suburban work centers generate less trips than similar locations without services (Cervero, 1986). In the United States, modifying the low-density suburban pattern that is dominant is a major challenge. Changes in land use toward higher density or mixed use are often resisted due to fear of reduction in property values. For new development, lower costs usually favor urban expansion rather than infill. In Europe and Japan, historic urban patterns and lack of space limit urban sprawl to some degree, but this could change if policies that subsidize agricultural land use are altered. Land use planning is especially significant for the LDCs, where cities are growing rapidly.

The major challenge in attaining reduced energy use (and emissions) in urban transportation lies in designing strategies that discourage use of single-occupancy automobiles and encourage less energy-intensive modes. The problem is that once automobiles are acquired, they tend to be driven. People can usually travel more rapidly and in greater comfort in automobiles than in collective modes, and the variable private cost of using cars is usually less than the cost of using other modes. Thus, strategies that make drivers bear the true social cost of operating the automobile are important to change behavior toward collective modes of transport. Congestion pricing (in which tolls are higher during peak periods), road tolls, and parking pricing are examples, although studies indicate that commuter traffic is highly price inelastic (Orski, 1990). More stringent methods of discouraging automobile use are also possible. Singapore has heavily restricted automobile use during peak hours through a combination of area licensing schemes, limitations on parking, steep road pricing, and mandatory carpooling (Ang, 1991). Discouraging automobile ownership (such as through high taxes) is a more controversial policy, but one that is still used in many developing and a few OECD countries.

Increase in the density of residential areas, work locations, and at other major destinations can encourage use of transit because it can be operated more efficiently and is easier for people to access when density is higher.

Curitaba, Brazil provides a good example of how land use planning can promote mass transit (Lerner, 1991). The city has five main arterial avenues that are mostly dedicated to express buses radiating from the city center. High density residential and commercial development is located along these avenues, and a system of local feeder buses connects to stations on the main lines. Automobile fuel use in Curitaba is significantly less than in other, similarly-sized cities in Brazil even though its automobile ownership rates are relatively high. Land use strategies are also important in the United States, where there is increasing interest in building fixed-rail transit systems but few changes in the conditions that might favor them over automobiles.

Other strategies to improve the service and viability of collective modes include pricing and improving the operation of transit and highway systems. Carpools and vanpools can be encouraged by providing free parking or other preferential treatment. Well-designed high-occupancy vehicle lanes can give these and other collective modes an important time advantage in peak periods. A similar advantage can be obtained if buses have designated lanes on major routes, or separate guideways in denser areas. Such systems have the flexibility of normal buses and can serve lower density neighborhoods than fixed-guideway transit while providing greater speed to major destinations. Transit services can also be improved by better integration of bus and rail systems through schedules and fares. Encouraging use of these systems during off-peak hours is difficult, but is important to improve their financial situation.

Greater use of bicycles and walking can be promoted through changes in building and street design, construction of separate bicycle paths, and allowing passengers to take bicycles onto transit. In several cities (e.g., Copenhagen and Davis, California) where bicycle paths have been integrated with the centers of activity and homes, the share of trips taken by bicycles is significant.

8.10 Conclusion

The level of energy-using activities is continuing to increase throughout the world, but the rates of likely growth differ among regions. Over the next 20 years, manufacturing production is expected to grow at a rapid pace in parts of the developing world, and moderately in the OECD countries. In the Former East Bloc, it seems likely to stagnate or decline for much of the 1990s, but could then grow at a moderate pace if the transition to a market economy is successfully managed. Domestic passenger travel seems likely to increase everywhere, and growth in international travel will be especially strong. Freight transport activity is difficult to evaluate in the aggregate, since the composition of goods changes over time, but increase is expected in all regions, especially in the developing countries.

Structural change within sectors will have significant impacts on energy use. In manufacturing, faster growth in light industry will lead to lower energy intensity in the OECD countries and especially in the Former East Bloc. The outlook in the LDCs suggests somewhat higher growth in energy-intensive industries, but this trend will vary among countries. In passenger travel, structural change is pointing toward higher energy intensity in most of the world as the role of automobiles and air travel continues to grow. Increase in the use of trucks is pushing in a similar direction in freight transport. In the residential sector, structural change will have only a moderate impact in the OECD countries, where per capita levels of home services are already high, but will push energy use significantly upward in the LDCs, and to a lesser extent, in the Former East Bloc.

References

Ambry, M. 1988. At home in the office. *American Demographics*, December:30–61.

Ang, B. 1991. The use of traffic management systems in Singapore. In *Driving New Directions: Transportation Experiences and Options in Developing Countries*. Washington, DC: International Institute for Energy Conservation.

Benard, M. 1989. Electricité de France. Private communication.

Cervero, R. 1986. Unlocking suburban gridlock. *Journal of the American Planning Association*. **52** (4), 389–406.

Grübler, A. & Nakićenović, N. 1990. *Evolution of Transport Systems: Past and Future*, Report to Shell Group Planning, UK. Laxenberg, Austria: International Institute for Applied Systems Analysis.

Kitamura, R. & Mokhtarian, P. 1991. Energy and air quality impacts of telecommuting. Paper presented at the *Conference on Transportation and Global Climate Change: Long-Run Options*, August 25–28, 1991. Pacific Grove, CA.

Korea Energy Economics Institute. 1989. *Sectoral Energy Demand in the Republic of Korea: Analysis and Outlook*. Seoul, South Korea.

Lerner, J. 1991. The Curitaba bus system. In *Driving New Directions: Transportation Experiences and Options in Developing Countries*. Washington, DC: International Institute for Energy Conservation.

Meyers, S., Tyler, S., Geller, H. S., Sathaye, J. & Schipper, L. 1990. *Energy Efficiency and Household Electric Appliances in Developing and Newly Industrialized Countries*. Berkeley, CA: Lawrence Berkeley Laboratory Report LBL-29678.

Orski, C. K. 1990. Can management of transportation demand help solve our growing traffic congestion and air pollution problems? *Transportation Quarterly*, **44** (4), 483–98.

Sathaye, J. & Tyler, S. 1991. Transitions in urban household energy use in Hong Kong, India, China, Thailand and the Philippines. *Annual Review of Energy*, **16**.

Schipper, L., Bartlett, S., Hawk, D. & Vine, E. 1989. Linking life-styles and energy use: A matter of time?, *Annual Review of Energy*, **14**, 272–320.

Sobey, 1987. Energy use in transportation: 2000 and beyond, *Summary of Presentations at the Workshop on Energy Efficiency and Structural Change: Implications for the Greenhouse Problem*. Oakland, CA, May 1–3, 1988. Berkeley, CA: Lawrence Berkeley Laboratory Report LBL-25716.

Terlecky, N. & Coleman, C. (eds.). 1989. *Regional Economic Growth in the US: Projection for 1989–2010. Vol. III, Projection by Age, Sex, and Race*. Washington, DC: National Planning Association.

US EIA (Energy Information Administration). 1991. *Annual Energy Outlook with Projections to 2010*, DOE/EIA-0383(91). Washington, DC: US Department of Energy.

Williams, R. H., Larson, E. D. & Ross, M. H. 1987. Materials, affluence, and industrial energy use. *Annual Review of Energy*, **12**, 99–144.

World Bank. 1990. *World Development Report 1990*. Washington, DC. 1991. *World Development Report 1991*. Washington, DC.

9

Energy intensities: prospects and potential

In the previous chapter, we described how rising activity levels and structural change are pushing toward higher energy use in many sectors and regions, especially in the developing countries. The extent to which more activity leads to greater energy use will depend on the energy intensity of end-use activities. In this chapter, we present an overview of the potential for intensity reductions in each sector over the next 10–20 years. It is not our intent to describe in detail the various technologies that could be employed to improve energy efficiency, which has been done by others (see, for example, Lovins & Lovins, 1991; Goldemberg et al., 1987). Rather, we discuss the key factors that will shape future energy intensities in different parts of the world, and give a sense for the changes that could be attained if greater attention were given to accelerate efficiency improvement.

It is important to have a conceptual framework when considering the potential for reducing energy intensities through technological change (Fig. 9.1). In many cases, the average new technology is less energy intensive than the stock average, and the best available technology is lower still. In addition, technologies which are relatively proven and understood with respect to their costs and benefits, but not yet commercialized, may be cost-effective from a societal perspective (especially if one adopts an energy price that incorporates environmental externalities). (We discuss what is meant by "cost-effective" below.) Their commercialization may be delayed because the producer believes that energy users would not pay the higher cost associated with the new technology. Beyond the current "optimal" level are technologies that are well understood and might be cost effective, but require further development or cost reduction before commercialization. Below this are techniques that are less proven, but could be cost-effective in the future. The physical limit for energy efficiency in various

250

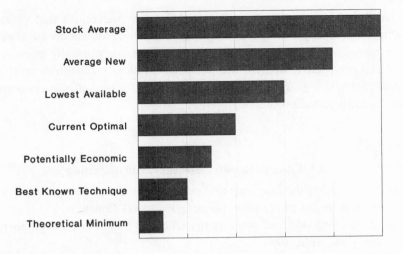

Adapted from Jochem (1990)

Fig. 9.1. Energy intensity potential

applications is defined by thermodynamic principles. The relationship between any particular efficiency level and its thermodynamic limit (its "2nd Law Efficiency") is important, since it defines how far that level is from what is physically possible (AIP, 1975). There are a few cases where cutting-edge technologies are actually nearing physical limits, but for the most part even the most advanced technologies are far from that point.

While the physical perspective is important, economic criteria are paramount for private and social decision making. An improvement in energy efficiency often involves a trade-off between additional investment and future reduction in energy costs (and sometimes other costs as well).[1] From an economic perspective, a "cost-effective" investment is one for which the present value of future energy and other savings over the life of the technology or measure is greater than the additional investment and other costs incurred to achieve the energy savings. The value given to the energy savings depends on expectations of future energy prices and the extent to which the future benefits are discounted. Energy users, particularly households, generally discount future savings (implicitly) at a much higher rate than is used when considering the societal perspective (typically 7–10%). Applying a lower discount rate usually increases the investment in efficiency that is cost-effective.

---

[1] There are some cases where energy efficiency improvement allows design changes that may result in a lower overall capital investment for a particular new technology or building.

In this chapter we describe potential intensity reductions that various sources have judged to be cost effective. While these sources do not always use common assumptions and methodologies, they usually present a societal perspective. Because this perspective differs – sometimes greatly – from that of energy users, there is a gap between the cost-effective potential and what is actually occurring.

9.1 Energy intensity, efficiency, and utilization

Before discussing the prospects and potential for energy intensities in each sector, it is useful to consider two important relationships: (1) between energy intensity and technical energy efficiency; and (2) between energy efficiency and utilization.

We illustrate the former with an example from Italy. Various new appliances in the late 1970s were more energy-intensive (in terms of kWh per year) than the stock average because they were larger or had more energy-using features (Fig. 9.2). By 1987, however, the energy intensity of new appliances had fallen, as technical efficiency improvements more than balanced increase in size. Thus, in the early 1980s, stock turnover was increasing energy intensity, but by the late 1980s it was decreasing it. New appliances planned for 1992–93 will have much lower energy intensity than 1987's new appliances in most cases, so stock turnover will depress average intensity considerably in the mid-1990s. Many developing countries are where Italy was earlier: refrigerators and other appliances are likely to become more energy-efficient from a technical viewpoint, but increase in size and features will push intensity upward.

Changes in utilization also affect energy intensity. Improving energy efficiency reduces the energy cost of particular activities, which in some cases can encourage increased use of equipment, thereby cancelling some of the energy savings. The magnitude of the "take-back effect" for different end uses has been debated. Khazzoom (1987) argued that the "take-back" resulting from appliance energy-efficiency standards was sizable, but Henly et al. (1988) estimated that less than 10% of the energy savings expected from appliances to which efficiency standards were applied would be lost due to increased use. For refrigerators and freezers, there is little, if any, scope for increased utilization. For clothes washers and dryers, it seems doubtful that households would use a more efficient device more frequently than an average one. For space conditioning, however, there is sometimes room for households to take back some of the

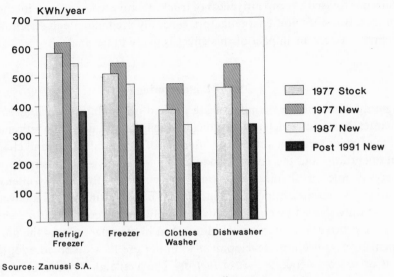

Source: Zanussi S.A.

Fig. 9.2. Italy: average appliance energy use

energy savings from efficiency improvement by increasing indoor comfort. There is evidence that utilization does increase for home heating after retrofit, at least when indoor temperatures are lower than desired. This occurred in the United Kingdom (Scott, 1980) and, to a lesser extent, in Japan. The data suggest that the reduction in energy use resulting from efficiency improvement was much larger than the increase due to greater comfort, however. The magnitude of the take-back effect for space conditioning depends on where indoor comfort levels are at a particular time and place. In countries where homes are already at a high level of comfort, there is little room for more utilization.

For automobiles, the size of the take-back effect is important for policy, since if automobile use is highly sensitive to the fuel cost per kilometer of travel, then fuel taxes might be a more effective way to reduce gasoline demand than fuel economy standards. A recent statistical analysis of vehicle-miles travelled and fuel cost per mile in the United States indicates that travel is rather insensitive to changes in fuel economy; the take-back effect was estimated to be 10–15% or less (Greene, 1991).[2] In freight transport, increase in truck energy efficiency reduces operating costs and

[2] In the framework of our analysis, the increased utilization of automobiles is a rise in activity. Increased use of space conditioning equipment manifests as a rise in energy intensity.

thus improves the competitiveness of trucks relative to less energy-intensive modes. But since non-energy factors generally predominate in selection of freight modes, the impact of this effect is likely to be small.

9.2 Manufacturing

Energy intensities in manufacturing depend on modifications to existing plants, the rate at which new production facilities are added, the retirement rate of older facilities, the energy intensity of new production, and change in operations and the mix of products. Expectations about energy prices play a role in shaping decisions, especially in the energy-intensive industries, but their importance should not be overestimated, as intensities are mainly shaped by concerns that are only loosely connected to energy.

The process of technological innovation in manufacturing – the movement from concept to development to initial use to widespread adoption – is driven by a complex set of factors. The availability and cost of new technologies, which depends on investment in R&D, is obviously critical. Adoption of new techniques, which often involves a certain amount of risk, depends on the economic environment in which industrial firms operate. A competitive environment is conducive to continuous innovation designed to produce goods at lower cost, while one that encourages maintenance of the status quo, either because incentives for innovation are weak or firms are shielded from competition, leads to much less innovation. The latter situation has characterized the Former East Bloc and many developing countries. As these regions move toward greater integration into the world market economy, the rate of adoption of new techniques will have to increase if their industries are to be viable. The transfer of advanced techniques will depend in part on the environment for outside investment, although encouraging indigenous innovation will also be important.

New processes are usually less energy-intensive than average practices. Thus, one might expect decline in energy intensity over a period of years to be greater when growth in output is high, since expansion of capacity implies addition of new production lines. The relationship between output growth and energy intensity over the 1973–87 period for eight OECD countries shows a mixed picture, however. The correlation is strong for the chemicals and paper and pulp industries, but less so in other industries. In the steel industry, for example, intensity declined by 30–40% in six of the eight countries even though there was little growth or even decline in output. This suggests that even where there is not significant capacity expansion, producers in a competitive environment attempt to enhance quality, improve productivity, and reduce costs by closing less-efficient plants and upgrading existing facilities through installation of advanced methods and technologies.

The steel industry provides a good example of how change in processes can lead to large energy savings. In the United States, 72% of the steelmaking capacity in 1965 used open hearth technology and only 18% used the basic oxygen furnace, which requires only about one-third as much energy per ton of raw steel produced (US EIA, 1990). By the mid-1970s, the basic oxygen process had increased its share to about 62%. In the following decade, there was a trend toward smaller mills utilizing electric arc furnaces, which make heavy use of recycled materials and are much less energy-intensive in terms of final energy. By the late 1980s, electric arc furnaces comprised about 40% of the total steelmaking capacity, and it is expected that the industry will move further toward this technology in the future. Various technologies promise to reduce energy intensity at every step of steel production: agglomeration, coking, ironmaking, steelmaking, primary finishing, secondary finishing, and heat treating.

Comparable changes in processes shape energy intensities in other industries as well.[3] The type of change will vary among countries depending on the nature of the local industry and its future development. In addition to change in unique industry processes, improvement in conventional technologies (motors, furnaces, boilers) will contribute to a decline in energy intensity across industries.

Along with technological innovation in manufacturing particular types of products, change in the product mix will affect the energy intensity of many industries. In the OECD countries, the trend in many industries toward products with higher value per unit will decrease energy intensity.

9.2.1 OECD countries

OECD manufacturing energy intensities will continue to decline with technological innovation. There are many new production techniques that result in substantially lower energy intensities, but their rate of penetration into the market when new facilities are constructed or when worn-out or obsolete equipment is replaced is uncertain. Further, it is difficult to gauge what the impact of explicit processes and technologies might be on energy intensities expressed in terms of value-added, or to translate specific

[3] For a discussion of technology prospects in various industries, see U.S. EIA (1990).

Table 9.1. *US manufacturing: potential energy intensity reductions*

Sector	State of the art	Advanced technology	Annual rate	Period
Iron and steel	40%	46%	3.0–3.6%	1983–2000
Paper and pulp	30%	49%	1.4–2.7%	1985–2010
Cement	31%	55%	1.5–3.2%	1985–2010
Glass	21%	44%	0.9–2.3%	1985–2010
Textiles	34%	53%	1.6–2.9%	1984–2010

Source: Energetics (1988).

physical energy intensities (e.g., MJ per tonne of steel) into industry-wide intensity measures, especially in very heterogeneous sectors such as chemicals and metal products. Even physical intensities such as "MJ per tonne of steel" are problematic, since the mix of steel products changes, and different products vary in their energy requirements.

Despite these problems, analysts have attempted to estimate future possibilities for manufacturing intensities. A series of engineering studies, for example, identified a substantial energy savings potential in several US industries (Energetics, 1988). In each industry, the analysts estimated the impact relative to the base year average with (1) demonstrated state-of-the-art technology, and (2) advanced technologies that are under development but not yet on the market. These advanced technologies were judged likely to be cost-effective in the future, based on interviews with manufacturers. In the steel industry, energy intensity could be reduced by 40% relative to its 1983 value using late-1980s state-of-the-art technology, and by 46% using advanced technologies (Table 9.1). The comparable values (relative to 1985) are 30% and 49% in the paper and pulp industry, and 31% and 55% in cement. The difference between state-of-the-art and advanced technologies is smaller in steel than in other industries.

The rate of intensity reduction depends on how quickly state-of-the-art or advanced technologies are adopted. The authors of the Energetics studies judged that the changes could occur in the steel industry by the year 2000, and by 2010 in the other industries. The rate of reduction achievable using state-of-the-art technology ranges from 0.9 to 3.0%/year; with advanced technology, it increases to 2.3 to 3.6%/year. (The higher figures apply to the steel industry and may be somewhat optimistic.) From the standpoint of technology, it appears that the historical trend of intensity decline can be maintained.

Industrial energy-efficiency improvement results from the implementation of myriad technologies. Basic process changes, such as the shift from the wet to the dry process in cement manufacture, constitute much of the potential. But the details of process management, waste heat recovery, and so forth are also important, as evidenced by the fact that the range of energy intensities of facilities using the dry cement production process is greater than the difference between the wet and dry techniques.

Manufacturing *electricity* intensities will decline much more slowly than *fuel* intensities, and may increase in some industries. Historically, growth in productivity in many industries has been associated with increasing application of electricity. The growing use of automated processes and industrial robots will contribute to this trend, as will various pollution control measures. At the same time, however, the electrical processes themselves will become more efficient. Electricity is being used more to control processes, which leads to greater overall efficiency, rather than as a direct substitute for fuels.

9.2.2 Former East Bloc

Manufacturing energy intensities in the former East Bloc, which are very high by international standards, will eventually be radically lower than they are now, but it is hard to say how rapidly they might fall. In the Soviet Union in the late 1980s, only half of the steel was made in more efficient oxygen-converted furnaces or electro-steel plants, which are the norm in Western Europe, and only about 15% of the cement was made by the more efficient dry process, versus over 50% in the West. The potential for intensity reduction is enormous, but as Dobozi (1991) points out: "The real issue is not the potential, but rather the future ability of the Soviet Union to realize it by instituting radical systemic changes in its economy. The poor record of Soviet energy conservation in the past two decades is an unambiguous indication that, despite the heavy emphasis planners have placed on energy conservation, the command economy methods of forcing conservation are not generally able to produce the desired level of savings, let alone to cure the heavy energy intensity bias of the economy."

The movement toward a more decentralized and market-oriented economy with a sizable private sector is clear, but great uncertainties surround the speed, comprehensiveness, and sequencing of market reforms, as well as the sustainability of a radical reform course in view of the serious obstacles it will have to face. Economic restructuring will eliminate the oldest and least efficient plants, most of which are quite inefficient indeed. Rapid introduction of market reforms might speed the scrapping of old factories, but concerns about unemployment may slow this development, as occurred in Poland in late 1991. Over time, the introduction of market pricing of energy will encourage energy-saving practices in those factories that survive restructuring. It is also probable that factories capitalized by foreign investors will be more energy efficient than even relatively modern Eastern factories, but it will take time before the new factories have a major impact on industry-wide average intensities.

9.2.3 Developing countries

Manufacturing energy intensities in the developing countries will likely decline as a result of improvement of existing facilities and addition of new plants embodying more modern technology. The extent of these forces will depend on the economic situation in particular places and the ability of LDCs to acquire state-of-the-art technologies. As in the former East Bloc, the market-oriented economic policies that are gaining favor in most developing countries will lead to greater competition, and probably more foreign investment as well, both of which should result in modernization and lower energy intensities. The impact of such a policy shift will likely be greatest in Latin America, since it represents a more marked break with the past than is the case in much of Asia. The recent moves toward liberalizing the economy in India should also lead to faster use of modern technologies.

In much of the developing world, many of the existing facilities utilize technology that is well below the state-of-the-art. Processes that were abandoned years ago in the OECD countries are commonly found in the developing countries. This is more the case in countries with relatively older industrial infrastructures, such as China, India, and Argentina, than in countries that have industrialized more recently. The situation also varies across industries. Those industries that are export-oriented have to be relatively efficient to compete internationally. The steel industries in South Korea and Brazil, for example, are relatively state-of-the-art, and are probably more energy-efficient than the steel industries in most OECD countries. In addition, facilities owned by transnational corporations are often equal in energy efficiency to factories of similar vintage in the OECD countries. On the other hand, industries that primarily serve the domestic market are often shielded from competition, utilize older technology, and invest little if anything in R&D.

While there have been relatively few industry-wide studies of conservation potential in developing countries, there have been thousands of energy audits of individual manufacturing plants that have demonstrated considerable potential for cost-effective energy management and retrofit investments. A recent World Bank study of China found numerous opportunities for improving energy efficiency in existing plants (World Bank, 1991). A key goal is to raise the efficiency of industrial boilers from an average of about 60% to 70–80%, which would involve better air control, air pre-heating and thermal insulation, and better boiler control features. Raising the inherent efficiency of new boilers will also result in important future energy savings, as will improvements in operations and maintenance. These findings for China apply in many other developing countries as well.

Along with improving technology and operations at existing plants, there are also opportunities to design new plants to be more energy-efficient. The extent to which new plants in developing countries are less energy-efficient than similar new plants in the OECD countries depends on who is building the plant, access to advanced technology, and the level of local energy prices. The energy intensity of particular new plants also depends on the product mix and the quality of inputs. Introduction of state-of-the-art (or nearly so) techniques depends on factors discussed earlier (e.g., access to technology, capital). Undoubtedly, there are many opportunities for cost-effective improvement in the energy efficiency of new plants in developing countries, but it is difficult to generalize.

Manufacturing energy intensities in developing countries will also be affected by change in the scale of industrial facilities. In many countries, small plants account for a substantial share of production in certain industries. This phenomenon, which is especially the case in the building materials industry, contributes to high energy intensity. In China, for example, 80–90% of cement and brick are produced in small plants, which are 50% or more energy intensive than larger modern plants. The extent to which small-scale industries, which may be inefficient but important for employment, give way to larger, more modern facilities will depend on market proximity as well as government policies. The ability of LDCs to acquire new technologies that allow for efficient production at a lower scale will also be a factor.

9.3 Passenger travel

Throughout the world, much of the growth in passenger travel is being provided by the two most energy-intensive modes, automobiles and air travel. For automobiles, the average fuel economy of new vehicles will depend on how much technical efficiency improvements are balanced by increase in size and power, and by how much emphasis auto manufacturers place on fuel economy. Both of these will be affected by government policies regarding gasoline taxation, vehicle import duties and registration fees, and fuel economy standards, as well as by consumer tastes. For air travel, technological innovation will continue to reduce fuel requirements per seat–kilometer, but the level of fuel prices in the 1990s may lead aircraft manufacturers to place greater emphasis on other features or to hold costs down.

9.3.1 OECD countries

Cars are probably the most visible of all energy users in the OECD countries, and they are accounting for a growing share of total energy – and especially oil – consumption. Improving automobile fuel efficiency is a hotly debated topic. The discussion is complicated because fuel economy can be increased by reducing acceleration capability and by making the vehicle smaller or lighter. Indeed, small, relatively low-performance cars that average (in tests) fuel use of around 5 1/100 km (50 mpg) are already on the market, but there is debate as to how much improvement is cost effective for the larger, higher-performance cars that OECD consumers are increasingly purchasing.

The foremost design changes available to manufacturers are: 1) smaller, more modern engines with the same maximum power, 2) electronically managed gears to reduce the average engine speed, 3) further streamlining the average vehicle, and 4) improving accessory efficiencies, especially air conditioning, power steering, and brakes.[4] General Motors' recently introduced Quad 4 engine, for example, is 50% more powerful per unit of displacement than the average. In a well-designed vehicle, a high-performance engine similar to the Quad 4 but with 1.3 liters displacement instead of 2.3 would provide the same acceleration capabilities as today's average kind of engine. With attention to friction reduction, such a power plant would enable a 20% increase in fuel economy (mpg) relative to the average 1990 new car. Transmission management can also provide major fuel benefits depending on how much reduction in engine speed is acceptable. A 10% improvement in fuel economy would not require technological virtuosity, for example, in controlling vibration. This improvement involves a slight loss of amenity because it causes more gear shifting and there is a slight delay associated with downshifting to access

---

[4] This section was contributed by Marc Ross, Professor of Physics at the University of Michigan. For further discussion, see Ledbetter and Ross (1990).

high power, though sophisticated electronic management would minimize this disbenefit. A 17% reduction in air drag, placing the average drag coefficient at 0.30 (which is well short of the best in a production vehicle – 0.26 with the Opel Calibra), with appropriate reductions in tyre resistance and transmission and accessories losses, would enable a 5 to 10% reduction in the load.

Assuming that the size and performance of an average new 1990 US car is maintained, the combination of the above improvements would result in an EPA test rating of about 40 mpg, around 40% higher than the average for 1990 new cars. This achievement would involve a retail price increase of perhaps $500, less after the changes became established. These changes could be implemented in new US cars over a period of 10 to 15 years. Much more could be achieved in perhaps 20 years, although the changes would involve greater technological ingenuity and/or higher costs. Among the most promising technologies are "stop–start," which turns the engine off when idling, recovery of energy in braking, and full control of valve timing, which would eliminate the need for a throttle as well as improve engine power. Altogether, such technologies would enable a further 30% increase in fuel economy. Another direction involves solving the diesel engine's emissions problems. If that could be done, a further 25% fuel economy improvement could be made. This level of improvement could also be achieved by switching to a high-octane fuel like methanol or methane and designing engines for that fuel. Together, advanced technologies and use of non-gasoline engines would enable a vehicle to have the average performance and size of current new cars with an EPA rating of around 65 mpg (3.6 l/100 km).

The above discussion refers to a vehicle with attributes in-between a US compact and intermediate-sized car. More striking reductions in fuel intensity can be achieved with very small cars appropriate for commuting and other in-town uses. With their much lower weight and air drag, such vehicles could achieve as much as a factor of two better fuel economy than an average size car with similar acceleration. Such vehicles are also more appropriate for electric propulsion.

Prototype vehicles that incorporate advanced technologies and designs have been developed by many of the leading car manufacturers (Bleviss, 1988). Among the vehicles designed for 4–5 passengers, fuel economy in city driving is in the 60–75 mpg range (3–4 l/100 km). The reduction in fuel economy achieved by these prototypes arises from a variety of features: extensive use of plastics and light materials, innovative engines and transmissions. Performance and size are generally slightly less than at present, but by no means poor. Lovins (1991) maintains that even higher fuel economy is possible if manufacturers "tunnel through cost barriers" by utilizing advances in materials, switched-reluctance electric motors in each wheel (in place of a driveshaft), and other weight-saving possibilities.

Bringing high-efficiency cars into production is the key issue. Most of the high-efficiency prototypes have been built to test performance and the compatibility of various experimental components, not to produce a car for the market. The number of years development of such cars would take is uncertain, as is their cost and consumer acceptance. Thus, technology development and incorporation in products involves a certain degree of risk. This idea is illustrated in an analysis by K. G. Duleep (1991), who used detailed information on vehicle technologies to model the impact of introducing various technologies into cars of different size. Assuming a 20-year period for introduction of technologies, he estimated the potential to improve fuel economy at different risk levels in each size class. Relative to the 1991 US sales-weighted new car average, he estimated that new cars could be 60% more fuel efficient (or 38% less fuel intensive) using low-risk technologies, and 60% more fuel efficient (or 49% less fuel intensive) using medium-risk technologies (assuming that the mix of sizes and other vehicle attributes in 2010 are equal to 1988 levels) (Table 9.2). What is lacking in the current environment are incentives for manufacturers to plan for production of highly fuel-efficient automobiles.

The future new car fleet averages in Table 9.2 refer to a fleet with the 1988 mix of vehicle sizes. If the trend toward larger cars continues, the values would be somewhat higher. If the opposite occurs, either through consumer choice or government policy, the values would be lower. It is worth noting, however, that the difference in average fuel economy between compact and intermediate cars, the two most popular size classes, is not very large.

In translating technical potential into real-world performance, it is important to bear in mind that the on-road fuel economy of new cars is falling further behind their rated fuel economy. The current gap between the two – estimated at 15–20% in the United States – is expected to increase with growing traffic congestion, higher highway speeds, and rise in the ratio of urban to highway miles (Westbrook & Patterson, 1989). The latter factor means that an increasing share of vehicle-km are driven on short trips; this increases fuel intensity, since a higher proportion of driving occurs when engines are not warmed up and running optimally. The fact that cars are designed and optimized not for city driving, but for the open road, contributes to higher intensity (An & Ross, 1991).

Table 9.2. *Potential automobile fuel economy at different risk levels*
(*average US new car test mpg in 2010*)[a]

	1991 Actual	Risk level I	Risk level II	Risk level III
Minicompact	—	68.5	83.4	110.0
Subcompact	31.2	51.5	63.4	86.6
Compact	29.3	46.4	57.0	74.0
Intermediate	25.8	42.2	51.3	69.7
Large	23.7	39.9	48.6	65.8
Luxury	—	37.2	46.2	62.1
Fleet	27.8	44.8[b]	54.9[b]	74.1[b]

Note: The attributes in each vehicle class are maintained at 1988 levels.
[a] To convert to 1/100 km, divide 236 by the mpg value.
[b] Assumes 1988 mix of sizes.
Source: Duleep (1991).

Air travel The evolution of air travel energy intensity is shaped by the technical energy efficiency of the aircraft fleet (seat-km per gallon), operational factors, and utilization of aircraft capacity (load factor). The addition of newer, more efficient aircraft will continue to increase fleet energy efficiency in the OECD countries. In the United States, the fleet fuel efficiency nearly doubled between 1970 and 1989 from 26 to about 49 seat–miles per gallon (smpg).[5] The best current generation aircraft deliver 50–70 smpg, while new aircraft scheduled for delivery in the early 1990s are expected to achieve 65–80 smpg. Increase in the number of seats per aircraft, which contributes to higher efficiency, is expected to continue as a necessary means of increasing passenger capacity without worsening airport traffic congestion. Technologically achievable efficiencies for post-2000 aircraft – using propfans as the standard engine – are estimated at between 110 and 150 smpg, but it is not clear whether manufacturers will actually introduce such aircraft or whether airlines will buy them.

Along with the characteristics of new aircraft, the future efficiency of the fleet will depend on the types of aircraft in operation, the rate of stock turnover, and the extent of engine retrofitting for older aircraft (aircraft lifetimes are typically 25–30 years). Using a government forecast of the US airline fleet composition in 2010, Greene created several scenarios of fleet efficiency, with results ranging from 65 smpg (no post-2000 generation models and no retrofit) to 82 smpg (post-2000 generation aircraft are

---

[5] The discussion draws on a study by David Greene of Oak Ridge National Laboratory (1990).

introduced via accelerated retirement and the existing stock improves due to engine retrofit). The latter would probably not occur unless fuel prices rise more than currently envisioned. Even so, an improvement of nearly 25% relative to 1989 (from 49 to 70 smpg) seems quite plausible.

Average load factor is expected to continue to rise. The US Federal Aviation Administration's 1989 forecast projected an increase from 62.3% in 1989 to 67% in 2010. Including this change, passenger-miles per gallon increase from 31 to 43–55 in Greene's scenarios. Improvements in air traffic control and airport operations are likely, but they are expected to be just sufficient at best to offset efficiency losses due to growing air traffic congestion.

While the above figures refer to the United States, roughly similar trends are likely for aircraft fleets in other large OECD countries. The degree of improvement will depend on the types of planes that are purchased by various airlines, which in turn will depend on the mix of flight lengths. The Swedish airline SAS, for example, has eliminated its DC-10s in favor of more efficient 767-ERs, and moved from the older DC-9s towards quieter and more fuel-efficient smaller planes (TPR, 1990). Other European airlines have indicated that their plans call for similar changes.

9.3.2 Former East Bloc

The most popular cars produced by Soviet manufacturers nominally require 7–9 liters/100 km (26–34 mpg), but they are relatively small, low-powered, and lacking in accessories. Over 10–15 years, it seems plausible that the average new car would be similar in technology, size, and performance to the current Italian level. Improving the quality of fuel, roads, parts, and vehicle maintenance would bring substantial reduction in the on-the-road fuel intensity of the fleet. If these two developments occur, the current fleet average of 11–12 liters/100 km (20–21 mpg) could drop to 7–8 (30–34 mpg) by 2010.[6] Similar factors apply in Eastern Europe. Future automobile intensity will also depend on the extent to which used cars from Western Europe are imported. These cars are technologically more advanced than the majority of cars in Eastern Europe, but are often larger and overall more fuel intensive.

Current new Soviet and Eastern European aircraft are roughly comparable to the technology level of 1975 in Western Europe, when fuel use per seat-mile was two-thirds higher than in 1988. If airlines can access hard

---

[6] This also assumes growth in the vehicle stock such that large cars (traditionally used by high officials) play a smaller role than in the recent past.

currency to acquire new aircraft or upgrade the engines in their own air frames, a significant improvement in efficiency is likely. Poland has already acquired Boeing 767s. As in manufacturing, infusion of capital from joint ventures will be an important factor. Operations also show promise for energy savings. In 1990, Soviet aircraft flying to North America made stops in out-of-the-way places such as Shannon, Ireland, or Gander, Newfoundland, where Aeroflot had its own fuel supplies purchased with rubles. With better planes, and capacity to purchase fuel with hard currency, the flights will go non-stop and thereby save fuel.

9.3.3 Developing countries

Automobiles Many of the technologies discussed above for the OECD countries can also be used to improve the fuel economy of new cars in developing countries. Given the growing internationalization of auto-mobile production, some convergence of vehicle characteristics seems likely, although the average new car in the LDCs will be smaller and lower powered than the OECD average. Duleep's analysis cited above suggests that a car fleet composed mainly of subcompact and compact cars, having somewhat less power and accessories than 1988 US cars of similar size, could achieve a test value of 4.3–4.7 l/100 km (50–55 mpg) fairly easily.

Gains from improved technical efficiency will in some cases be offset by increases in size and power. The extent to which new cars are taxed based on engine size, which is now the case in many countries, will affect the mix of vehicles and hence the average fuel economy of the fleet. The price of fuel will also play a role in shaping consumer decisions. Regulations regarding exhaust emissions may be a factor. On the production side, a key factor is the nature of technology transfer from the major world auto manufacturers to LDC subsidiaries or partners. Trade in second-hand cars will also play a role.

Apart from vehicle characteristics, in-use automobile energy intensity will be shaped by real-world operating conditions. The prospects for worsened urban traffic congestion are even greater in the developing countries than in the OECD countries. If actions are not taken to ameliorate the situation, actual automobile (and bus) energy intensity may be far worse than one would expect on the basis of vehicle characteristics.

Air travel The average fuel use per seat-km of jet aircraft in most developing countries is much higher than in the OECD countries, since the aircraft tend to be older. The potential for improvement as new aircraft enter the fleet will depend on whether the airline, which is a state-owned

company in most cases, is able to purchase relatively state-of-the-art
aircraft, or is forced to purchase planes second hand from Western airlines
due to lack of capital. In this context, privatization of national airlines and
encouragement of joint ventures could allow more rapid modernization of
aircraft fleets.

9.4 Freight transport

Energy intensities of freight transport modes will be shaped by changes in
vehicles and in operations. As in manufacturing, transport companies
operating in a competitive environment have an incentive to reduce costs
overall. Energy costs are a consideration, but other factors may be more
important. Minimizing transport time, for example, is a primary objective
that encourages truckers to travel at high speeds that are suboptimal with
respect to fuel economy.

For the main freight transport modes, truck and rail, change in the type
of engine used affects energy intensity. In the OECD countries, most
freight trucks use diesel engines, which are more energy-efficient than
gasoline engines. In the Former East Bloc and many developing countries,
a large share of the truck fleet still uses gasoline engines, so a transition to
diesel engines will reduce fuel intensity. For rail, transition from steam to
diesel or electric locomotion will reduce energy intensity considerably in a
few developing countries, especially China. The transition from diesel to
electric locomotion greatly reduces final energy intensity, but has little
effect on primary energy intensity (assuming fossil-fuel electricity generation).

The technological potential to improve the fuel economy of diesel trucks
is less than that for automobiles. The diesel engine is already a rather fuel-
efficient power plant, and weight reduction is a less promising strategy for
improving fuel economy in heavy trucks, since the load is a large part of the
weight that needs to be moved. Still, there are a number of available or
near-commercial technologies that could provide fuel savings, as we
describe below. There is also potential to reduce truck fuel intensity
through improvements in operations. This includes better matching of
truck specifications to missions, reducing empty or low-load return trips,
improved maintenance, and changes in driver behavior. The operating
environment is also important, especially in the Former East Bloc and
developing countries, where poor road conditions cause inefficient operation.

9.4.1 OECD countries

Heavy trucks account for the bulk of long-distance truck freight transport in the OECD countries. The fuel economy of these trucks can be improved considerably through implementation of commercially available technologies, changes in driver behavior, and introduction and widespread use of near-commercial technologies. A recent US study of tractor–trailer ("combination") trucks evaluated the cost and potential savings of a wide range of possible improvements and estimated a savings potential (reduction in fuel per km relative to the average 1982 truck) of 40% from technical improvements (Sachs et al., 1991). The savings derive from improvements in tractor and trailer aerodynamics, engine efficiency and engine control, drive train, and use of radial tires. The estimated cost of conserved energy for most measures (using a 7% discount rate) is approximately $0.40/gallon, half the current diesel fuel price. Some of the technologies have already penetrated into the truck fleet, so the potential improvement relative to 1990 would be somewhat less than 40%.

An additional 5% reduction in fuel intensity could be obtained from changes in driver behavior. Use of "progressive shifting," in which the driver shifts from lower gears at lower rpm, can be encouraged with electronic engine controls. Reduced road speed can also improve fuel economy considerably, but is not cost-effective given the value of reducing transport time. (The 5% figure assumes that road speed is not reduced.)

Use of engine technologies that are not yet commercial could reduce fuel intensity another 5%. These include adiabatic engines designed to operate at higher temperatures with lower heat rejection, and use of Rankine cycle technology to utilize exhaust gas heat. If some combination of these technologies was implemented, the overall reduction in fuel intensity relative to the 1982 truck would be 50%[7]

9.4.2 Former East Bloc

The transition to a market economy should bring significant improvement in the fuel economy of truck freight in the Former East Bloc. When fleet operators gain an incentive for reducing fuel consumption, they will be more careful in their operations and maintenance, as well as in their selection of vehicles. As fuel prices rise, vehicle producers may attach

[7]   The reduction in fuel intensity is from 45 1/100 km (5.2 mpg) in the base case to 27 1/100 km (8.6 mpg) using available technology, to 25 1/100 km (9.4 mpg) with driver behavior change, to 22 1/100 km (10.4 mpg) using advanced engine technology.

greater importance to fuel economy as an attribute of their vehicles. Diesel trucks accounted for only 50–60% of Soviet truck freight tonne–km in 1990, so average intensity will decline as their share increases. In addition, many of the technologies that have been incorporated in heavy diesel trucks in the OECD countries are not yet in widespread use in the Former East Bloc.

9.4.3 Developing countries

The transition from gasoline to diesel trucks will also be important in many developing countries. This is especially the case in China, where 75–80% of trucks use gasoline. The Chinese government plans to increase production of diesel trucks, which are larger than gasoline-fueled trucks and require only about half as much fuel per tonne-km (Liu et al., 1991). Improving the fuel economy of diesel trucks is hindered by the conditions that prevail in most developing countries. Turbocharging engines and aerodynamic improvements bring relatively less savings at the low speeds and stop-and-go movement commonly found on highways, and improved drivetrain matching is less effective because of overloading (Energy and Environmental Analysis, Inc., 1990). Improved lubricants and switching to radial tyres can yield some savings, and other technologies will have an impact in the long run.

Better matching of vehicles with missions could also reduce energy use. In China, most trucks are medium sized with maximum load of 4–5 tonnes. Use of more heavy trucks would reduce the overall energy intensity of truck freight. Another source of inefficiency in China is the common use of tractors designed for agricultural activities for transporting products. The current policy is to substitute diesel vehicles for these tractors, but efficient light trucks suited to the road conditions of rural areas need to be designed and manufacturing capability developed.

Improving road conditions would allow trucks to operate more efficiently in most countries. Trucks often have to share the road with slower-moving vehicles, including animal-drawn carts. In addition, the poor physical condition of roads causes trucks to be heavier than they would otherwise need to be for purposes of durability (as is the case in India).

Rationalization of operations could reduce energy use per tonne–km in most freight modes. In China, for example, half of truck travel is with empty loads; in rail freight transport, goods are often unloaded and reloaded before reaching their destination. Modernizing management systems with computer-aided dispatch would help the situation in China and elsewhere. Modern control technologies would also enable higher speeds of trains. For trucks, driver training has potential for modest energy savings, especially if it is combined with an incentive system.

9.5 The residential sector

The potential for improvements in energy efficiency in the residential sector is high in all regions and end uses. The increasingly international market for appliances and heating and cooling equipment means that changes in one region could spread quickly. Actions by government or industry to improve the efficiency of equipment could have positive side effects in other countries, and increasing competition for markets could encourage improvements in efficiency. In contrast, reducing heat losses and gains through building shells depends largely on local building traditions and materials, which are much less subject to international or even regional competition. Thus, achieving significant changes in this area may require greater involvement of governments and utilities.

9.5.1 OECD countries

Space heating In the United States and Western Europe, the energy intensity of heating will continue to decline as improvements are made on existing homes, and as new, more energy-efficient homes enter the stock. In Japan, intensity is likely to rise as indoor comfort improves, but space heating will remain a relatively minor end use. Since new housing is being added slowly, especially in Western Europe, the extent of overall decline in heating intensity over the next two decades will depend mainly on retrofit activity. Without major increase in energy prices, we expect relatively little change in indoor temperatures, except in Japan and perhaps the United Kingdom, where they are likely to rise. Growing use of sophisticated control systems may allow more efficient operation of space conditioning equipment, however.

While there has been considerable retrofit activity since 1973, a large fraction of most housing stocks still has poorly insulated walls and leaky windows, and additional ceiling insulation is cost-effective in some cases. A recent West German study estimated costs of conserving energy for buildings of different vintage in five building types (Ebel et al., 1990). As a rough average over the different building types, investments saving 40% of baseline heating energy would be cost effective even in a low future energy price case. Achieving this level of savings would require a considerable investment, however – approximately 55 DEM/m² ($33/m², or around

$2600 for the average West German home). Going to 50% energy savings costs considerably more (around 80 DEM/m²), but is cost-effective in the high energy price case.

In the United States, a recent government study estimated that energy savings of 30–35% could be attained over the next 20 years through technically feasible retrofits in dwellings built before 1975, but only about half of this was estimated to be cost effective (US EIA, 1990). The estimated potential in dwellings built between 1975 and 1987 is somewhat less. The smaller cost-effective potential reduction in the United States relative to West Germany partially reflects lower residential energy prices.

The cost effectiveness of retrofits depends on whether energy-saving improvements are made on their own or as part of general renovation. A recent study by the Danish Ministry of Energy estimated that for homes built before 1979, the average simple payback period (using current energy prices, including taxes) to achieve a 25% reduction in energy use was 17 years for single-family homes and 11 years for apartments (Energistyrelsen, 1990). Assuming the work is carried out when renovation and other improvements are made anyway (i.e., counting only material costs), the payback period decreases to 11 and 7 years, respectively.

The retrofit potential also depends on the current state of the housing stock. In Sweden, for example, building thermal integrity is already quite high, so the cost-effective conservation potential is modest. A recent study of electrically heated single-family houses, which account for about half of the housing stock (and are particularly energy efficient), found that the most attractive measure, additional attic insulation, would save about 10% of heating energy, and cost between 105 and 130 SEK/GJ saved (Vattenfall, 1990). (Swedes paid approximately 140 SEK/GJ for electric heating in 1990.) Once this were done, it is unlikely that extra wall insulation or adding a third window pane would be cost effective (relative to the current price) in most cases, although for homes that already have adequate attic insulation these measures might be economic.

We estimate that space heating intensity in new homes in the United States and Western Europe in 1987 was 70–80% of the intensity of the existing stock. Thus, even without significant further improvement in the thermal integrity of new homes, their penetration will gradually reduce average intensity. Adoption of Scandinavian levels of insulation in the rest of Western Europe and North America would bring a reduction of at least 25% in the space heating requirements of new dwellings relative to those built in the late 1980s. Analysis for Great Britain indicates such levels of insulation (and other measures to reduce heat losses) would be cost effective (Pilkington, 1990). An obstacle to such improvement in Britain and continental Europe as well is the tradition of using masonry, which is more difficult and costly to insulate than wood. Insulating new homes to the levels of Sweden could be difficult without effort by builders and material suppliers to improve insulation techniques and reduce costs.

Water heating and cooking Energy use per capita for water heating may increase somewhat in Western Europe and Japan due to growing use of larger storage tanks, often in combination with central heating. On the other hand, improvement in efficiency of new equipment, and better insulation of existing water heaters and pipes will reduce energy intensity. Increased insulation of electric water heaters, electronic ignition of gas water heaters, higher efficiency gas burners, and separation of water heating from space heating functions in large boilers all promise significant savings over conventional technology. Heat-pump water heaters, which offer great savings in warm climates, and exhaust-air heat pumps connected to ventilation systems may come into more widespread use.

Cooking energy use per capita is likely to decline somewhat due to changes in cooking equipment and in cooking habits, as well as improved energy efficiency of cooking devices. The gradual shift from single oven/ranges to independent cooktops, small ovens, microwaves, and other small devices will tend to reduce energy use for typical meals. Halogen-element cooktops and microwave and convection ovens are penetrating the market for reasons of speed and convenience, and offer energy saving as a side benefit. At the same time, continued increase in participation of women in the work force and greater use of prepared foods is likely to reduce cooking time.

Electric-specific appliances Since most home appliances have short life-times (10–20 years), changes in new appliances shape the intensity of the stock rather quickly. The energy intensity of new appliances is declining in most cases, but for some appliances improvement in technical energy efficiency is being balanced by growth in size and features. This is especially true for refrigeration equipment in Western Europe and Japan. Two-door refrigerator-freezers with full freezing capability are gradually replacing smaller refrigerators or single-door refrigerators with small freezer compartments, and automatic defrost is just being introduced (Schipper & Hawk, 1991).

A series of engineering-economic analyses by researchers at Lawrence Berkeley Laboratory (LBL) – conducted to support setting of US energy efficiency standards – have established that the potential for cost-effective efficiency improvement in major home appliances is considerable. For the most popular type of refrigerator, incorporation of cost-effective design options would reduce electricity use by 28% relative to the average model produced in 1989 (this level was chosen for the 1993 standard) (Turiel et al., 1991). The analysis suggests that use of evacuated-panel insulation for walls, which would result in an additional 12% savings, is also cost-effective, but their long-term reliability is not yet clear. A recent study carried out by the Agence Français pour la Matrise d'Energie (the French Energy Conservation Agency) found similar results for European refrigerators and freezers (Lebot et al., 1991). Application of cost-effective and available technology would result in a reduction in energy use (relative to the average consumption of new models in 1990) of around 33% for refrigerators and upright freezers, and 44% for chest freezers. Both the US and French studies found that use of other technologies (such as a two-compressor system) could reduce energy use further, but these were not cost-effective at expected electricity prices

For clothes washers and dishwashers, the LBL analysis found that design options that reduce energy use (including energy to heat water) by 30% are cost effective (standards at this level will take effect in 1994). For clothes washers, a change from vertical axis to horizontal axis clothes washers (which can be designed to load from the top) would reduce energy use by about two-thirds relative to the baseline (mainly due to much lower use of hot water). Such washers are common in Europe, but not in the United States. For clothes dryers, the cost-effective reduction is only 15%, but much greater savings (about 70% relative to the baseline) are possible through use of a heat pump dryer. The latter has significantly higher cost, but a prototype has been developed and successfully tested. Increasing the spin speed during the spin/dry cycle of a clothes washer can also reduce drying energy use considerably (because much less energy is expended in mechanical water removal than in thermal water removal). Many European washers already utilize higher spin speeds than are standard in the US.

In the United States, appliance efficiency improvement is being strongly pushed by national standards and utility incentives. The standards for refrigerators and freezers due to take effect in 1993 set a minimum efficiency level that almost none of the models produced in 1989 could meet. In Western Europe, momentum is building for standardized energy efficiency labelling, which could help to improve efficiency somewhat. Sweden is considering efficiency standards, and a new Nordic Secretariat has invited other nations to participate in their development (Karbo et al., 1991).

The residential end use with perhaps the largest potential for efficiency improvement is lighting. Expanded use of compact fluorescent lamps (CFLs), which use 20–25% of the electricity of standard incandescent lamps to produce the same light output, could have a major impact. Incentives are generally needed to lower the much higher first cost of CFLs, but utility programs that encourage CFLs have become popular in the United States and Western Europe (Nadel et al., 1991; Mills, 1991). Ultimate market penetration will also depend on design of lighting fixtures that can accommodate CFLs.

9.5.2 Former East Bloc

Both technology and changes in behavior of occupants and institutions can contribute to reduction in heating intensity, particularly if energy prices rise to conform to production costs.[8] Building thermal integrity is low by Western standards, and lack of metering and controls discourages household conservation efforts. Direct metering of heat actually consumed at the building apartment level, combined with provision of thermostats and control valves in each apartment. Comparisons of energy consumption show that unmetered homes use 10–50% more heat than metered ones. Improvements in virtually every component of the building shell, preferably at the time of maintenance or rehabilitation, will also yield a reduction of perhaps 25% in heating intensity. Separating provision of domestic hot water from space heating, so as to reduce the enormous summertime losses in large-scale heat systems providing only domestic hot water, will also contribute energy savings. Improvements in the distribution of heat and hot water from both large boilers (which heat large apartment buildings or entire blocks) and smaller boilers found in homes in rural areas will also reduce energy use. Overall, reducing space and water heating intensity in existing buildings by 25% should be relatively easy to achieve. Over a longer time period, a 50% reduction appears feasible, particularly if energy prices rise to reflect costs.

The large demand for new housing increases the opportunities to save energy in heating. Apartments can be designed with individual heat management controls. There is great room for improving building shell components and heating equipment, but the building design process,

---

[8]   As reform of energy prices proceeds, residential heating prices will probably be the slowest to reach market levels, since many households would be unable to maintain a reasonable level of comfort if prices rose significantly in the near future.

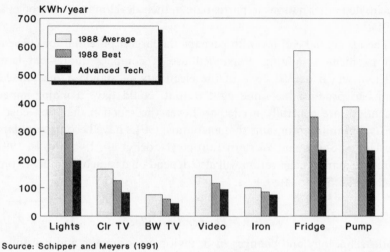

Source: Schipper and Meyers (1991)

Fig. 9.3. Potential for reducing appliance electricity intensity in Indonesia

building materials fabrication and provision, and construction methods will have to undergo change to succeed in this area.

Electric appliances are another area where great improvements in efficiency can occur at low cost. Components currently being used in most domestic appliances are very old and inefficient. The entrance of Western companies and the convertability of currency will make it easier for manufacturers to use better wiring, compressors, motors, and lights. Efficiency gains will be partly counteracted by increase in size and features, however. Washing machines will be increasingly automated, and refrigerators will get bigger and probably have larger freezer compartments. On balance, we estimate that a reduction of at least 25% in the unit consumption of new refrigerators, TV sets, and washing machines is feasible at very little increase in equipment costs. Introduction of more advanced improvements may take longer because more sophisticated manufacturing processes are involved.

9.5.3 Developing countries

The energy efficiency of most existing and new residential equipment in the developing countries is quite low relative to Western levels. Some improvement will occur as households are able to afford better-quality devices. At the same time, households will demand higher levels of service, so the average energy use per device may grow even as the technical energy efficiency improves.

In cooking, currently the most important residential end use in most developing countries, the ongoing transition from biomass to modern fuels will reduce energy intensity in most cases, since kerosene and LPG stoves are much more energy-efficient than biomass stoves. There is also potential to produce kerosene and LPG stoves that are more efficient than the typical devices now used. In rural areas, many households will continue to rely on biomass fuels. The technical potential to improve biomass stove efficiency is great, but transferring designs from the laboratory to the field has proven difficult. There has been considerable effort to improve the efficiency of stoves in rural areas, but in general results have been disappointing (Manibog, 1984). If the lessons of the past are applied to future efforts, however, and more fuel-efficient stoves satisfy other criteria important to users, some improvement seems likely. In urban areas, where commercial markets already exist for biomass fuels and stoves, the prospects for improving efficiency are much greater. While many households will move "up the ladder" to modern fuels, the influx of new urban dwellers and continued presence of poverty means that biomass fuels will continue to be used in many cities.

Lighting is the most common electric end use in developing countries, and could be made much more energy-efficient through use of circular fluorescent lamps and CFLs. Recent studies of the economics and potential for widespread use of CFLs in India and Brazil have shown that such use would be very cost-effective from national and utility perspectives (Gadgil & Jannuzzi, 1991). The cost of avoided peak installed electric capacity from use of CFLs rather than incandescent lamps is only 15% and 10% of the cost of new installed capacity, respectively, but subsidized electricity prices and the high first cost of CFLs limit their attractiveness to households. The extent to which these lamps spread will depend on utility or other programs to overcome the first-cost barrier.

For other electric appliances, many cost-effective options exist to improve their energy efficiency, and changes are in fact occurring. At the same time, size and features are increasing for many appliances, which will push upward on energy intensity (measured as average consumption per device). Lack of test data for new appliances makes quantification of the degree of electricity saving that could be achieved with particular measures difficult, but reasonable estimates have been made in Brazil (Geller, 1991), Egypt (Turiel et al., 1990), and other countries. Figure 9.3 illustrates estimates made in a study of Indonesia (Schipper & Meyers, 1991). It shows the change relative to the 1988 stock average energy intensity that might be achieved if all appliances used best-available cost-effective technology (i.e., the technology used in the most efficient appliances available in the OECD countries), and if all appliances incorporate advanced technology that is beginning to be adopted, or could soon be adopted, by OECD manufacturers. The percentage reduction relative to 1988 is 20–30% in the "Best" case, and 30–60% in the "Advanced Technology" scenario.[9]

Air-conditioning is currently a minor end use in the developing countries, but it could grow significantly. Considerable improvement in equipment efficiency is possible. In Thailand, for example, the average home air conditioner draws 1.6 kW, while the best units require less than 1 kW to provide the same cooling capacity. Designing new residential buildings with higher thermal integrity and less solar gain could also reduce intensity in those buildings where air-conditioning is likely to be used. A recent study for Thailand found that installing 7.5 cm of insulation in the attic of a typical single-family house would permit the use of an air conditioner 30% smaller to meet cooling needs (Parker, 1991).

Water-heating equipment is becoming more common in many developing countries. Storage water heaters can be made less energy intensive by improving insulation and reducing thermostat setpoints. Hot water heat pumps are very energy efficient and may be cost effective in many countries, but their high first cost may limit their market penetration.

Space heating is not a major end use in most developing countries. China is an important exception, however. Space heating energy use is constrained by regulations and mandated coal allocations, so indoor temperatures are typically very low. From a technical standpoint, there is considerable potential to improve building thermal integrity (especially in new buildings) and heating equipment efficiency. Since indoor temperatures in most homes are lower than desired, however, some of the savings from efficiency will go toward greater indoor comfort rather than reduced energy use, at least until comfort reaches a reasonably satisfactory level. Even so, computer simulations show that efficiency improvement can reduce energy use by 40% relative to mid-1980s practice, while allowing considerable increase in indoor temperatures (Huang, 1989). The government has introduced a standard that calls for new buildings in cities to be

[9] These estimates were made by comparing the average energy use of appliances in Indonesia with those in other countries, notably Japan, West Germany, and the United States, and by consideration of technical and economic studies of conservation potentials for home appliances. We did not account for the likely increases in the level of service demanded by future consumers (larger refrigerators and water heaters, hot water in washing machines, larger TV sets). Taking these changes into account would raise the base-line of electricity use, and raise the benefits of greater efficiency as well.

designed to use 30% less heating energy relative to 1980 practice, but implementation is proceeding slowly.

9.6 The service sector

The energy intensities of different service subsectors are shaped by the levels of energy services present in buildings (heating, cooling, ventilation, lighting, and others), building operation, and the technical energy efficiency of buildings and their equipment. As in the residential sector, there are signs that some saturation of service levels has occurred in upper-income countries, and the marginal equipment (such as PCs and fax machines) are relatively non-energy-intensive for the value they provide. An important difference from the residential sector is that commercial buildings are more likely to require space cooling (due to the greater density of people and equipment); growth in this end use is likely to increase electricity intensity in much of the developing world.

In the United States and Japan, electricity use in this sector is a major source of the peak demand that utilities must meet, and a similar situation is evolving in many developing countries. Since peak electricity is most expensive for utilities to deliver, promoting electric end-use efficiency in this sector is particularly attractive. There are many cost-effective opportunities for saving electricity through retrofit of existing buildings and especially in design of new buildings. There are also synergies between measures. For example, more energy-efficient lighting systems reduce the amount of heat that needs to be removed by the air-conditioning system, which allows use of a smaller and less expensive system.

9.6.1 OECD countries

While electric end uses are becoming increasingly important, there are still opportunities to save fuel used in space and water heating. The energy intensity of these end uses is likely to decline at a slower rate than between 1973 and 1988, however. Improvements in lighting energy efficiency will reduce the free heat from lights, though this phenomenon will be partially balanced by growth in use of office equipment. New ventilation systems, often designed to improve indoor air quality, make it easier to recuperate heat from exhaust air, or provide heat from warmer to cooler parts of a building. Improvements in windows will reduce heat losses in winter and heat gains in cooling months.

The situation for electricity intensity is more complicated. On the one hand, the penetration of computing and other office equipment is increasing. One study estimated that computing electricity use in the US service sector would at least double between 1990 and 2000 without innovations to save electricity (Norford et al., 1990). These devices not only consume electricity directly; they also contribute to internal heat that must be removed by air-conditioning systems. On the other hand, there are many opportunities to improve electricity efficiency in the service sector with new lighting technologies, optical coatings on windows to reduce undesirable heating gains, more efficient motors and compressors, and computer-controlled energy management systems (Geller, 1988). US utilities are increasingly working with building owners or operators to reduce electricity use, and utilities in Japan are beginning to show interest because of summer cooling problems.

In Western Europe, lighting and cooling in commercial buildings has begun to show up in summer loads, raising interest for more careful management of electricity use. A number of studies show a considerable economic potential for electricity saving in existing buildings. A Swedish study of 14 commercial buildings that examined electric heating, ventilation, lighting, and other measures aimed at electricity found that a savings of 20% would be cost-effective at the current electricity price (Bodlund et al., 1989). A major Danish project found a large number of electricity saving measures in commercial buildings (Gjelstrup et al., 1989). One study, which considered lighting, ventilation, pumping, cooling, refrigeration, and other technologies, found that 24% of public sector (schools, hospitals, etc.) electricity could be saved by behavior and management changes, and an additional 18% reduction was possible with available technologies (Johansson & Pedersen, 1988). A study of the private service sector found a somewhat lower conservation potential: 10–18% through behavior and management changes, and an additional 15% reduction through retrofit (Nielsen, 1987). The difference between these findings indicates that the private sector buildings are more carefully managed.

A factor that is likely to raise energy use in retail buildings is increase in opening hours. This trend will have a larger impact in Western Europe than in the United States, where the phenomenon is already rather advanced. Relative to building new space, however, increased utilization of existing space should reduce energy use per unit area, since there is still heat (or cool) in a building at closing hours.

9.6.2 Former East Bloc

The level of services in buildings is likely to increase considerably due to changes in existing buildings and addition of new, more modern buildings. Indoor comfort levels for lighting, ventilation, and heating should increase, particularly as stores and other buildings for personal services compete with each other for business. Proliferation of office communication and computing equipment will be rapid. With time, comfort levels in schools and hospitals may increase. Growth in refrigeration is likely as the retail food and restaurant sectors expand.

While rising levels of service will increase energy intensities, there is much room to improve efficiency. The intensity of space heating in the Soviet Union, which is very high compared with Northern Europe, could be reduced by at least 30–35% through retrofit. Studies of Poland and Czechoslovakia show similar potential savings (Chandler, 1990). Efficiency can also be improved considerably for ventilation, lighting, and refrigeration.

In contrast to the situation in the OECD countries, the addition of new buildings will likely raise average energy intensity, since many such buildings – especially those built with Western investment – will have higher amenity levels than existing ones. Unless energy-efficient building practices are encouraged through codes and other measures, the significant growth in area for private offices and hotels that is likely to occur could increase average intensity substantially.

9.6.3 Developing countries

As in the former East Bloc, rise in the level of building services will push upward on energy intensity. The use of air-conditioning and office equipment will increase. Comfort levels in schools and hospitals should increase, as will refrigeration in stores and restaurants. In many countries, growth in construction of modern office, retail, and tourism-related buildings will be considerable, and these will have higher amenity levels than most existing buildings.

The areas with the greatest potential for efficiency improvement are lighting and air-conditioning. Many developing countries have tropical climates, and the cooling load in modern-style buildings is often very high. Improving the energy efficiency of lighting (and other equipment) reduces heat gains inside the building, thereby reducing the cooling load, which may allow downsizing of HVAC equipment. A study conducted in

Thailand found that a combination of commercially available high-efficiency fixtures, lamps, and ballasts can save 70% of typical lighting electricity consumption in fluorescent systems at a cost of conserved energy less than half the retail rate for electricity (Busch, du Pont & Chirarat-tananon, 1991). With the drop in cooling demand that comes from reducing the amount of heat given off from the lights, the savings for the building as a whole from the lighting measures alone can reach 30%. There is also a high savings potential for air-conditioning through application of commercially available technologies. These include variable air volume systems, high-efficiency centrifugal chillers, and high-efficiency fans equipped with efficient motors and variable-speed drives. A study of large commercial buildings in Thailand found that each of these bring electricity savings in the range of 10% (Busch, 1990). Operational choices are also important: more careful management of thermostat settings and ventilation can significantly reduce cooling energy use. Busch estimated that use of energy-efficient lighting could reduce energy use by 20% to 35%, and use of daylighting could provide an additional 6% to 15% savings. Roughly comparable energy savings have been estimated in Pakistan (Almeida & Geller, 1986) and Jamaica (Resource Engineering, 1991).

In addition to the work in Thailand, studies of the efficiency improvement potential in large commercial buildings have been conducted in other Southeast Asian countries (Levine et al., 1989). Key findings are: (1) energy savings in new commercial buildings of 30–40% can be achieved with present technology; (2) savings of 10–20% are possible in most existing buildings; (3) these reductions in energy use are economically beneficial, with payback periods of one to three years; and (4) an overall 20% savings for commercial buildings would yield an annual benefit on the order of $400 to $500 million in the ASEAN region. As a result of these studies, energy-efficiency standards for new buildings are being implemented or formulated in several countries.

In contrast to the OECD and former East Bloc countries, in most developing countries there is little energy use for space heating. China is an important exception, and Korea also has significant space heating demand. The situation in these two countries is very different. Space conditioning standards in South Korea are approaching those in Japan, especially in the cities. In China, on the other hand, the thermal integrity of most buildings is low, but improving efficiency is constrained by several barriers. Prevailing construction techniques (brick) make addition of wall insulation difficult, as does shortage of insulation itself. Fuel is subsidized and allocated by square meter of heated area, but supplies are uneven.

Buildings with ample supply can be overheated, requiring opening windows to cool off, while those with not enough make do with cold rooms. For many buildings, then, retrofit improvements may be used to improve indoor comfort rather than reduce energy use. New buildings could certainly be constructed with higher levels of thermal integrity. A design standard for tourist hotels is expected to be approved by the end of 1991.

9.7 Adding up the energy efficiency potential: an example

The overall energy-efficiency potential within a given sector or in an entire country depends on the potential in many end uses, and the relative importance of them. The energy-efficiency "resource" is like fossil fuel resources in that the total amount deemed "available" depends on the price of energy and the cost of saving a unit of energy (which may change over time due to technological innovation). A way of expressing this relationship that has been used in many analyses of energy-efficiency potential is with a supply curve of conserved energy.

Figure 9.4 shows a supply curve of conserved electricity for the United States residential sector estimated by a recent comprehensive study performed at Lawrence Berkeley Laboratory (Koomey et al., 1991). Each step of the curve represents the total electricity savings that would be gained if the particular conservation measure (or package of measures) was implemented in all equipment and homes in 2010. The savings are relative to a "baseline" that assumes energy intensities "frozen" at 1990 levels.[10] The magnitude of savings depends on both the future number of households and the market penetration of each type of equipment (both of which were forecast by the model used in the analysis).

Each step has an estimated cost of conserved energy (CCE), which is calculated independently of energy price but can be compared to the price of electricity (or an estimate of the societal electricity cost). Using the 1989 US average residential price of electricity, measures that would save 40% of "baseline" electricity consumption would be cost-effective. This assumes a discount rate of 7%. Reducing the discount rate to 3% would decrease the CCEs by about 30%, but since the CCE of the more expensive measures rises steeply past the 40% level of reduction, using a lower

[10] The baseline case assumes that buildings and appliances existing in 1990 remain at 1990 efficiency levels (no retrofits), and all new homes and appliances that enter the stock remain at the efficiency of new systems in 1990. Thus, stock turnover reduces average energy intensities in the baseline case even though new and existing devices are frozen at their 1990 efficiencies.

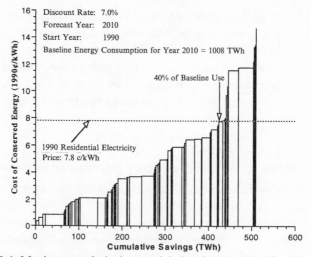

Fig. 9.4. Maximum technical potential electricity savings for US residences

discount rate would not substantially change the size of the cost-effective potential. If electricity prices rise and/or the costs of the more expensive efficiency measures decline, however, the cost-effective savings potential would increase.

9.8 Conclusion

The prospects for energy intensities, and the potential for reduction, vary among sectors and parts of the world. In the majority of cases, intensities are tending to decline as new equipment and facilities come into use and improvements are made on existing stocks. The effect of stock turnover will be especially strong in the developing countries, where stocks are growing at a rapid pace, and the Former East Bloc, where much of the existing industrial plant will eventually be retired and replaced with more modern facilities. While reductions in energy intensity are likely in most areas, there is a large divergence between the technical and economic potential for reducing energy intensities and the direction in which present trends are moving. In the next chapter, we present scenarios that illustrate where trends are pointing, and what could be achieved if improving energy efficiency were a focus of public policies.

References

AIP (American Institute of Physics). 1975. Chapter 2. Second-Law Efficiency, *AIP Conference Proceedings: Efficient Use of Energy*, New York, NY.
Akbari, H., Rosenfeld, A. H. & Taha, H. 1990. Summer heat islands, urban trees, and white surfaces, *ASHRAE Transactions*, **96** (1), 1381–8.

Almeida, A. & Geller, H. 1986. The potential for electricity conservation in the industrial and commercial sector in Pakistan. Prepared for ENERCON: The National Energy Conservation Center, Islamabad, Pakistan: ENERCON Task Order 86–051-DEL 2.

An, F. & Ross, M. 1991. *Automobile Energy Efficiency at Part-Load: The Opportunity for Improvement.* Washington, DC: American Council for an Energy Efficient Economy.

Bleviss, D. 1988. *The New Oil Crisis and Fuel Economy Technologies: Preparing the Light Transportation Industry for the 1990s.* New York, NY: Quorum Press.

Bodlund, B., Mills, E., Karlsson, T. & Johansson, T. 1989. The challenges of choices: Technology options for the Swedish electricity sector. In T. B. Johansson et al., eds. *Electricity. Efficient End Use and New Generation Technologies, and Their Planning Implications.* Lund, Sweden: Lund University Press.

Busch, J. F. 1990. *From Comfort to Kilowatts: An Integrated Assessment of Electricity Conservation in Thailand's Commercial Sector,* Ph.D Thesis. Berkeley, CA: Lawrence Berkeley Laboratory Report LBL-29478.

Busch, J. F., du Pont, P. & Chirarattananon, S. 1991. Conserving electricity for lighting in Thai commercial buildings: A review of current status, potential, and policies. To be published in *Proceedings of Right Lite Bright Lite,* 1st *European Conference on Energy-Efficient Lighting.* Stockholm, Sweden, May 28–30, 1991.

Chandler, W. U. (ed.). 1990. *Carbon Emissions Control Strategies: Case Studies in International Cooperation.* Washington, DC: World Wildlife Fund and the Conservation Foundation.

Cumper, J. & Marston, S. 1991. Energy and Economic Analyses for the Jamaica Energy Efficiency Building Code: Draft Report. Kingston, Jamaica: Resource Engineering, Enertech Ltd.

Dobozi, I. 1991. Impact of market reforms on USSR energy consumption. *Energy Policy,* **19** (4).

Duleep, K. 1991. *Potential Fuel Economy in 2010.* Arlington,VA: Energy and Environmental Analysis, Inc.

Ebel, W. et al. 1990. *Energiesparpotentiale im Gebaüedebestand.* Darmstadt, Germany: Institut für Wohnen und Umwelt.

Energetics Incorporated. 1988. *The U.S. Cement Industry: An Energy Perspective.* Report to the US Department of Energy. Columbia, MD.
   1988. *The U.S. Glass Industry: An Energy Perspective.* Report to the US Department of Energy. Columbia, MD.
   1988. *The U.S. Pulp and Paper Industry: An Energy Perspective.* Report to the US Department of Energy. Columbia, MD.
   1988. *The U.S. Steel Industry: An Energy Perspective.* Report to the US Department of Energy. Columbia, MD.
   1988. *The U.S. Textile Industry: An Energy Perspective.* Report to the US Department of Energy. Columbia, MD.

Energistyrelsen. 1990. Miljoe 2000. *Arbejdsgruppen om energiforbrug i bygninger.* Energyforbrug i bygninger. Copenhagen, Denmark.

Energy and Environmental Analysis, Inc. 1990. *Policy Options for Improving Transportation Energy Efficiency in Developing Countries.* Prepared for the US Congress, Office of Technology Assessment. Arlington,VA.

Gadgil, A. & Jannuzzi, G. 1991. Conservation potential of compact fluorescent lamps in India and Brazil. *Energy Policy,* **19** (5), 449–63.

Geller, H. S. 1991. *Efficient Energy Use: A Development Strategy for Brazil.* Washington, DC and Berkeley, CA: American Council for an Energy-Efficient Economy.

1988. *Update on Electricity Use in the Service Sector.* Report to the US Office of Technology Assessment. Washington, DC: American Council for an Energy-Efficient Economy.

Gjelstrup, G., Larsen, A., Nielsen, L., Oksbjerg, K. & Togeby, M. 1989. *Elbesparelser i Danmark.* AKF Forlaget, Copenhagen, Denmark.

Goldemberg, J., Johansson, T. B., Reddy, A. K. N. & Williams, R. H. 1987. *Energy for Development,* Washington, DC: World Resources Institute.

Greene, D. L. 1991. Vehicle use and fuel economy: How big is the rebound effect? *Energy Journal,* **13** (1), 117–143.

1990. Commercial aircraft fuel efficiency potential through 2010. *Proceedings of the Intersociety Energy Conversion Engineering Conference,* **4**, 106–11. Piscataway, NJ: IEEE.

Henly, J., Ruderman, H. & Levine, M. D. 1988. Energy saving resulting from the adoption of more efficient appliances: A follow-up. *Energy Journal,* **9** (2), 163–70.

Huang, Y. 1989. *Potentials for and Barriers to Building Conservation in China.* Berkeley, CA: Lawrence Berkeley Laboratory Report LBL-27644.

Johansson, M. & Pedersen, T. 1988. *Tekniske elbesparelser.* AKF Forlag, Copenhagen, Denmark.

Karbo, P., Østergaard, V., Lorentzen, J. & Hammar, T. (eds.). 1991. *NORDNORM Workshop on Energy Labelling and Efficiency of Household Appliances,* Stockholm, Sweden, May 6, 1991. Copenhagen, Denmark: Energistyrelsen.

Khazzoom, J. D. 1987. Energy saving resulting from the adoption of more efficient appliances. *Energy Journal,* **8** (4), 85–9.

Koomey, J., Atkinson, C., Meier, A., McMahon, J. E., Boghosian, S., Atkinson, B., Turiel, I., Levine, M. D., Nordman, B. & Chan, P. 1991. *The Potential for Electricity Efficiency Improvements in the U.S. Residential Sector,* Berkeley, CA: Lawrence Berkeley Laboratory Report LBL-30477.

Lebot, B., Szabo, A. & Despretz, H. 1991. *Gisement des Economies d'Energie du Parc Européen des Appareils Electroménagers obtenues par une Reglementation des Performances Energétiques.* Report to the European Economic Community, Director General for Energy. Sofi Antopolis: Agence Français pour la Matrîse de L'Energie.

Ledbetter, M. & Ross, M. 1990. A supply curve of conserved energy for automobiles. *Proceedings of the 25th Intersociety Energy Conversion Engineering Conference.* New York: American Institute of Chemical Engineers.

Levine, M. D., Busch J. F. & Deringer, J. J. 1989. Overview of building energy conservation activities in ASEAN, *Proceedings of the ASEAN Special Sessions of the ASHRAE Far East Conference on Air Conditioning in Hot Climates.* Kuala Lumpur, Malaysia, October 26–28. Berkeley, CA: Lawrence Berkeley Laboratory Report LBL-28639.

Liu, F., Davis, B. & Levine, M. 1991. *Energy in China.* Berkeley, CA: Lawrence Berkeley Laboratory draft report.

Lovins, A. 1991. *Advanced Light Vehicle Concepts.* Notes for the Workshop on An Evaluation of the Potential for Improving the Fuel Economy of New Automobiles and Light Trucks in the United States. Snowmass, CO: Rocky Mountain Institute.

Lovins, A. & Lovins, H. 1991. Least cost climate stabilization. *Annual Review of Energy*, **16**.

Manibog, F. 1984. Improved cooking stoves in developing countries: problems and opportunities. *Annual Review of Energy*, **9**, 199–277.

Mills, E. 1991. Evaluation of European lighting programmes: Utilities finance energy efficiency. *Energy Policy*, **19** (3), 266–79.

Nadel, S., Atkinson, B. A. & McMahon, J. E. 1991. A review of U.S. and Canadian lighting programs for the residential, commercial and industrial sectors. In *Proceedings of the 1st European Conference on Energy-Efficient Lighting*, Stockholm, Sweden, May 28–30, 1991.

Nielsen, B. 1987. *Vurdering af elbesparlsespotentiale. Handel og service.* Lynbgy, Denmark: Dansk Elvaerkers Forenings Udredningsinstitute (DEFU), Teknisk rapport 257.

Norford, L., Hatcher, A., Harris, J., Roturier, J. & Yu, O. 1990. Electricity use in information technologies. *Annual Review of Energy*, **15**, 423–53.

Parker, D. 1991. *Residential Demand Side Management for Thailand.* Bangkok, Thailand: International Institute for Energy Conservation.

Pilkington. 1990. *Environmental Policy and Energy Demand.* Prepared in association with Oxford Economic Research Associates. St Helens and Oxford, UK.

Sachs, H. M., DeCicco, J. M., Ledbetter, M. & Mengelberg, U. 1991. *Heavy Truck Fuel Economy: A Review of Technologies and the Potential for Improvement.* Washington, DC: American Council for an Energy-Efficient Economy.

Schipper, L. J. & Hawk, D. 1991. More efficient household electricity use: An international perspective. *Energy Policy*, **19**, 244–65.

Schipper, L. J. & Meyers, S. 1991. Improving appliance efficiency in Indonesia. *Energy Policy*, **19** (6), 578–88.

Scott, A. 1980. Economics of house heating. *Energy Economics*, **2** (3), 130–41.

TPR (Transportrådet). 1990. *Trafik, Energi och Koldioxid: Strategier för att reducera bränsle-förbrukning och koldioxidutsläpp*, [In Swedish, Traffic, Energy, and Carbon Dioxide: Strategies for Reducing Fuel Use and Carbon Dioxide Emissions.] Rapport 1990:11, Stockholm, Sweden.

Turiel, I., Berman, D., Chan, P., Chan, T., Koomey, J., Lebot, B., Levine, M. D., McMahon, J. E., Rosenquist, G. & Stoft, S. 1991. U.S. residential appliance energy efficiency: Present status and future policy directions. In Vine, E. and D. Crawley (eds.) *State of the Art of Energy Efficiency: Future Directions*, Washington, DC: American Council for an Energy-Efficient Economy.

Turiel, I., Lebot, B., Nadel, S., Pietsch, J. & Wethje, L. 1990. *Electricity End Use Demand Study for Egypt*, Berkeley, CA: Lawrence Berkeley Laboratory Report LBL-29595.

US EIA (Energy Information Administration). 1990. *Energy Consumption and Conservation Potential: Supporting Analysis for the National Energy Strategy*, SR/NES/90–02. Washington, DC: US Department of Energy.

Vattenfall, 1990. *Electricity Conservation: Problems and Potentials.* Vällingby, Sweden.

Westbrook, F. & Patterson, P. 1989. Changing driving patterns and their effect on fuel economy. Paper presented at the 1989 SAE Government/Industry Meeting, May 2, Washington, DC.

World Bank. 1991. *China: Efficiency and Environmental Impact of Coal. Volume 1: Main Report* Washington, DC.

10

Scenarios of future energy intensities

In this chapter, we present scenarios of potential change in energy intensities in the OECD countries and in the Soviet Union. These scenarios are meant to illustrate how intensities might evolve over the next 20 years given different conditions with respect to energy prices, energy-efficiency policies, and other key factors. Changes in intensity will also be affected by the rates of growth and stock turnover in each sector. We have not tried to forecast how activity levels and structure will evolve. However, the OECD scenarios assume a world in which GDP averages growth in the 2–3%/year range, with some differences among countries. For the Soviet Union, the degree and pace of intensity decline will be highly dependent on the success of the transition to a market economy; each scenario explicitly envisions a different degree of success.

We have not constructed comparable scenarios for the developing countries. The stock of equipment and facilities in most developing countries in 2010 will consist heavily of equipment and facilities that will be added between the present and then. This factor, along with uncertainty about current stock-average intensities, makes it difficult to predict future intensity levels. However, we present scenarios of electricity use in the residential sector in Indonesia as an illustration of how intensities could decline in the future.

10.1 Scenarios for the OECD countries

Over the next 20 years, most energy intensities in the OECD countries will decline as new capital stock replaces the old and as improvements are made to buildings and industrial facilities. The extent of decline will depend on many factors, including economic growth, energy prices, technological developments, and government policies. The scenarios presented here reflect our judgement of how producers and users of end-use systems might respond to different "boundary conditions." We use historical rates of decline in different periods as a guide as to what might occur in the future, along with consideration of the intensities of current new systems and techniques that are near-commercial.

The intensities are built from patterns of energy use in 1985 for approximately 25 different end uses. We estimated intensities separately for the United States, Japan, and a European aggregate of seven countries. The 1985 energy intensities and end-use patterns are based on our analysis of nine OECD countries (presented in the sectoral chapters), extrapolated to the entire OECD using national energy balances and data provided by Shell International Petroleum Company. We use 1985 as the base year in order to utilize past estimates of final energy use by sector for the entire OECD (Schipper, 1987). The results in the text are presented in terms of all-OECD average intensities. Appendix B contains a more detailed discussion of the assumptions used in constructing the scenarios. Below we describe each scenario briefly, and then discuss the differences among the scenarios for each sector.

Trends This scenario reflects a world in which only modest attention is given to energy efficiency. The world oil price increases by around 50% between 1990 and 2010; natural gas prices increase somewhat more and coal prices somewhat less; electricity prices rise only slightly, since the fuel component is only part of the total price.[1] The role of policies is limited to present automobile and appliance standards in the United States, existing building codes, and a slow increase in the participation of US and European utilities in demand-side management (DSM) programs.

The declines in energy intensities reflect our judgment about energy-efficiency improvements that are likely to occur, given the above conditions and trends. In most cases, the rates of change are similar to those experienced between 1985 and 1988 or 1989. Exceptions include electricity intensity in the service sector, which is expected to fall slowly after rising in recent years, and residential electricity intensities, which decline after being more or less flat in recent years. The main reason for these changes is that DSM programs will affect trends even without major changes in energy pricing. Many stock-average intensities in 2010 are near the average values for *new* systems in 1990.

---

[1] These assumptions roughly correspond to our interpretation of the current consensus. See, for example, the US Department of Energy's *Annual Energy Outlook* 1991 (US EIA 1991).

Moderate effort In this scenario, adoption of marginal-cost pricing, removal of subsidies, and internalization of some environmental externalities boosts real energy prices to users by 25–35% relative to "Trends". (These increases should occur by 1995 if their full effect is to be felt by 2010.) Governments also step up the role of energy-efficiency policies, with particular emphasis on automobiles and retrofit of existing buildings. No major technological'breakthroughs are assumed, however.

Most intensities in "Moderate effort" decrease between 1985 and 2010 at approximately the same rate as they did between 1972 and 1985. Exceptions are air travel, which decreases more slowly, and truck freight, which decreases faster than in the past. For many end-uses, the average intensity levels in 2010 are near the lowest of new systems in 1989.

Vigorous effort This scenario embodies what might be attained if energy prices rose significantly, if a very strong effort was made to improve new energy-using technologies, particularly for motor vehicles, and if a comprehensive program of retrofitting buildings and factories were undertaken as well. Energy prices are 50–100% higher than in "Trends", as there is more aggressive internalization of externalities associated with local environmental problems related to energy production and use, as well as those associated with greenhouse-gas emissions. National and international consensus to greatly improve the efficiency of new and existing systems is reached in the mid-1990s, leading to rapid response from transnational manufacturers, national and local energy supply authorities, and other actors who will market energy efficiency. Energy-efficiency R&D is pursued in earnest, leading to cost reductions for efficiency options.

The rate of decline in intensities resembles that which occurred in the 1979–1983 period, when energy prices and programs stimulated much action to save energy. Average intensities in 2010 lie below the lowest of new systems on the market in 1989, but are close to the levels achieved by the best products expected to be available by the early 1990s, or at levels represented by prototypes. Some technological breakthroughs are important, especially those leading to lower cost for materials that permit weight reductions in motor vehicles, higher temperature combustion in boilers and engines, and highly efficient insulating materials and windows.

10.1.1 Sectoral trends

The average annual rates of change in energy intensities in each scenario are shown in Table 10.1. For comparison, historical rates achieved between

1972 and 1985 are also shown. For most sectors, the rates shown were calculated based on the intensities derived for 2010. For manufacturing, the estimated rates of change were the basis of intensity decline, and the intensity values for 2010 were calculated based on those rates.

Manufacturing Energy intensities in manufacturing, measured as energy per unit of value-added, have declined for many decades due to technological change. *Fuel* intensity has declined much faster than *electricity* intensity, and this difference is very likely to continue (and perhaps increase). The rate of decline in fuel intensity in "Trends" (3.0%/year) is only somewhat slower than what occurred in the 1972–85 period (3.7%/year). This situation reflects continuation of the trend of technological innovation. Fuel intensity falls by 4.0%/year in "Moderate effort" as manufacturers respond to higher energy prices; the pace is more rapid (4.5%/year) in "Vigorous effort", since the price rise is much greater.[2]

Electricity intensity declines by 0.3%/year in "Trends." There is increasing use of electricity in many applications, but higher efficiency still results in a net decline in intensity. The drop of 0.5%/year in "Moderate effort" is comparable to the historic trend from 1972–1985. In "Vigorous effort", the rate of decline is twice that of 1972–85, reflecting the impact of higher prices and DSM programs.

There is relatively modest decline in intensity relative to "Trends" in the "Moderate effort" and "Vigorous effort" scenarios (Fig. 10.1). The reason is that the primary force of intensity reduction is ongoing technological change that is relatively independent of energy prices. There is also more limited scope for efficiency policy and programs in manufacturing, relative to other sectors. Stimulating R&D may be the most important ingredient in accelerating efficiency improvements.

Transportation We focus on three subsectors – automobiles, air travel and truck freight – which together account for over 75% of OECD transportation energy use. While reductions in intensities of other travel and freight modes are likely to occur, they are not very significant in the overall picture. Modal shifts which could save energy are not considered here.

In "Trends," OECD average automobile fuel intensity declines from

---

[2] The intensity changes do not include the effects of structural change among basic manufacturing subsectors, which would reduce *aggregate* manufacturing energy intensity by an additional 0.8%/year were historic trends to persist.

Table 10.1. *Average annual rate of decline in OECD energy intensities*
(percent)

Sector	1972–1985 Actual	1985–2010 Scenarios		
		Trends	Moderate effort	Vigorous effort
Industry (energy/VA)				
Fuels	3.7	3.0	4.0	4.5
Electricity	0.5	0.3	0.5	1.0
Travel (energy/p–km)				
Cars[a]	2.2	1.0	2.0	4.1
Air	3.9	1.2	2.0	2.8
Freight transport (energy/t-km)				
Trucks	(−)0.4	0.4	1.1	2.7
Residential (energy/capita)	2.2	0.7	2.1	4.8
Fuels: heat, HW, cooking	2.0	0.6	2.0	5.4
Electricity: appliances	1.3	0.9	1.7	2.8
heat, HW, cooking	1.7	1.2	2.8	4.2
lights	0.8	1.2	3.7	5.5
Services (energy/VA)	1.8	0.6	1.9	3.0
Fuels	3.6	0.6	2.0	3.1
Electricity	(−)1.3	0.4	1.7	2.7
Weighted total[b]				
Energy	2.3	1.2	2.2	3.6
Fuels	3.1	1.3	2.3	3.9
Electricity	(−)0.9	0.6	1.5	2.4

[a] Energy/km.
[b] Intensities are weighted by end-use shares of 1985 final energy use.

10 1/100 km (24 mpg) in 1985 to about 8.5 1/100 km (28 mpg) in 2010. Most of the change occurs in the United States (Fig. 10.2).[3] The slow improvement reflects trends in recent years in new car fuel economy, as well as the effect of increased traffic congestion. In "Moderate effort", intensity declines to 6.9 1/100 km (35 mpg). Higher fuel prices play a role, as do regulations or agreements that cause automobile producers to focus more effort on fuel efficiency. This level is somewhat better than the on-road figure for Italy in 1988, which implies that cars will be somewhat smaller and less powerful than those being sold today, or that the share of diesels will increase. In "Vigorous effort", automobiles use only 4–5 1/100 km (48–60 mpg), roughly the level of the most efficient small cars sold in

[3] Changes in the US figure significantly in the OECD average, since the US accounts for about 57% of OECD automobile fuel use. In considering the scenario intensity levels, one must bear in mind that actual on-the-road intensity is some 20% higher than test intensity.

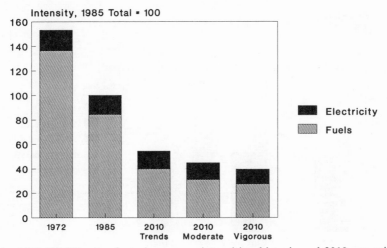

Fig. 10.1. OECD manufacturing energy intensities, historic and 2010 scenarios

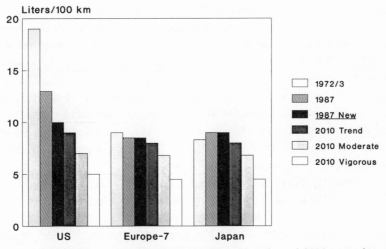

Fig. 10.2. OECD automobile fuel intensities, historic and 2010 scenarios

1990. In this scenario we assume a shift toward smaller cars as a result of higher gasoline prices and policies, as well as technological improvements that help commercialize the fuel-saving features now found only on lightweight prototypes (see Chapter 9, Section 9.3.1). Reaching this level of fuel economy would require clear agreement among political leaders, citizen groups, and automobile manufacturers so that efficiency goals have wide support.

The intensity of air travel declines at 1.1%/year in "Trends," much less

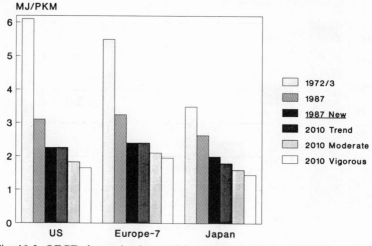

Fig. 10.3. OECD domestic air travel intensities, historic and 2010 scenarios

than the 3.9% rate of decline during the 1972–1985 period. This reflects the slowing of intensity reduction in recent years, a slower lower rate of improvement in load factors, and somewhat less turnover and retrofit of aircraft. In "Moderate effort", intensity declines more rapidly as airlines respond to higher prices; the rate is still only half of its historical rate, but intensity in 2010 is nonetheless 40% lower than in 1985 (Fig. 10.3). This scenario reflects replacement of almost all existing aircraft by models with the lowest intensities available in 1990. In "Vigorous effort," the intensity declines at close to 3%/year as a result of significant penetration of propfan aircraft.

The intensity of truck freight declines at 0.4%/year in "Trends". The historic upward trend in intensity in the United States reverses as various technologies penetrate the fleet and operations improve. In "Moderate effort", intensity declines at 1.1%/year as higher fuel prices provoke careful effort to improve both trucks and freight operations. In "Vigorous effort", there is much faster decline (2.7%/year), reflecting widespread use of advanced technologies made available through accelerated R&D.

Residential sector The "Trends" scenario envisions a continued but slow pace of retrofitting of existing homes. Entrance of new homes into the stock also contributes to decline in average heating intensity in 2010, although it remains above the 1987 average for new homes (Fig. 10.4).[4]

---

[4]   In Japan, indoor comfort levels are increasing; consequently, space heating intensity is higher in 2010 than in 1987 in "Trends".

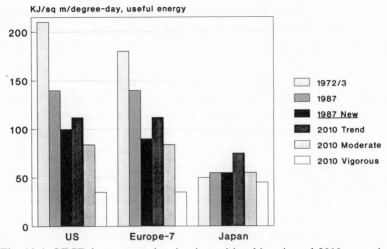

Fig. 10.4. OECD home space heating intensities, historic and 2010 scenarios

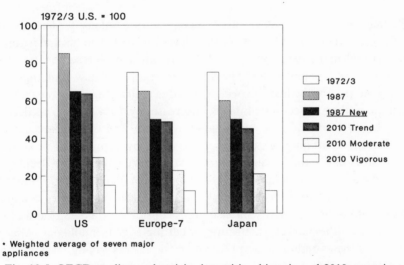

• Weighted average of seven major
appliances

Fig. 10.5. OECD appliance electricity intensities, historic and 2010 scenarios

Overall, intensity of fuel use for space heating, water heating, and cooking declines by 0.6%/year. In "Moderate effort", stepped-up retrofit activity and programs that target new homes cause fuel intensity to fall by 2.0%/year, about the same rate as between 1972 and 1985. In "Vigorous effort", much higher prices and strong programs cause a large reduction in heating intensity relative to "Moderate effort" in Europe and the US. The average heating intensity in 2010 is far below the level of new homes in 1987 in the US and Europe. Space heating intensity in new homes falls close to the levels now found in Scandinavia. More careful energy management in the home, induced by higher fuel prices, is facilitated by advanced control systems.

For electric appliances, the average intensity in 2010 in "Trends" is about the same as that of new appliances in 1987 (Fig. 10.5). The decline is driven mainly by the current US efficiency standards and other programs in place in Europe in 1990. In "Moderate effort", stronger policies and programs, especially in Europe, cause the average intensity to fall to near the level of the present market lowest (which is about half the level of the average new appliance in 1987). "Vigorous effort" includes programs to push the efficiency frontier in new appliances, and to encourage market uptake of these devices. The average intensities of appliances in 2010 is considerably below the lowest intensity of products available in 1990. "Vigorous effort" also embodies considerable reduction in energy intensity of lighting as compact fluorescent lamps come into widespread use.

Service sector In the "Trends" scenario, fuel intensity declines by 0.6%/year, reaching 75% of its 1985 level by 2010. The decline is much slower than in the 1972–1985 period (3.6%), which reflects some saturation of the retrofit potential and the slow penetration of new buildings. (In addition, part of the historic decline was due to growth in market share of electric heating.) Electricity intensity decreases by 0.4%/year. This decline is a break from the past, when intensity rose. The decline reflects the increasing attention given to the service sector by utility DSM programs. In "Moderate effort", higher prices cause fuel intensity to decline at 2.0%/year. Electricity intensity falls at a rate of 1.7%/year. In "Vigorous effort," fuel intensity declines at 3.2%/year as retrofit activity becomes more ambitious. Electricity intensity falls considerably as DSM and other efforts push the market penetration of cutting-edge technologies. A large share of new buildings soon incorporate energy-saving designs and technologies that are found in the most efficient new buildings today.

10.1.2 Summary

To express the combined effect of the various intensity changes, we weight each intensity according to 1985 sectoral consumption patterns. The average decline in the overall weighted energy intensity is 1.2%/year in "Trends", well below the 2.2%/year achieved between 1972 and 1985. Fuel intensity declines at 1.3%/year in "Trends" (compared to 3.1%

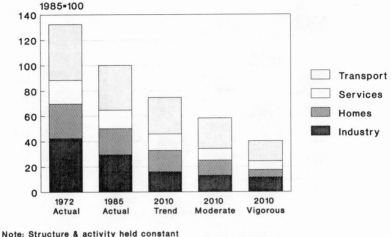

Note: Structure & activity held constant
at 1985 levels.

Fig. 10.6. OECD final energy demand in 2010, effect of intensity reductions

historically), and electricity intensity falls at 0.6%/year (compared to an historic rise of 0.9%/year). In "Moderate effort", the average energy intensity declines by 2.2%/year, about the same as the historic rate. Fuel intensity falls by 2.3%/year, and electricity intensity by 1.5%/year. In "Vigorous effort", average energy intensity declines by 3.6%/year, close to the rate experienced during the period 1979–1982 for many countries. Fuel intensity falls at 3.9%/year, electricity intensity by 2.4%/year.

Figure 10.6 illustrates the effect of the different intensity scenarios, assuming that activity levels and sectoral structure are held constant at 1985 levels.[5] The intensity reductions in the "Trends" scenario would reduce future demand by 25% relative to 1985. The demand in "Moderate effort" is 22% below that in "Trends", while "Vigorous effort" is 47% lower. Relative to 1985, "Moderate effort" is 42% lower, while "Vigorous effort" is 60% lower. Put differently, the level of activity supported in 1985 is achieved with only 40% of the energy. This would not require an enormous amount of technological innovation, but would require great effort to accelerate the market penetration of highly energy-efficient technologies and techniques between now and 2010. At present, the OECD nations are clearly not on a path towards either the "Moderate effort" or "Vigorous effort" end points.

[5] In actuality, changes in activity, structure, and intensities will probably reduce the importance of the manufacturing, freight, and residential sectors relative to that of services and travel. This introduces a small distortion into the results, because the effect of intensity declines by 2010 will depend on the shares of different sectors in final energy use.

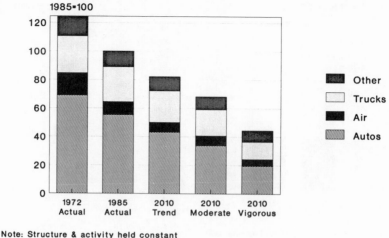

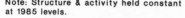

Note: Structure & activity held constant
at 1985 levels.

Fig. 10.7. OECD transport energy demand in 2010, effect of intensity reductions

Figure 10.7 illustrates the impact of intensity changes on OECD transportation energy demand, using the same method as for Fig. 10.6. The reduction in transportation energy demand from "Trends" to "Moderate effort" is somewhat smaller than for *total* final energy because changing current trends in transport is more difficult than in other sectors, in our judgment. With "Vigorous effort", however, considerable reduction could be achieved: demand in 2010 is only 45% of the 1985 level.

10.2 Scenarios for the Soviet Union[6]

Constructing a "Base Case" or "Trends" future scenario for the Soviet Union requires a considerable amount of conjecture. Events following the failed coup of August 1991 have made the political and economic future of the Union and the Republics very uncertain. While the outlook at present is rather bleak, it is conceivable that slow reform within an economic confederation could hold the economy together, albeit with little economic growth over the next few years. The reorganization now taking place seems likely to hasten the speed of reform, but the response of the economy will not be clear for many years. Consequently, the scenarios we present here represent end-points. Given the outlook as of end-of-1991, it seems unlikely that their attainment would occur before 2010.

[6] The Soviet scenarios are based on work originally published in Cooper and Schipper (1991). They were modified for this book after the disintegration of the USSR in 1991.

The "Slow reform" scenario illustrates what might happen if the dissolution of the Union and public opposition were to hinder the momentum of reform, and if energy was deemed a plentiful resource (in Russia) relative to important ingredients in energy efficiency such as capital and technical know-how. Prices are allowed to rise to internal replacement levels, roughly 2–3 times the 1985 levels in most cases (these still lie below Western levels). However, no special effort is undertaken to carefully meter and charge for energy as used.[7] Thus, energy prices rise but the full costs are generally not seen by ultimate consumers. Under these circumstances, it is reasonable to expect a slow decline in the energy intensities of industry and household and service-sector heating through gradual replacement of old facilities and equipment. Relatively little effort and scarce capital is focused on energy efficiency, however, and only a modest amount of Western technology penetrates into the Soviet economy.

In the "Rapid reform" scenario, development of a market economy proceeds at a relatively fast pace, and there is substantial assistance and investment from the West. Energy intensities decline to around the average levels of Western Europe in 1985. Implicit in such a change is that today's least productive factories are closed, and there is considerable investment in new factories that utilize state-of-the-art technology. A rapid upgrading of the quality and durability of components of industrial and consumer goods takes place, which helps to boost energy efficiency. The energy efficiency of new homes increases, bolstered in part by improved quality of design and components. Energy users face prices that reflect actual consumption, not just allocations.

The "Extra effort" scenario is similar to "Rapid reform", except it assumes that energy efficiency is given higher priority by authorities and energy users. Considerable assistance is provided by countries with particular sectoral expertise: Scandinavian countries for space heating, district heating, and large heat pumps; Japan and Germany for industrial energy uses and cogeneration; W. Europe in general for small, efficient automobiles; the US for many activities in the energy sector. Energy prices reach Western levels to reflect opportunity costs of exporting oil and gas. Healthy growth in the economy hastens the processes of rebuilding and refurbishment, and encourages even more Western participation in an increasingly open economy. Rapid reduction in the level of resources devoted to defense frees not only capital, but know-how as well, which

---

[7] Very few apartments heated by central boilers or district heat in market-oriented Sweden are individually metered, so continuation of this tradition in the Soviet Union is plausible. Large industrial complexes can still be run without submetering.

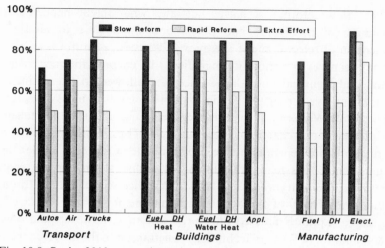

Fig. 10.8. Soviet 2010 energy intensity scenarios, reductions relative to 1085

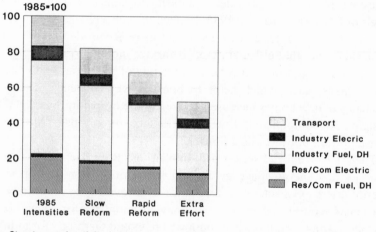

Note: Structure and activity held
constant at 1985 levels

Fig. 10.9. Soviet final energy demand in 2010, effect of intensity reductions

further propels technological improvements in the efficiency of energy utilization. Above all, however, individuals, private companies, and public authorities become engaged in energy use as never before. The result is that average intensities fall to around the level of current *new* Western European intensities.

Figure 10.8 shows the reductions in key energy intensities relative to 1985 for each scenario. In "Slow reform," the reductions are less than

20%, except for automobiles, air travel, and manufacturing fuel intensity. Even with only slow reform, many of today's least efficient factories are likely to close before 2010, and market forces will enhance energy productivity. In the "Rapid reform" scenario, the reductions are mostly in the 20–35% range, with a larger drop in manufacturing fuel intensity as new factories replace the old at a higher rate. The reductions in the "Extra effort" scenario are mostly in the 40–50% range; here too the decline is larger in manufacturing.

Figure 10.9 shows final energy demand with 1985 intensity levels and with the levels in each of the scenarios. To illustrate the aggregate impact of the intensity reductions, we hold activity and structure at 1985 levels. (This assumption is not a reflection of what will actually occur, but is necessary to illustrate the point.) The reduction in final energy consumption is 18% in "Slow reform", 32% in the "Rapid reform" scenario, and 48% in "Extra effort". The values in the "Rapid reform" scenario are in the range deemed feasible by Makarov and Bashmakov (1991). Sinyak (1990) found similar potentials in his study of the Soviet Union.

10.3 Developing countries

Scenarios of energy intensities and energy demand in 2025 have been developed for major LDCs in a series of studies performed and coordinated by the International Energy Studies (IES) Group as part of assessment of future CO_2 emissions from energy use. Initial work by Sathaye et al. (1989) was followed by studies performed by country experts within a framework developed by the IES Group; for an overview of this work, see Sathaye and Ketoff (1991). Country experts developed a "High emissions" case and a "Low emissions" case that incorporates greater effort to improve energy efficiency. For 12 countries combined, total primary energy requirements in 2025 in the "Low" case are approximately 20% lower than in "High." As these scenarios differ in structure and time-frame from those presented for the OECD and Soviet Union, we do not present them here in detail. Instead, we illustrate the potential for energy intensity reduction with scenarios developed by the authors in a study of residential electricity use in Java, the main island of Indonesia (Schipper & Meyers, 1991).

We developed two scenarios of future energy intensity for the various electric end uses, and compared these to a "Frozen intensity" case. As discussed in Chapter 9, the "Best 1988" scenario assumes that the average appliance in 2010 incorporates the technology used in the most efficient appliances available today in the OECD countries. The percentage

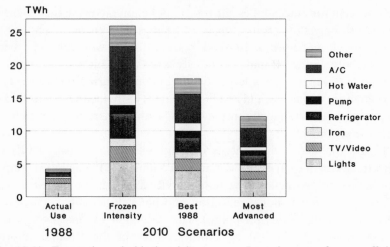

Fig. 10.10. Future household electricity use on Java, impact of more efficient devices

reduction relative to the energy intensities (UECs) in 1988 is in the 20–30% range for most appliances. The "Advanced technology" scenario assumes that the average appliance in 2008 incorporates technology that is beginning to be or could soon be adopted by manufacturers in the OECD countries. In this scenario the percentage reduction relative to the UECs in 1988 is 30–50% for most appliances (and 60% for refrigerators and air conditioners).

In these scenarios, we also estimated future levels of appliance saturation (based on discussions with appliance manufacturers on the levels of saturation found in upper-income households in Java and in wealthier Asian developing countries). Applying the assumed levels of saturation to the projected number of households in urban Java in 2008, we estimated total residential electricity consumption in each scenario. Compared to the consumption that would obtain if the average UECs in 2008 were the same as those of today, the results show a reduction in total consumption of 30% in the "Best 1988" scenario, and of about 50% in the "Advanced technology" scenario (Fig. 10.10). Total electricity use still grows substantially in these two scenarios relative to 1988, since appliance ownership grows so much, but the growth is much less than if intensity remains unchanged.

Although the information available did not allow a thorough analysis of the costs of saving electricity through improved appliance efficiency vs. the cost of electricity supply, it is very likely that the improvements incorporated in the "Best 1988" scenario and many of those in the "Advanced technology" scenario are cost-effective compared to the marginal cost of electricity production of approximately $0.09/kWh. The estimated "savings" of 8.0 TWh in the "Best 1988" scenario is equivalent to the output of around 1500 MW of power plants (operating at 60% capacity factor). Using the approximate cost of new coal-fired capacity of $1500/kW, the savings from avoided power plants resulting from improved efficiency would amount to $2.2 billion by 2010.

10.4 Conclusion

The scenarios presented in this chapter do not predict what will happen in the future. We believe, however, that they illustrate a plausible set of outcomes if energy prices, policies, programs, and other factors evolve as described in each case. With higher energy prices and vigorous policies and programs, intensities in the OECD countries in 2010 could be nearly 50% less on average than the level where trends seem to be pointing. In the former Soviet Union, a combination of rapid, successful economic reform and "extra effort" to improve energy efficiency might result in average intensity being nearly 40% less than in a "slow reform" case. And in the LDCs, a mixture of sound policies, programs, and energy pricing reform could also lead to intensities being far lower than they would be otherwise.

We have alluded to the changes that need to occur if energy intensities are to decline as described in each scenario. In the next part, we describe the kinds of strategies that could bring such changes about.

References

Cooper, R. C. & Schipper, L. J. 1991. The Soviet energy conservation dilemma. *Energy Policy*, **19** (4), 344–63.

Makarov, A. A. & Bashmakov, I. 1991. The Soviet Union: an energy development strategy with minimum emissions of greenhouse gases. *Energy Policy* (December).

Sathaye, J. & Ketoff, A. 1991. CO_2 emissions from major developing countries: Better understanding the role of energy in the long term. *Energy Journal*, **12** (1),161–96.

Sathaye, J., Ketoff, A., Schipper, L. & Lele, S. 1989. *An End-Use Approach to Development of Long-Term Energy Demand Scenarios for Developing Countries*. Berkeley, CA: Lawrence Berkeley Laboratory Report LBL-25611.

Schipper, L. J. 1987. Energy conservation policies in OECD countries: did they make a difference? *Energy Policy*, **15** (6), 538–48.

Schipper, L. J. & Meyers, S. 1991. Improving appliance efficiency in Indonesia. *Energy Policy*, **19** (6), 578–88.

Sinyak, Y. 1990. *Energy Efficiency and Prospects for the USSR and Eastern Europe*. (CRIEPE Report EY90001). Laxenburg, Austria: International Institute for Applied Systems Analysis. A shorter version was published under a slightly different title in *Energy*, **16** (5), 791–815.

US EIA (Energy Information Administration). 1991. *Annual Energy Outlook with Projections to* 2010, DOE/EIA-0383(91). Washington, DC: US Department of Energy.

Part III
Shaping the future

11

Encouraging energy efficiency: policies and programs

Based on past experience and current trends, it is safe to say that some reductions in energy intensities will occur in the future as a result of moderate rise in energy prices, policies and programs currently in place or planned, and the course of technological change and stock turnover. As we illustrated in Chapters 9 and 10, however, there is a large gap between where trends are pointing and the intensity levels that are technically possible and appear to be economically attractive in terms of life-cycle costs. Even assuming only a modest rise in energy prices due to market forces, reductions in energy intensities of 30–40%, or more in some cases, appear to be cost effective and achievable over the next 20 years. But the direction of current trends makes it unlikely that this degree of progress will in fact occur. Narrowing the "efficiency gap," the difference between the average energy efficiency resulting in the market (in new equipment and buildings as well as in existing stocks) and the level of efficiency that would be economically beneficial from a societal point of view, is an important role for public policy.

11.1 Barriers to improving energy efficiency

In part, the pace of energy-efficiency improvement reflects the typical situation with diffusion of new technologies and techniques. It takes time for their advantages to be understood and accepted, and to outweigh the risks associated with the unfamiliar. But there are also a number of specific factors that tend to inhibit the market's adoption of higher energy efficiency.

A basic problem is the pricing of energy below its true cost. Prices affect initial purchase decisions and also how systems are operated. Direct and indirect subsidies that create lower prices than what the market would otherwise deliver are common, particularly in the LDCs and Former East Bloc (Kosmo, 1987). Even in the OECD countries, marginal-cost pricing of electricity is not used in some jurisdictions. Only gasoline is priced substantially above its market price (due to taxation), and that not in all countries. In addition to the above problems, market prices do not reflect the environmental and sociopolitical externalities associated with energy supply and use. Among economists, pricing below true social cost is a well-recognized market failure which leads to underinvestment in energy efficiency.

Other obstacles to energy efficiency may be institutional or behavioral in nature (Hirst & Brown, 1990). Institutional barriers are primarily beyond the control of individual energy users. Government fiscal and regulatory policies and certain codes and standards may inhibit energy efficiency. For example, the United Kingdom does not tax value-added on gas, electricity and coal, but taxes nearly every other consumer product or service. This uneven treatment raises the cost of energy-efficiency investments relative to the cost of energy. Limited access to or high cost of capital often limits investment in energy efficiency, especially in developing and transitional economies. Lack of supply of energy-efficient products and services is also a problem in these countries. High import duties often contribute to this problem, and raise the cost of energy-efficient products that are not locally produced.

Behavioral barriers concern the decision-making of energy users. Lack of reliable information about the financial impacts of different investments or actions, or the difficulty of obtaining it, is especially a problem for households, small and medium-sized companies, and small public administrations. Even if relevant information is available, few energy users perform an economic calculation of costs and benefits, or take a long-range perspective. Studies of actual choices of appliances and household conservation investments have found that consumers often exhibit very high implicit discount rates in trading off between investment in efficiency and future energy savings (Train, 1985). Private firms also tend to require a relatively rapid payback for energy-efficiency investments – often a higher rate of return than for other types of investments. The perceived risk of energy-efficiency investments (will benefits be as claimed?) is one reason for this. Uncertainty about future energy prices is also a factor.

The fact that energy costs are usually a small part of household or business expenditures is part of the problem. People may simply not be interested in them, or do not want to bother to take the necessary steps. Cost-effectiveness is not the only criterion; "cost relevance" is also required (Grubb, 1990). Consumers are usually more concerned with the initial cost of a technology than with life-cycle costs. Thus, the higher first cost of energy-efficient technologies often inhibits their adoption, especially for low-income consumers.

A barrier that is common in the buildings sector is "misplaced incentives", or the separation of expenditure and benefit. For example, in a rented building in which tenants pay the energy costs, they either cannot renovate the building, or may be unsure whether they will be there long enough to obtain the full benefit. A similar situation exists for new homes, for which builders select the characteristics of the building shell, the type of heating equipment, and often even the major appliances. Being very conscious of their own costs, they often select levels of energy efficiency well below those that are cost-effective for households. In principle, renters and homebuyers should place some value on energy efficiency, which would give landlords and builders an incentive for improving efficiency, but lack of information or interest as well as other barriers often get in the way.

Barriers to energy efficiency apply to varying degree in different sectors, and within sectors as well. Energy users differ with respect to their circumstances, perspectives, and the criteria that they use in making decisions that affect energy efficiency. For each type of energy user, there is often a "group-specific" set of obstacles that needs to be addressed with a matching set of measures (Jochem & Gruber, 1990). Barriers are usually greater among households and small companies than among large companies, who generally have more knowledge about energy-efficiency options and the technical staff to evaluate and implement them. Even among large companies, however, capital scarcity can be a problem, and investments with payback of more than three years are usually not made unless they are part of a larger change.

While barriers to energy efficiency have been widely discussed, there is some controversy about whether they constitute "market failures" that policy should address. It has been argued that many (or even all) of the behavioral barriers mentioned above are normal features of a competitive market, and that high implicit consumer discount rates reflect real investment costs, including risk and the non-liquid nature of energy-efficiency investments (Sutherland, 1991). The implication of this argument is that, aside from providing information to consumers, which can lower uncertainty and risk, and supporting long-range R&D, the appropriate government role is to address "true" market failures, the most serious of which is the pricing of energy at less than its social cost.

Even if energy prices fully reflected social costs, however, there would still be a discrepancy between the criteria used for energy-efficiency investments and those used for investment in energy supply. While implicit discount rates for energy-efficiency investments are typically 20–40% or higher, utilities make supply-side investments on the basis of a discount rate of about 5%. Providing energy users with better information might lower their discount rates somewhat, but the evidence suggests the impact would be relatively small. Moreover, energy suppliers can generally access capital at lower cost than most energy users. Policies and programs to encourage the market's adoption of energy efficiency can help to "balance the scales" between supply and energy-efficiency investments. They can often reduce the risk of investment in energy efficiency at a low social cost, and can provide capital at lower cost as well.

11.2 Targets of energy-efficiency policies

Broadly speaking, energy-efficiency policies may target either new products and capital goods or existing capital stocks. The impact of policies in each area depends on the rate of stock growth and turnover. In countries that are growing slowly, affecting the existing stock is important, whereas in countries where the stocks are increasing rapidly, policies affecting new equipment and buildings can have a large impact. Policies for new systems are especially important because it is much harder and more costly – and sometimes impossible – to upgrade later than to build in high levels of energy efficiency from the start.

11.2.1 New products and capital stock

The main actors that shape the energy efficiency of new products, buildings, and other capital stock are the companies that produce them and the households, businesses, and other institutions that purchase them. Producers of energy-using technologies are usually large or medium-sized companies. In some cases, the product is the result of several parties working together (e.g., a large building). For most products, but especially for buildings and factories, many components affect the energy use of the overall system.

The market's adoption of higher energy efficiency in new products has two aspects. One is the extent to which users select from among the commercially available products those that are relatively more energy efficient. The other is the degree to which producers push the "efficiency frontier" through development, commercialization, and incorporation of new technologies. The latter is of course affected by the expected buying habits of purchasers. Stimulating the demand for higher energy efficiency is important to gain the attention of producers. One of the virtues of a well-functioning market economy (and drawback of centrally planned economies) is the tight feedback between producers and buyers. The importance of energy costs to buyers affects the degree of emphasis that producers give to energy efficiency relative to other features. If producers perceive that energy costs are important to buyers, they allocate the resources to design products that are more energy efficient. For example, the decisions of aircraft manufacturers that affect the energy efficiency of new aircraft are shaped by their perception of the importance of fuel economy to airlines. That importance is in turn influenced by the airlines' expections of future fuel prices, the role of energy costs in total operating costs, and the value attached to future reductions in fuel costs.

The product planning process typically involves long lead times between concept and commercialization. Having seen energy prices go up sharply but then decline, producers are wary of putting too many eggs in the energy-efficiency basket. They may have technologies ready to commercialize that incorporate higher energy efficiency, but judge that the market is not yet ripe. They may even plan to commercialize such technologies, but pull back upon discovering that the market is not sufficiently interested.[1] Thus, targeting producers *and* buyers will usually be more effective than addressing one but not the other.

Strategies to overcome the reluctance of producers to push energy efficiency are especially attractive in cases where there are relatively few of them. For certain types of equipment, large companies make (or control through license) technologies that are found throughout much of the world. This is especially true for transport vehicles, home appliances and other electrical equipment, and some industrial equipment. Components may be manufactured in various places, but the design and specifications are produced in corporate headquarters. This centralization of design, which one could argue is on the rise in the world today, could mean that innovative ideas may have a hard time reaching the places where design

---

[1] For example, the efficiencies of 1990s generation aircraft could have been approximately 20% higher than they seem likely to be if manufacturers had been successful in marketing propfan engines to airlines. As late as 1987–88, Boeing, McDonnell, and Airbus all planned to offer propfans on their 1990s aircraft, but the drop in jet fuel prices shifted the focus of aircraft buyers from fuel costs to purchase costs (Greene, 1990). The manufacturers could not interest buyers in the more expensive propfans, and dropped plans to introduce them.

decisions are made; but it also presents important opportunities for strategic intervention.

In considering strategies to encourage adoption of higher energy efficiency, it is essential to bear in mind a simple but important idea: *Those innovations spread which simultaneously meet several demands* (Berg, 1979). Or, as Amulya Reddy recently put it: "Those technologies are successful which potential purchasers cannot afford to reject because they are superior on several counts compared to existing technologies" (Reddy, 1991). This idea implies that changes that only improve energy efficiency will be greeted less favorably than those that also provide other benefits. For automobiles, for example, improving aerodynamics or reducing vehicle weight improves both acceleration performance and fuel efficiency. Energy efficiency may improve due to changes that are implemented only partly with that goal in mind. For example, refrigerator manufacturers moved to smaller, more energy-efficient compressors to allow more room for food storage and provide quieter operation as well as to improve energy efficiency. In Sweden, several factors contributed to the building of very energy-efficient housing (Schipper et al., 1985). Occupants wanted a comfortable indoor environment free from drafts and moisture, with control over the quality of indoor air, and owners of apartment buildings sought to reduce the maintenance requirements caused by use of poor components. On the producer side, energy efficiency was partly the result of the desire to industrialize and streamline the housing planning and production process, and to guarantee a market for suppliers of innovative and reliable technologies. These goals were largely met by the 1960s and early 1970s, and the result was the most energy-efficient housing stock in the world. Only after 1975 was energy added to the building regulations as an explicit concern.

By the same token, improvement in energy efficiency may be constrained by other goals that take precedence. Automobile manufacturers have made great strides in increasing the energy efficiency of engines, for example, but much of the gain has gone to increase acceleration/top-speed performance rather than to improve vehicle fuel economy. An example that is important in the LDC context is the need to make products rugged enough to withstand the often harsh operating conditions. These include large voltage fluctuation in electricity supply in the case of electrical equipment, which limits the usability of some energy-saving technologies (or adds to their cost), and poor road and fuel quality in the case of motor vehicles, which leads to use of rugged, heavy vehicles and engines that can tolerate a range of fuel quality.

11.2.2 Retrofit and operation

Influencing the energy efficiency of new equipment and buildings is important because the choices that are made affect energy use for many years or even decades. However, the in-use energy intensity of all technologies is also shaped by user management and operation. Management includes decisions to make improvements that increase energy efficiency. In some cases, changes may be undertaken solely to reduce energy costs, but energy-saving retrofits or changes in operations are more likely to occur where there are other benefits as well. When these changes interfere with normal business operations, there may be reluctance to undertake them.

Large companies generally have better information about their energy use and the benefits of improving energy efficiency, and are usually more able to implement retrofit measures and manage their facilities. A German study of eight industries found a direct correlation between company size and implementation of technical energy-saving measures (Gruber & Brand, 1991). Even large companies may require a rapid payback (usually less than three years) for retrofit investments, however. Top management often places higher priority on projects that open up new markets as opposed to those that improve existing plants without clear market implications.

In the OECD and some other countries, an "energy management" industry has developed to serve the manufacturing and service sectors, while businesses such as insulation contractors serve households. In the United States and, to a lesser extent, elsewhere, utilities have taken an active role in providing energy management assistance to their customers.

11.3 Policies for accelerating energy-efficiency improvement

Below, we discuss basic policy methods for encouraging energy efficiency. Since this topic has been the subject of a great deal of writing, we do not discuss in depth the various issues related to particular polices, but instead attempt to provide a context and overview.

11.3.1 Energy pricing: sending the right signal

Governments throughout the world are increasingly recognizing the need to remove controls that prevent energy prices from reflecting market realities. Changes have been made, but there is still strong resistance to removing subsidies from users who have grown accustomed to cheap energy. Where survival needs are at issue, a gradual approach seems reasonable, but subsidies often benefit energy users who do not really need the subsidy. They inhibit all users from taking measures to use energy more efficiently, and drain national resources that could be used to address the needs of the poor more effectively. In addition to removing subsidies, policies can promote marginal-cost pricing where it is not currently used, and time-of-day pricing where it is practical. Both of these provide clearer signals of the true cost of energy use.

The other key issue in energy pricing is internalization of environmental and social externalities. One way of doing this is to require the actors involved in energy supply to reduce the environmental damage (or the risk of damage) caused along the supply pathway (in waste disposal), which will presumably lead to higher production costs and prices. Another method is taxation of energy products, with the tax based on the estimated environmental cost per unit. An indirect method involves requiring utilities to incorporate environmental externalities in planning and resource selection. Approximately half the state electric regulatory commissions in the United States have begun to do this in one form or another (Ottinger, 1991).

11.3.2 Information: increasing awareness

Information relating to energy efficiency can assist energy users in making decisions regarding new equipment and buildings and retrofit of existing facilities. Reliable information can reduce the risk of energy-efficiency investments. Providing information is especially important for households and small businesses, for whom the cost of obtaining information is typically high (and usually cannot be spread over many investments, as is the case for a large business).

For new equipment and buildings, standardized testing or measuring by an independent party such as a government agency conveys more reliable information than claims from individual manufacturers, though government programs may build on efforts by private sector associations. The energy-efficiency labelling that such testing permits can supply more information at lower cost than private markets (Sutherland, 1991). Energy-efficiency labelling has been used for household appliances in the United States, Canada, and a number of other countries, and is beginning to be used for new homes in the United States. It has been used for many years for automobiles in the United States, but not in most of Europe. The experience to date with appliance labelling in several countries suggests that it has relatively little impact on consumers' decisions (Marbek Resource Consultants, 1990). Information is sometimes not presented in a way that is useful for consumers, or they may simply lack the interest to make use of it.

For existing buildings and factories, information can be either generic (e.g., advice on managing particular types of equipment or on generic conservation opportunities) or site-specific, such as results from an energy audit. The latter is more likely to result in investment, especially if followed up by further assistance, including financing and incentives.

11.3.3 Efficiency regulations or agreements: pushing producers

The most common type of energy-efficiency regulation is one that sets a minimum efficiency level for particular types of new equipment or buildings. Such regulation currently applies to a number of household appliances in the United States, and is embodied in building energy codes in most northern European countries and US States. The effect of minimum efficiency standards depends on how strict the standard is. Standards initially may only eliminate the least efficient products, but then be gradually tightened, as has occurred with the US appliance standards (Turiel et al., 1991).

In the United States, several state and local governments have begun to adopt minimum efficiency standards for lamps, lighting fixtures, and motors used in commercial buildings (Nadel, 1991). Standards can also be applied to common types of motors and commercial HVAC equipment. Standards primarily apply to new equipment and buildings, but it is also possible to require that existing buildings meet minimum criteria at point of sale, as is the case for homes and commercial buildings in San Francisco, California.

While most efficiency standards specify a miniumum acceptable level, they may also specify an *average* efficiency level to be attained. An example of this approach is the US fuel economy standards on automobiles and light trucks, which set a level for each manufacturer's sales-weighted average of cars sold in a given year, and established a penalty system for failure to meet the standard. Unlike a minimum efficiency standard, meeting an average efficiency level partly depends on what manufacturers are able to sell. The US standards have been criticized for being biased against manufacturers whose sales are weighted toward larger cars; standards that take vehicle size into account are an alternative regulatory form.

Voluntary efficiency improvement targets negotiated between government and equipment producers are another option. This has been the preferred approach in Europe and Japan for automobiles and appliances, and did result in significant improvement. In Brazil, the government and utilities have proposed protocols to improve the efficiency of refrigerators, freezers, motors, and lamps (Geller, 1991). Voluntary targets are usually based on the average efficiency of products sold. They seem to have greater effect if it is believed that nonattainment might lead to binding regulations. Regulations provide greater certainty than voluntary targets, but may take considerable time to implement due to the process required to decide on a particular level, as well as opposition from the affected industry.

11.3.4 Financial incentives: enticing users

Financial incentives to encourage energy efficiency include low-interest loans, direct payments (grants and rebates), and tax incentives. Incentives may be provided by governments or utilities, and may target new equipment as well as retrofit. Financial incentives are most commonly given to energy users, but incentives for equipment dealers and payments to manufacturers (to subsidize the price of more efficient equipment or assist in retooling) are also possible.

Incentives have been primarily used in the buildings sector. These have included grants, low-interest loans, and tax credits for residential retrofits; rebates for higher-efficiency appliances; and various types of payments to encourage retrofit and energy-efficient new construction in the service sector. Tax incentives can be useful in sectors that are hard to affect through other means. Accelerated depreciation of energy-efficient equipment may entice manufacturers, airlines, and truck owners to make choices that they otherwise would not. Experience has shown that non-financial aspects of incentive programs, especially marketing strategy, can be even more critical to participation than the level of incentives.

Disincentives may also be used. An example is the US "Gas Guzzler Tax," which raises the price of cars whose rated fuel economy is below a certain level. Increase in the threshold for this tax had a clear effect on the fuel economy at the lower end of the market in the mid-1980s, even as real gasoline prices were declining (Ledbetter, 1991). A similar effect results from variable import duties that place higher taxes on cars with larger engines or greater overall weight; basing the tax on measured fuel economy would be even more effective. Another disincentive that has been used by utilities with some success to encourage energy-efficient new homes is charging builders a connection fee if homes do not meet a minimum standard. Connection charges can be an effective way of encouraging builders to meet (or exceed, if the fee varies with the level of efficiency) voluntary efficiency standards.

A potentially effective way to influence consumer decisions at the point of purchase is to combine incentives and disincentives. For new cars, for example, a system of fees and rebates could vary according to their rated fuel economy (Gordon & Levenson, 1990). If a car had above-average fuel economy, its buyer would receive a rebate whose amount could vary depending on how far above average it is. The reverse would apply for cars of below-average fuel economy. A program of this nature is being implemented in the Canadian Province of Ontario. Vehicle emissions could also be brought into the fee/rebate framework, as has been proposed in legislation introduced in California.

An objection that is sometimes raised to financial incentives is that some of the money is given to users that would have made a particular efficiency investment anyway. (This does not represent an economic welfare loss, but rather a transfer payment.) Even in these cases, however, the incentive may accelerate investment, which usually has some value. Moreover, it is possible to minimize the "free rider" problem by limiting incentives to high-efficiency measures that would probably not be implemented (except by a few) without the incentive. Similarly, the size of the incentive may be varied according to the level of efficiency, as is the case with a number of rebate programs for electric appliances in North America.

Providing financing for energy-efficiency investments may be helpful in cases where users lack access to capital or are deterred by high initial costs. Spreading costs over time (such as including them as part of the utility bill) and utilizing lower-cost capital than that available to consumers can be especially effective for lower-income households.

11.3.5 Research, development, and demonstration

Research, development, and demonstration (RD&D) is essential to continue technical progress that contributes to more efficient energy use. The evolution of efficiency-enhancing technologies and techniques depends on both private sector and publicly-sponsored RD&D. Private sector RD&D mainly focuses on technologies that appear commercially viable in the near or medium term. Publicly sponsored RD&D is important for basic research that may lead to commercially viable results, as well as for ideas that are promising but too risky for the private sector to invest in. Demonstrations of cutting-edge technologies and techniques can also help speed their market penetration.

We have not studied RD&D explicitly, yet certain needs are clear from our review of the successes – and failures – of the past. We offer a list of areas where RD&D could have considerable payoff in Box 11.A. In most cases, a key issue is how to reduce the time it takes for energy-saving technologies to reach the marketplace. For example, the existence of a large number of low-emission, high fuel economy prototypes suggests automakers are familiar with many technical options (Bleviss, 1988). But if fuel economy is not high on the list of features that make a car marketable, they are not likely to incorporate such options in their products unless these features have other benefits, or the government acts to stimulate the market.

One important need is demonstration and testing of optimized, integrated packages of advanced energy-saving measures. Many projections of energy savings are based on the performance of individual energy-efficiency measures. Field tests of packages of advanced technologies are needed to measure the effects of component interactions on energy performance, economics, reliability, and end-user acceptance. The Advanced Customer Technology Test for Maximum Energy Efficiency project of Pacific Gas and Electric Co. in northern California is a recently begun program of field experiments designed to explore these issues in real buildings (Brown, 1991). In the initial pilot retrofit project, four of the five firms involved created designs that are expected to save more than 70% of the current gas and electricity consumption and meet cost-effectiveness criteria. One lesson learned so far is that even relatively advanced designers need help in identifying and sorting through the available technology options.

In many areas, RD&D in the OECD countries will have important benefits elsewhere, since many of the markets for energy-using equipment are worldwide. The degree of benefit will depend on technology transfer from the OECD countries where much of the advanced equipment is produced or designed. In many cases, joint efforts to adapt new technologies to local climates, capabilities, and other conditions is very important.

Box 11.A. Priorities for energy-efficiency RD&D

AUTOMOBILES AND TRUCKS The effect of various technical efficiency improvements on actual fuel economy has been offset by increases in weight, speed, and performance. RD&D is needed on design options that provide the features that people want while also improving fuel economy. One example is light-weight materials that provide safety but are not prohibitively expensive. Improving the emissions characteristics of lean-combustion and diesel engines is another. Advanced engines for trucks offer considerable fuel savings over current designs. Attention should also be given to improving truck operations. Since the share of truck freight is rising in almost every country, this effort would have a large payoff.

EFFICIENT ELECTRICITY USE The most important uses of electricity are for lighting, motor power (including compressors), heat pumps for heating and cooling applications, electro-technologies for industry (e.g., electrolysis, electro-plating), and other applications, some of which compete directly with fuel-using processes. Improving the performance of motors would permit significant electricity savings in all applications. Improving compressors has become a high priority because of the impending switch away from CFCs in refrigeration applications. Further development of electronic controls for lighting, appliances, and motors affords additional electricity savings as well as likely increases in productivity. Reducing the cost of these technologies is an important R&D priority. Improving the design of lighting systems permits a significant reduction in lighting needs, just as lowering the heat gain through windows using special optical coatings or glazing whose transparency can be regulated reduces requirements for cooling in buildings.

BUILDING MATERIALS FOR COLD CLIMATES Almost 1.5 billion people live in climates where space heating represents a significant energy use. There remains a great need to develop building materials and construction methods that reduce heat losses and reduce the cost of doing so in new homes and buildings and. more important, in existing ones. Private firms have some incentives to pursue these developments, but only if they see that housing policies, particularly financing, will support inclusion of the technologies in new homes. Many energy-saving strategies can only be tested at the whole-building level, and the makers of individual components have little incentive to undertake such tests. Government-sponsored demonstration can accelerate market adoption of new products and building techniques.

ADVANCED INDUSTRIAL PROCESSES AND MATERIALS Much work is needed to develop more energy-efficient industrial processes, particularly those related to combustion, materials forming, automation, electricity use, and pollution control. While individual firms have an interest in reducing their own energy costs, few are interested in developing generic energy-saving techniques. Since it is difficult for government to pick which processes are the most important or promising, a coordinated effort with manufacturers is called for.

TRANSPORTATION NETWORK Research is needed on technologies and policies that allow vehicles to operate more efficiently on existing highway systems to reduce congestion, pollution, and fuel use. Priorities include improved signal timing on major local roads and ramp metering on limited-access highways; and pilot testing of "smart highways" (offering improved accident detection and clearing and information on traffic conditions) and "smart cars" (e.g., having on-board mapping, or automatic braking to allow reduced headway). Better understanding of the impact of road pricing and other policies to internalize the cost of using automobiles is also needed.

11.4 Implementing energy-efficiency strategies

Policies and programs to encourage energy efficiency may be implemented by a variety of actors. These include national, state, and local governments; energy suppliers, especially electric utilities; and professional associations for people who design and manage energy-using systems. In the developing and Former East Bloc countries, international lending organizations and assistance agencies can play an important role as well.

11.4.1 Utility programs to improve end-use efficiency

Utility companies are potentially an important vehicle for implementing strategies that are tailored for different types of energy users. Utilities have access to funds, established relations with customers, and can benefit financially from well-planned and executed programs. Utilities have been involved in so-called "demand-side management" (DSM) for over a decade in the United States, but it is less advanced in the rest of the world.[2] DSM has become widely accepted in the United States, and plays a major role in the strategy of some utilities.

DSM encompasses a wide range of programs that provide general information, energy audits, financial incentives, low- or zero-interest financing, direct installation of energy-efficient equipment at zero or low-cost to the customer, and technical assistance. Some utilities run comprehensive programs that combine several of the above elements. For example, a program to encourage energy efficiency in new commercial buildings may offer design assistance and rebates for incorporation of specific measures. Utilities may also contract with energy service companies and even customers to "deliver" specified energy savings at a certain price per kW or kWh saved, and solicit competitive bids from these parties (Goldman & Wolcott, 1990).

[2] Demand-side management includes load management (clipping peak demand and/or shifting demand from one time period to another) and load building (increasing loads during certain periods) as well as programs to improve end-use energy efficiency. In this discussion, however, we refer to the latter.

An important target for utility programs is "lost opportunity resources" – situations such as new construction where there is a one-time opportunity to improve energy efficiency at a low cost. Other important situations for intervention include cases when major equipment is being replaced and when buildings or process lines are remodelled. Intervening in these situations requires good communication and outreach on the part of utilities.

DSM activity has primarily affected the residential and commercial sectors. While most utilities design programs to serve both commercial and industrial customers, many have found it difficult to encourage industrial process improvements. A recent comprehensive review of US utility conservation programs (Nadel, 1991) made a number of key findings:

- Energy audits alone have relatively little impact. Implementation of audit recommendations is increased by the provision of follow-up services and financial incentives.

- Rebate programs can promote efficient equipment at a moderate cost to the utility. However, they generally only reach a minority of customers and have not been very effective at promoting improvements involving the complex interactions of multiple pieces of equipment.

- Loan and performance contracting programs can be useful for customers who lack capital to finance conservation improvements.[3]

- Direct installation programs can achieve high penetration rates and savings per customer, but generally at a higher cost to the utility (though not necessarily to society) than rebate and loan programs. This approach is particularly suitable for serving hard-to-reach customers (e.g., low-income households and small commercial customers).

- Marketing strategies and technical support services have a large impact on program participation and savings. It is important to keep customers' needs in mind and tailor programs for particular market segments. Personal one-on-one and community-based marketing strategies can be especially effective. Equipment dealers, contractors, and design professionals can be important allies in promoting programs.

11.4.2 Pushing the efficiency frontier

Coordinated strategies that target technology producers and users can accelerate the movement of new technologies from the laboratory to the market. To do this, it is necessary to stimulate the innovative impulses of

---

[3] In performance contracting, a third party provides the customer with assistance – often including financing – in implementation of energy-efficiency improvements.

producers while reducing their risk, which means encouraging a market for the new products.

An example of such an approach is the strategy of "innovative procurement" being used in Sweden. The government is seeking to stimulate development of more energy-efficient products "upstream", where they are designed and manufactured, *and* to cultivate a market for these technologies "midstream", where they are bought in quantity. The strategy is focused on a few key actors who account for a large share of the market for a given product, such as owners or administrators of large numbers of residential or commercial buildings. Working with these groups, technical specifications for the product(s) desired – such as a refrigerator, lighting system, or ventilation system – are clearly spelled out so that other attributes are not sacrificed in order to save energy. Having organized a market, the government then opens a competition among suppliers for the technologies desired. A small subsidy is guaranteed so that the first production will not be too costly to the buyers. In the longer run, of course, the winning technologies should be self-supporting. Two examples of the Swedish approach to "innovative procurement" are described in Box 11.B.

Box 11.B. Innovative procurement in Sweden

In 1990, the National Energy Administration (STEV) announced an international competition for a fridge–freezer that should undercut the electricity consumption per liter of present "lowest" models on the market by at least 15%, with a bonus of SEK 500 ($80) per unit payable if 25% reduction were achieved. The winning company was guaranteed an order for at least 500 units, and buyers were offered an incentive of SEK 1000 ($160) per unit. Major purchasers of such applicances – public, cooperative, and commercial housing corporations – were involved in the program and even helped to formulate the specifications. Three companies entered the competition (two from outside Sweden), and Electrolux was the winner with a model that will use 33% less than the current lowest. Production of the winning refrigerator, which may initially cost 10–12% more than current models, is expected in the fall of 1991.

STEV has also organized major property owners to work with lighting system companies to implement advanced methods for reducing electricity use (Stillesjoe, 1991). The objective is not simply to replace existing equipment with more energy-efficient devices, but to incorporate new and improved lighting systems. In three pilot projects, reductions in electricity requirements for lighting of 50–65% were achieved. The simple payback period (including the effect of the bonus) was over 20 years in two of the buildings, but only seven years in the third. Given the overhead costs of these start-up projects, it is likely that future projects will yield more rapid paybacks. Similar efforts in industrial equipment and building shell retrofits have been announced. Unfortunately, the current low price of electricity in Sweden inhibits interest in these programs.

A program similar to the Swedish competition for refrigerators is being organized in the United States by a consortium of utilities, government agencies, refrigerator manufacturers, and other interested parties (ACEEE, 1991). The "Golden Carrot" program will provide incentives to manufacturers who develop and market refrigerators that use at least 25–30% less energy than is required by the 1993 federal efficiency standard (which is considered rather strict), and use no CFCs (HCFCs are acceptable). Manufacturers will compete for a winner-take-all pot of money on the basis of energy efficiency and the amount of incentive requested per unit. Funds will come from a consortium of electric utilities. Utilities will also offer rebates to customers, dealers, or manufacturers for the super-efficient devices, as well as incentives for replacement of inefficient units with super-efficient ones. Products that do not win the competition, but are as energy-efficient as the winning bid, would be eligible for the rebates. The consortium is also trying to incorporate the purchasing power of federal agencies and public housing authorities. Refrigerator manufacturers who have participated in organizing meetings believe the efficiency target is ambitious, but are inclined to participate in the program.

11.4.3 *Targeting large energy users*

Another strategic approach for pushing the market toward higher energy efficiency is to target large companies that are responsible for significant amounts of energy use. Not only can substantial energy savings result; the demand for energy-efficient products resulting from large-scale programs can have an important effect on the market, potentially lowering costs for all purchasers. The ratio of benefits to program costs may be low if the program can influence key decisionmakers at corporate headquarters to implement company-wide policies. For the companies, the public relations benefit from acting as an environmentally responsible 'corporate citizen' can be a key motivation to act.

An example of targeting large energy users is the Green Lights Program of the US Environmental Protection Agency (EPA). 'Green Lights' is a voluntary program aimed at large corporations for the purpose of creating corporate-wide installations of energy-efficient lighting (Friedman & Hoffman, 1991). A participating corporation agrees to survey the lighting in all of its US facilities, to retrofit 90% of its floor area with cost-effective energy-efficient lighting packages, and to re-analyze profitable lighting options no later than five years after completing the retrofits. In return, EPA provides several types of program support, including a software package to help analyze options and a registry of utility rebates and financing options for energy-efficient lighting technologies. Perhaps most important, the program also gives participating corporations an opportunity to "green" their public image. EPA hopes to enlist 200–500 major US corporations in the program by the end of 1991.

EPA is also involving manufacturers of energy-efficient lighting technologies. Participants agree to help establish and take part in an independent product testing and information program to validate vendor claims about product performance and generally enhance consumer confidence in the technologies. EPA is also seeking participation of electric utilities. Utility "allies" agree to work with EPA to publicize the program and recruit new corporate partners, to assist in the product information program, and to meet the same lighting standards as the corporate partners in its own offices. For utilities, the program will lower the cost of marketing lighting programs, which usually focus on a facility-by-facility selling strategy. It will also help provide technical support, assist in evaluation of new technologies, and increase awareness of rebate programs at appropriate corporate levels.

The government itself is often the largest user of energy, and has a particular responsibility to take a long-term perspective in decisions that affect the energy efficiency of public property. In many countries, buildings and vehicles owned by government at different levels comprise a significant share of the total stock. Thus, policies and programs to encourage energy efficiency in new buildings and vehicles, building retrofit, and sound energy management practices can yield important energy savings, set an example, and exert considerable "market pull". Policies can also impact public housing, or buildings constructed with public financing.

11.4.4 Considerations for LDCs and the former East Bloc

The policies and programs discussed in the preceding section can be implemented in all countries, although selection and design must match local conditions. Barriers to the success of a particular policy must be foreseen, and complementary policies implemented to deal with them. Removing energy-price subsidies is an important step, but where managers have historically paid little attention to energy, or are unfamiliar with concepts such as life-cycle cost, the impact may be slow to develop. Similarly, policies that require products that are not widely available and/or skills and services that are not well developed may have little impact at first. Policies can create incentives for the local market to deliver those products and services, but technical or financial assistance may be needed to accelerate the process.

The barriers to energy-efficiency improvement are much greater in the former East Bloc and the LDCs than in the OECD countries. Some of them reflect general problems these countries face in their economic development, such as scarcity of capital. Political instability or high rates of inflation may lead to a very short-term focus that works against efficiency improvement. Many countries suffer from lack of modern components and assembly methods, low quality of much equipment, difficulty of obtaining equipment, and lack of skilled personnel. Where energy-efficient technologies are not locally produced, import duties often make their use prohibitively expensive. To make matters worse, energy prices have often been heavily subsidized, so the development of skills and technologies to improve energy efficiency has been artificially stifled. Other results of past policies, such as the lack of metering of heat and hot water in the majority of homes in the former East Bloc, also hinder conservation efforts.

Technical assistance and training are needed in many areas. As demand for energy efficiency grows, so does the need for energy auditors (for buildings and factories), energy management specialists, and architects and other designers who understand how to effectively incorporate energy efficiency in new buildings and industrial processes. There is a need to develop skills at a basic level (i.e., at universities and technical institutes) and also build them among those already practising in the field. Governments and/or utilities can provide specialized types of technical assistance as well, such as providing building designers with workbooks, educational seminars, and free computer simulations of building energy use. There is also a need to develop skills in program design, implementation, and evaluation.

Improving the link between indigenous R&D and commercialization of energy-efficient technologies is an important goal. There is a need to increase collaboration among research institutions, commercial and industrial enterprises, and end-users, thereby increasing the responsiveness of the R&D community to market forces. One promising model designed to further this aim is the Program for Commercial Energy Research (PACER) being implemented in India. PACER seeks to promote development of goal-oriented and market-responsive technological innovations in the Indian power sector through financial assistance to consortia of manufacturers, research institutions, and end-users (Jhirad, 1990). Encouraging joint-venture research linked to eventual commercialization between indigenous and foreign firms is another attractive approach.

11.5 Other policies that affect energy efficiency

The focus in this chapter has been on direct methods to encourage improvement in the energy efficiency of particular technologies and systems. However, there are many non-energy policies, or policies of which energy is only one component, that also impact energy efficiency, sometimes in significant ways. For example, policies that reduce traffic congestion have a positive effect on vehicle fuel economy. In LDCs, improving road conditions will contribute to lowering vehicle fuel intensity, since road quality affects the efficiency with which vehicles operate and the choice of vehicles. Improving the quality of fuel also can boost vehicle energy efficiency. A related "infrastructure policy" that is important in many LDCs is improving the quality of electric power supply. Because of the large voltage fluctuations that are common in many LDCs, certain technologies that could increase the efficiency of motors, compressors, lighting systems, and other electrical equipment are not used, or equipment is oversized so that it will provide adequate power when voltage is low. Voltage fluctuation also decreases the lifetime of many types of equipment, reducing the period over which energy savings can pay back the initial extra investment in energy efficiency.

In the manufacturing sector, policies that encourage use of 'post-consumer' scrap rather than virgin material can reduce the energy intensity of making certain products. For example, producing aluminum using scrap requires only about 10% as much energy as using bauxite. The relative savings are less in other areas, but still significant. Regulations affecting air pollution may impact energy intensity. In some cases, the equipment to control emissions may increase energy intensity, but regulations can also create an incentive for more efficient energy use. This is especially the case if the regulations are based on total emissions from a company rather than emissions per unit.

The policies discussed above will not be implemented primarily to save energy, but that can be a factor in their favor. If energy considerations are to play a role in formulating more general policies, however, it will be necessary to better understand the impact on energy use of broader policies.

11.6 Conclusion: balancing policies

Successfully overcoming the barriers to higher energy efficiency requires development of policies designed for specific users and locations. Reform of energy pricing, which entails removing subsidies and beginning internalization of externalities, is critical to give technology producers and users proper signals for investment and management decisions. But while a rise in energy prices increases the amount of energy-efficiency improvement that is cost-effective, it does not remove other barriers that deter investment. Minimum efficiency standards or agreements can "raise the market floor," and are important because they affect the entire market in the near-term. But they may not "raise the ceiling" very much, and do little to push the efficiency frontier. To accomplish these goals, incentives and other market-development strategies are needed. Utility programs in particular can play a key role in pushing energy efficiency beyond the level where users are likely to invest on their own.

Policies, programs, and pricing should complement one another. Pricing reform alone will not overcome the many entrenched barriers to higher energy efficiency, but trying to accelerate energy efficiency improvement without addressing energy pricing problems will lead to limited success. Whether targeting new equipment or management of existing systems, policies must reflect a thorough understanding of the particular system and an awareness of the motivations of the actors.

References

ACEEE (American Council for an Energy-Efficient Economy). 1991. *The Golden Carrot News*. Washington, DC.

Berg, C. A. 1979. Energy Conservation in Industry: *The Present Approach, The Future Opportunities*. Washington, DC: Council on Environmental Quality.

Bleviss, D. 1988. *The New Oil Crisis and Fuel Economy Technologies: Preparing the Light Transportation Industry for the 1990s*. New York, NY: Quorum Press.

Brown, M. 1991. The ACT² Project: Demonstration of Maximum Energy Efficiency in Real Buildings. San Francisco, CA: Pacific Gas and Electric Co.

Friedman, T. L. & Hoffman, J. S. 1991. EPA's green lights program and electric utilities: How they can work together to promote demand reduction programs. *Proceedings of the Conference on Demand-Side Management and the Global Environment*, April 22–23, 1991. Arlington, VA.

Geller, H. S. 1991. *Efficient Energy Use: A Development Strategy for Brazil.* Washington DC and Berkeley, CA: American Council for an Energy-Efficient Economy.

Goldman, C. & Wolcott, D. 1990. Demand-side bidding: Assessing current experience. *Proceedings of the ACEEE 1990 Summer Study on Energy Efficiency in Buildings*, Vol. 8. Washington, DC: American Council for an Energy-Efficient Economy.

Gordon, D. & Levenson, L. 1990. Drive + : Promoting cleaner and more fuel-efficient motor vehicles through a self-financing system of state sales tax incentives. *Journal of Policy Analysis and Management*, **9** (3), 409–15.

Greene, D. L. 1990. Commercial aircraft fuel efficiency potential through 2010. *Proceedings of the Intersociety Energy Conversion Engineering Conference*, **4**, 106–11. Piscataway, NJ: USA IEEE.

Grubb, M. 1990. *Energy Policies and the Greenhouse Effect. Volume One: Policy Appraisal.* Aldershot, Hants, UK: The Royal Institute of International Affairs, Dartmouth Publishing Co.

Gruber, E. & Brand, M. 1991. Promoting energy conservation in small and medium sized companies, *Energy Policy*, **19** (3), 279–87.

Hirst, E. & Brown, M. 1990. Closing the efficiency gap: Barriers to the efficient use of energy. *Resources, Conservation and Recycling*, **3**, 267–81.

Jhirad, D. 1990. Power sector innovation in developing countries: implementing multifaceted solutions. *Annual Review of Energy*, **15**, 365–98.

Jochem, E. & Gruber, E. 1990. Obstacles to rational electricity use and measures to alleviate them. *Energy Policy*, **18** (4), 340–50.

Kosmo, M. 1987. *Money to Burn? The High Costs of Energy Subsidies.* Washington, DC: World Resources Institute.

Ledbetter, M. 1991. Overcoming barriers to fuel-efficient automobiles. *Proceedings of the Conference on Tomorrow's Clean and Fuel-Efficient Automobile: Opportunities for East-West Cooperation*, Berlin, Germany, March 25–27, 1991.

Marbek Resource Consultants. 1990. *Survey of Appliance Labelling Programs.* Ottawa, Canada.

Nadel, S. 1991. Electric utility conservation programs: A review of the lessons taught by a decade of program experience. In Vine, E. and Crawley, D. (eds.) *State of the Art of Energy Efficiency: Future Directions*, Washington, DC: American Council for an Energy-Efficient Economy.

Ottinger, R. L. 1991. Consideration of environmental externality costs in electric utility resource selections and regulations. In Vine, E. and Crawley, D. (eds.) *State of the Art of Energy Efficiency: Future Directions*, Washington, DC: American Council for an Energy-Efficient Economy.

Reddy, A. 1991. Barriers to improvements in energy efficiency. *Energy Policy*, **19** (10), 953–61.

Schipper, L. J., Meyers, S. & Kelly, H. 1985. *Coming In from the Cold: Energy-Wise Housing in Sweden.* Cabin John, MD: Seven Locks Press.

Stillesjoe, S. 1991. Using Innovative Procurement Mechanisms to Help Commercialize New Energy Efficient Lighting Products. Stockholm, Sweden: Proceedings of Stockholm Lighting Conference.

Sutherland, R. J. 1991. Market barriers to energy-efficient investments. *Energy Journal*, **12** (3), 15–33.

Train, K. 1985. Discount rates in consumers' energy-related decisions: a review of the literature. *Energy*, **10** (12), 1243–53.

Turiel, I., Berman, D., Chan, P., Chan, T., Koomey, J., Lebot, B., Levine, M. D., McMahon, J. E., Rosenquist, G. & Stoft, S. 1991. U.S. residential appliance energy efficiency: Present status and future policy directions. In Vine, E. and Crawley, D. (eds.) *State of the Art of Energy Efficiency: Future Directions*. Washington, DC: American Council for an Energy-Efficient Economy.

12

Energy and human activity: steps toward a sustainable future

In this closing chapter, we suggest a path for developing a relationship between energy and human activity that is sustainable over the long run. As John Holdren described in the Prologue of this book, a sustainable supply and use of energy is one that does not reduce, over time, the quantity and quality of goods and services that the planet's environmental, economic, and social systems are able to provide. The size of the human population, the per capita levels of various activities, and the sophistication of the technology with which energy resources are obtained and transformed to support those activities are the three most basic determinants of the global impacts of civilization's energy use, as well as the impacts in a given country. As we have shown, the efficiency of energy-using technologies is improving, but not fast enough to balance the growth in global population and increase in per capita activity levels. In addition, the spread of advanced technology around the world is very uneven. If the growing economies of the developing world rely heavily on old technology, and are able to adopt more sophisticated technology at only a slow rate, the impacts of their energy use will hamper their own development prospects, as well as international efforts to reverse the degradation of the global environment.

12.1 Trends in energy use: cause for concern

The record of improved energy efficiency in the OECD countries is in some respects encouraging. Between 1973 and 1988, the weighted average of sectoral energy intensities in Japan, West Germany, and the United States declined by close to 25%. However, structural change within sectors – mainly in transportation and the residential sector – increased energy use

328

somewhat in most countries, and growth in activity pushed up demand in all sectors. The net result was that total final energy use in the OECD countries grew by an average of 0.5%/year between 1973 and 1988. Since electricity use rose faster than final consumption of fuels, the average growth in primary energy use was higher – 0.9%/year.

Between 1973 and the mid-1980s, higher energy prices and concerns over the security of oil supplies played an important role in stimulating improvement in energy efficiency. While oil security has not faded entirely as a concern, the issue does not have the same appeal to policy makers as it did in the early 1980s, in part because governments have promoted oil storage programs to mitigate the impact of potential supply problems. Real energy prices today are somewhat higher than they were in 1970, but their current impact on OECD economies has been largely mitigated by the improvements in energy efficiency. The levelling off of energy prices has not made significant improvement in efficiency uneconomic, but has slowed momentum toward such improvement, resulting in an "efficiency plateau" in many sectors. This plateau is especially problematic for automobiles, as trends toward larger size and higher performance are contributing to higher energy intensity, and ownership is rising in much of the world. Most observers expect that market forces will result in only modest increase in international energy prices during the next 20 years. Thus, one of the factors that contributed to improvement in energy efficiency in the past will probably be less strong in the future.

Even with only moderate rise in energy prices, OECD intensities will likely continue to decline in most sectors, especially in manufacturing, where technological progress is relatively independent of change in energy prices. In addition, historic and current energy efficiency policies and programs are having an impact in some areas. Averaged over all sectors, however, the net decline will likely be much smaller than that which occurred before 1985. Our "Trends" scenarios in Chapter 10 shows an average reduction in OECD sectoral energy intensities of 1.2%/year between 1985 and 2010, well below the rate of 2.3%/year achieved between 1972 and 1985.

The decline in intensities is unlikely to keep pace with the pressure on energy demand from rising activity. The current general consensus is that GDP growth in the OECD countries will average between 2.5% and 3.0% per year between 1990 and 2010 (it averaged about 3% per year in the 1980s). Certain types of energy-intensive activity, especially travel, may grow faster than GDP. Structural change at a macroeconomic level and in the manufacturing sector will contribute to lower aggregate energy intensity, but the shift toward energy-intensive modes is increasing energy use in transportation. Energy use will grow much less than GDP, but some absolute increase seems likely. Recent projections from the US Department of Energy (DOE) show average growth in OECD primary energy consumption of about 1.2% per year between 1989 and 2010, which would result in an absolute increase of 28% (US EIA, 1991).

In the developing countries, energy efficiency will improve with stock turnover and use of more modern technologies; but population growth, increase in per capita GDP, transition away from biomass fuels, and rise in various energy-intensive activities will lead to major increase in energy use. Economic growth will be particularly strong in much of Asia, and is likely to be higher than in the 1980s in Latin America, and perhaps in Africa as well. Along with continued urbanization, rise in income will bring considerable increase in demand for consumer goods. Moreover, the slackening in the development of more energy-efficient technologies in the OECD countries could slow the progress toward higher efficiency in the LDCs.

The outlook for energy use in the Former East Bloc countries is very uncertain. Economic restructuring involving changes in the quantity and mix of goods produced and closing of many outdated facilities could greatly reduce the overall energy intensity of the economies, even without major efforts to encourage energy efficiency. Pricing reform should also have a significant effect, although the high cost of capital could hinder the response of energy users. On the other hand, successful economic reform should eventually bring a surge in the demand for personal mobility, home comfort, and greater activity in the service sector, all of which would push energy use upward. The pace and nature of reform and investment is obviously the critical factor. It seems certain that growth in energy demand over the next 20 years will be less than in the past, however.

How do the various forces affecting world energy use add up? We have not attempted to construct scenarios of world energy use. Projections are routinely made by various international and national agencies, major energy companies, and other institutions, but none of them are very sophisticated at capturing the complex interactions among factors and regions. And while the art of energy forecasting has become more sophisticated, embodying more analysis and thought, there remains the inherent uncertainty about the future.

Despite these caveats, it is useful to consider some recent assessments of the global energy outlook. The US DOE projections cited above show an average growth rate in world total primary energy over the next 20 years of about 1.3% per year, which would result in consumption being one-third higher in 2010 than in 1989. The 1989 World Energy Conference (WEC) projections envisioned growth through 2020 of between 1.2% and 1.6% per year, which would result in world energy use being 40–60% higher than in 1989 (WEC, 1989). Recent scenarios by sector and region done for the US Working Group on Global Energy Efficiency show (in the base case) demand growing through 2025 at about the same rate as the WEC's high projection (Levine et al., 1991). These projections are by no means destiny, but neither are they implausible. A world in which energy use grows by 50% over the next 30 years may not be desirable, but it is not improbable unless actions are taken to restrain demand.

If future energy demand could be largely met with sources whose environmental impacts were relatively benign, increase in energy use would be less cause for concern. But few analysts expect such an outcome. The main reason is that supply of fossil fuels at prices not greatly higher than today's seems quite possible over the next two decades. The oil reserves of the Middle East are huge; gas resources are thought to be larger than was once believed; and the resources of relatively low-price coal are large. From an environmental perspective, the abundance of fossil fuels is troubling because their use places an increasing burden on the environment, particularly with respect to the greenhouse effect. The nuclear alternative has problems of its own, and only substitutes for fossil fuels in electricity production.

To the extent that fossil fuel prices grow only moderately, it will be hard to achieve restraint in consumption; and unless the energy from renewable sources comes down substantially in price, they will face difficulty in achieving very high market penetration. The problem is that the costs of fossil fuel use are rising faster than *prices*. The environment is absorbing more damage, and costs are also being imposed on future generations. Some of the costs are not yet known, especially those connected with the greenhouse effect. Nor do we understand the capacity of the environment to continue to absorb insults without very damaging effect. That capacity could turn out to be greater than many believe, but to act as if that were in fact the case is a risky experiment indeed.

All countries share in the concern over environmental problems associated with rising energy demand. The developing countries are increasingly aware of the local environmental impacts of energy production and use, and are also recognizing that the effects of global climate change could seriously impact them. But with rising populations and so many people living in poverty, governments understandably give priority to needs for economic and social development.[1] At the same time, supplying energy for domestic use is an enormous drain on the resources of many developing countries (and for much of Eastern Europe as well). Oil imports contribute heavily to the foreign debt under which many countries suffer. And for the more populous oil-exporting countries, such as Mexico, Indonesia, and Nigeria, high growth in domestic oil demand limits export earnings in the long run. In addition, rapid increase in electricity use is leading to rising capital demands for power supply. The power sector already claims a large share of total capital investment, yet electricity shortages plague much of the developing world, hampering economic development. One recent study estimated that total LDC power sector capital requirements might average around $140 billion (1990 dollars) per year in the first quarter of the next century (Levine et al., 1991). Yet the current investment level of about $50 billion per year is already jeopardized by overall national indebtedness and poor performance of many electric utilities. With capital scarce in many countries, even partially meeting the rising demands of the power sector would mean shifting resources from other critical areas.

12.1.1 Signs of hope

Rising energy demand threatens both the health of the environment and prospects for development for much of the world's population. In the long run, environment and development are closely linked: the problems of one cannot be solved without addressing the problems of the other. There is growing recognition of this reality, as expressed in the landmark report of the Brundtland Commission (WCED, 1987).

Concern over the environmental impacts of energy use is growing. Environmental issues have become central to energy policy considerations in the OECD countries. The threat that emissions of CO_2 and other greenhouse gases will cause problematic changes in the world's climate is increasingly gaining credence among scientists, as evidenced by the report of the Intergovernmental Panel on Climate Change (IPCC, 1990). To be sure, some of the crucial parameters of these changes are uncertain, and the costs are difficult to quantify, but restraint in the increase of fossil fuel use, or indeed an absolute decline, has been identified in every study as a key

---

[1] For example, the Statement of the International Conference on Global Warming and Climate Change: Perspective from Developing Countries, held in New Delhi in 1989 says that when responses to the greenhouse challenge conflict with development needs, priority should be given to development. "Only in this way can the poorest of the populations be brought to the minimal level of health and resilience needed to cope with environmental stress and stabilize population sizes." (Gupta & Pachauri, 1989).

element in dealing with the threat of climate change. And many governments are beginning to heed the call of the scientific community. About a dozen industrial countries[2] have officially pledged to stabilize or reduce CO_2 emissions by 2005 (US OTA, 1991). In addition, global concerns are being buttressed by the local and regional environmental impacts of energy supply and use.

Complementing the increasing awareness of the need to restrain energy use for environmental reasons is a new understanding of the relationship between growth in energy use and development. Supported by the experience of the OECD countries in decoupling GDP growth from growth in energy use, voices from the developing world have called for shifting the focus from increasing energy supply to meeting demand for energy services at least cost (Goldemberg et al., 1987). The idea that improving energy efficiency should be an important part of the energy strategy of developing countries is being increasingly accepted by public officials, multilateral development banks (such as the World Bank), and bilateral donors. Implementation of energy-efficiency strategies is well behind the recognition of their importance, but there are signs of progress (Levine et al., 1991).

12.2 Restraining energy use

At global, regional, and local levels, restraining growth in energy demand, or even reducing demand in the industrial countries, can contribute greatly to easing the environmental and sociopolitical problems associated with supplying and using energy. At a global level, restraining demand will allow more time for improvement and greatly expanded use of the renewable energy technologies that could support a sustainable energy future in the long run. For all countries, but especially for developing and restructuring countries, restraint in energy use will allow transfer of scarce resources from energy supply to other pressing needs.

What is involved in restraining energy use? The most important component is improving the efficiency with which energy is used to support various human activities. Such improvement has technological and behavioral aspects. Restraint in energy use can also be accomplished by moving toward less energy-intensive activities, or by reducing certain types of activities. This dimension of restraint goes well beyond the domain of energy policy. It is connected to the evolution of the built environment,

---

[2]   As of January 1991, they were Austria, Australia, Canada, Denmark, France, Germany, Italy, Japan, The Netherlands, New Zealand, Norway, Sweden, United Kingdom. Notably absent from this list is the United States.

lifestyles, and culture. While the evolution of human activities certainly has a momentum of its own, it is also shaped by public policies. Implementing a strategy of energy restraint requires understanding the impacts on energy use of policies in many areas, and bringing the benefits of energy restraint into the policy discussion.

12.2.1 *Energy restraint and global climate change: how much? what cost?*

The recognition that restraint in global energy use may be necessary has arisen from analysis of the greenhouse problem. The question of how much restraint may be called for is difficult and much debated. Addressing it involves economics, environmental science, and ethics. The value of reducing CO_2 and other emissions depends on estimates of how high emissions will be, how they will change the climate, and how that climate change will affect our economies and lives. It also depends on how society values the future and generations not yet born. The IPCC and the US Environmental Protection Agency have estimated that at least a 50–80% worldwide reduction in CO_2 emissions is needed to keep the atmosphere at today's already altered level. Given the inevitable increase in energy use in the LDCs, achieving such a goal is difficult to imagine, nor is it necessarily needed. But it gives a rough illustration of the extent to which it may be necessary to reduce fossil fuel use. Such reduction has many aspects, including switching to other energy sources and improving the efficiency of electricity supply, but restraint in energy use would have to play a major role. The more energy use rises with population growth and per capita activity, the larger will be the use of less desirable energy sources, and the overall level of impacts. The higher the level of impacts, the greater the severity with which restraint might need to be applied to hold damage to an acceptable level.

Despite the increasing calls to restrain or reduce fossil fuel use, some economists are skeptical about whether policies can lead to meaningful emissions restraint without significant costs.[3] Their views collide with those of many energy efficiency analysts (e.g., Williams, 1990) who insist that large reductions in CO_2 emissions are available at zero and even negative cost (i.e., the CO_2 reductions are a "bonus" from actions that are cost-effective anyway). As discussed in Box 12.A, the economists' results depend greatly on assumptions for key parameters. Of course, the results of energy-efficiency potential studies also depend on assumptions about

---

[3] For a review of economic studies of the cost of slowing climate change, see Nordhaus (1991).

technology performance and costs, as well as how easily the barriers to market penetration can be overcome. Who is right? To argue that energy users are fully taking advantage of the opportunities for energy saving that are socially cost-effective is to ignore all the evidence. But to argue that all or most of the efficiency gap can be eliminated may understate some of the real costs and barriers that cannot be overcome by even the most creative energy efficiency programs. What is needed to resolve this controversy is a hard and rigorous look at the nature of the efficiency gap, as well as better understanding of the extent to which policies and programs are economically beneficial in fact, as opposed to on paper. While some strategies cost very little to implement or administer, others have direct or hidden costs that are potentially significant in relation to benefits. Unfortunately, there was little careful analysis of the impacts of the energy efficiency policies introduced in OECD countries in the 1970s and 1980s, so it is difficult to judge with precision which policies worked the best. To do so requires more thorough evaluation and analysis of trends in energy use than is currently occurring.

Box 12.A. Macroeconomic impacts of reducing CO_2 emissions

Some analysts have suggested that significantly restraining energy use to reduce greenhouse gas emissions would be relatively expensive, resulting in reduction in economic growth. Such concerns are based on macroeconomic estimates of the energy price level (or carbon tax) required to reduce fossil fuel use and CO_2 emissions to a specified level relative to an expected baseline (e.g., Manne & Richels, 1990). Baseline trends depend on a range of model parameters related to the degree of substitutability between energy and other inputs, the rate of autonomous (price-independent) energy efficiency improvement, and the structural evolution of the economy. Variations of these parameters within reasonable bounds change the estimated costs of CO_2 emissions restraint from a few percent of national income to almost zero (Manne & Richels, 1990a).

The estimation of the "correct" parameter values is problematic since the past behavior of the energy system, captured in a particular way in econometric studies, does not explicitly account for current knowledge about the technical potential to save energy. Such studies are often carried out at a high level of aggregation, abstracting from the details that are revealed only in disaggregated studies. While the models might capture the approximate price elasticity of energy demand, they are inadequate to assess the potential costs and benefits of applicance efficiency standards or programs to install insulation in older homes. In this sense, the macroeconomic studies reflect the potential costs of one policy measure – a carbon tax – rather than the costs of emissions reduction per se.

While the debate goes on, there is increasing agreement that considerable emissions reduction is indeed available at negative or zero net cost. A recent study conducted under the auspices of the US National Academy of Sciences (NAS) estimated that end-use efficiency improvement in the United States could reduce its CO_2 emissions by around 35% of the 1988 level at negative or zero net cost (Mitigation Panel, 1991).[4] Strategies to encourage adoption of negative and zero net cost measures have been dubbed "no regrets", since they are beneficial for reasons other than greenhouse concerns. Indeed, the NAS study found that demand-side efficiency measures were the most cost-effective options for reducing greenhouse gas emissions. Other policies that restrain energy use while serving economic and social goals (such as building a well-functioning transportation system) are also "no regrets" actions. For the developing countries, a "no regrets" strategy means that responses to the greenhouse challenge should be those that enhance development prospects, such as seeking to provide energy services at least cost. Going beyond the actions that constitute "no regrets" would be a form of insurance against too much future damage from events or changes that are uncertain but may have dangerous consequences.

12.3 Energy efficiency: potential and realization

In the past decade, a variety of studies have estimated how much end-use efficiency improvement may be technically achievable and cost-effective in many countries. The methods used have become more sophisticated, and the analysis has become more rigorous, especially with respect to electricity end-use efficiency. Yet it is difficult for such studies to capture the dynamics of the real world, with real people choosing how to live or how to run factories. They indicate the rough magnitude of energy savings available in different areas, and give rough estimates of costs and benefits, but the paucity of hard information and rigorous analysis makes it difficult to predict the actual performance and economics or the take up of the potential. The fact that efficiency improvements appear cost-effective on paper does not mean that they will penetrate widely in the marketplace. The difficulty and cost of overcoming the barriers to efficiency improvement must be taken into account.

In considering policies, it is important to distinguish between potential energy savings and practical realization. There are few hard data on the

---

[4] This reduction assumes 100% implementation of measures to the 1988 stock, but the conclusion applies across the range of estimates of cost per tonne of CO_2 avoided. The reduction in total greenhouse gas emissions (CO_2 equivalent) is 23%.

marginal impacts of energy efficiency programs or policies in the past, so it is difficult to predict how much of the projected savings in any given plan are likely to be realized. Moreover, changes in energy efficiency that will occur in the absence of any policies affect the economics of further change. If reducing environmental impacts is a goal of restraining energy demand, a realistic assessment of the outcome of efforts is needed, lest society face significantly higher impacts than expected. Assessing the experiences around the world to better understand past achievements in energy efficiency will help reveal how far energy efficiency can be pushed, and where the push needs to be hard. There is also a need to improve understanding of the human processes that lead to – or hinder – improvement of efficiency. Important areas include:

- *How small energy users assess energy-efficiency options*: how individuals and firms use information on potential energy savings, and how they value present investments against future savings.

- *How producers incorporate energy efficiency in new products*: understanding how companies invent, incorporate, and market energy efficiency in their products is important in designing effective programs to accelerate that process.

- *How large users manage energy*: better understanding of how manufacturers have reduced energy intensities over time, and of the role that energy considerations play in choosing new techniques. Are energy efficiency investments allocated a lower priority than other investment decisions? If so, why?

Clearly, optimism about *potential* should be tempered with realism about *achievement*. The efficiency levels that appear to be cost-effective even at present energy prices will probably not be realized except at somewhat higher prices and with active energy efficiency policies. Recall that our "Trends" scenario foresees a reduction in OECD energy intensities (relative to 1985 levels) of about 25% by 2010. In "Moderate effort", which reflects a world in which policies, programs, and higher prices encourage investments in more efficient technologies otherwise hindered by market barriers, intensities are on average about 40% lower than 1985. The "Vigorous effort" scenario, with yet higher energy prices, and policies and programs that are more forceful, results in a 60% reduction. This case roughly embodies pushing efficiency to the edge of cost-effectiveness measured against likely market prices, but probably within the bounds assuming some internalization of environmental costs. It is by no means easily achievable, but with determination, well-designed policies, and serious pricing reform, it might be attained. Other argue that even greater savings are possible if the dissemination of new technologies can be accelerated (Lovins & Lovins, 1991).

Considerations about the gap between potential and realization apply even more in the former East Bloc and LDCs, where the barriers to efficiency are greater than in the OECD countries, and the human and financial resources to overcome them are smaller. Estimates of the efficiency potential in these countries are less firm due to lack of information and other factors. Subsidy of energy prices and uncertainty about costs of efficiency improvement make it difficult to develop reliable estimates of the economic potential in many cases. Even so, the estimates of technical potential suggest that considerable reduction in intensities is attractive.[5] Indeed, much efficiency improvement will be needed if the economies of these countries are to compete in the world market. Achieving major reductions will require serious pricing reform and strong government policies and programs, as well as effective assistance and investment from the OECD countries.

Lastly, it is perhaps dangerous to expect policies to work too far against consumer behavior or reach too far beyond the message contained in energy prices. The present price environment is clearly not reinforcing energy-saving policies as was the case in the early 1980s. Moreover, in much of the world, people are demanding greater amenity and, in some cases, employing more energy-intensive technologies to provide services. A prime example of this trend is the desire for greater comfort and performance in automobiles, which counteracts the effect of technical improvements on fuel economy. Changing tastes and lifestyles may increase energy use even as technological change improves efficiency. The evolution of energy use as populations age, as people emigrate from one country to another, and as new technologies bring new behavior is an important part of the context in which energy-efficiency efforts must work (Schipper et al., 1989). Few analyses of technical potential have confronted the implications of change in lifestyles. One must beware the narrow view that focuses only on the savings from efficiency gains, while excluding changes that may offset the impact of those savings.

---

[5] The December 1991 issue of *Energy Policy* contains a collection of articles describing a number of recent efficiency potential studies for LDCs and former East Bloc countries.

12.4 The role of government

The problems of energy and the environment demand thoughtful government action and leadership. While government intervention in the workings of the market is looked on with more suspicion than in the past, it remains true that collective concerns about environment, security, and long-term sustainability are not adequately addressed by the private transactions that govern the design, purchase, and use of energy-using equipment. Governments have a responsibility to deal with long-term problems whose impacts lie outside the time horizons of present consumers or private firms. Many environmentalists and policy makers have called for a *sustainability criterion* that ensures that the life opportunities of future generations are no worse than those of today (WCED, 1987). An argument may be made that the sustainability criterion requires strong policy action to avoid catastrophic future outcomes even if their probabilities are small (Howarth, 1991).

Government actions can take several forms. Only governments can undertake pricing reform, sending appropriate signals to individuals and firms to implement efficiency and behavioral measures to restrain energy use. In many cases, government action can effectively address barriers and accelerate investment in energy efficiency at low societal cost. Only governments can assume the risks of vigorous R&D that may not pay off to individual firms or even present generations. Governments can encourage utilities to play a larger in promoting energy efficiency. Governments can also provide leadership to express clearly the best scientific knowledge about the collective problems of environment and energy security, and to negotiate among conflicting interests. Many of these ideas were recently supported in a paper by oil giant British Petroleum Co. (Box 12.B).

Box 12.B. The government role: a business perspective

In a paper prepared for the Business and Industry Advisory Committee of the OECD, British Petroleum recognizes the important role of government in establishing boundary conditions and sending clear signals to producers (energy and equipment suppliers) and consumers as well. BP recognizes the many imperfections in the market for energy, as well as some of the present subsidies that encourage greater use. BP states that there is no certainty over what private firms or governments should do about climate change, but that no action is a form of action, too. They advocate investments now to expand options for more efficient equipment, cleaner fuels, and coordination of incentives to consumers and suppliers of equipment. A step-wise approach is needed to rank options by costs and benefits, so that those with the least cost and greatest benefits can be taken first. Various measures that provide "level playing fields" (i.e., rationalization of taxes on energy and non-energy goods, cost recovery for producers who comply with environmental goals) are necessary. The choice between fiscal instruments (i.e., taxes) and regulations depends on circumstances. Energy consumers should be involved in decision making to develop priorities.

Source: BP (1990).

12.4.1 Institute rational energy pricing

Rational energy pricing begins with removal of subsidies that give energy users improper signals. In addition, governments can use energy taxes and related policy instruments to stimulate socially desired improvements in energy efficiency and reductions in the activities that drive energy use. One goal of pricing policy is the incorporation of external costs into energy prices to improve the economic efficiency of energy markets. Moving to internalize the external costs of energy supply and use into prices will not only help to restrain demand. It will also accelerate the development and market penetration of alternative energy sources that have higher private costs but lower social and environmental costs than fossil fuels or nuclear power. Estimates of externalities are not easy to develop, particularly those that result from emissions of greenhouse gases, but that is not a reason to not undertake the process.

Some governments, notably Sweden, with other Nordic countries and Germany not far behind, have begun the process of internalizing externalities in energy prices through taxation.[6] Sweden imposed pollution taxes for CO_2, SO_2, and NO_x in 1990 and 1991 on various fuels. Such taxes were not held up because of the uncertainty over exactly how high the tax should be; quite the contrary, these taxes are part of the insurance policy deemed necessary *because* estimates of potential risk and damage are uncertain.

A common objection to energy taxation or removing subsidies is the burden placed on the poor. If prices are raised slowly, and other taxes are reduced, this problem can be mollified. Socio-economic groups hardest hit during the transition to higher energy costs can be helped through various

---

[6] For a review of taxes and alternative or complementary methods, such as emissions targets or tradable permits, see Pearce et al. (1991).

forms of assistance, including programs to improve their efficiency of energy use.[7] Another consideration is the impact that significant energy taxes might have on the productive sector and its international competitiveness, especially if countries act unilaterally. Higher energy prices would raise costs in the short run, but their effects could be offset by improvements in energy efficiency. A comparison of the industrial performance of Japan, which has faced expensive energy for decades, with the United States, which has had lower growth but paid less for energy, suggests that low energy prices are no guarantee of economic success, while high prices need not retard growth. This is particularly true in the advanced economies, where energy-intensive industries are shrinking in importance, while high-tech products, for which energy is a minor cost input, are on the advance. The growth of the chemicals industries in Japan and West Germany after the first oil crisis suggests that know-how can overcome the impact of higher energy prices. One strategy is to lower other taxes, particularly those on capital inputs that enhance energy efficiency. There is some evidence that such a shift might stimulate economic growth (Bye et al., 1989).

While internalization improves the economic efficiency of the market mechanism, it is generally blind to the distribution of benefits between members of society or between present and future generations (Howarth & Norgaard, 1990). Discounting future costs and benefits is fundamental to the measurement and characterization of externalities, yet the use of a positive discount rate implies that little or no weight is attached to events that occur a generation or more into the future. To the extent that governments are concerned about the challenge of sustainable development – balancing the interests of the present and future people – they may choose to raise energy prices beyond the level implied by internalization alone to reduce the risks that current energy use pose to future welfare.

12.4.2 Address barriers to more efficient energy use

Rational energy pricing, while important, will not by itself overcome other barriers that hinder improvement in energy efficiency. Consumers have poor information about economic tradeoffs between investment and energy use, and generally do not spend the time to sort out the economics

---

[7] That the impact of higher energy prices should continually arise as an argument against proper energy pricing in some countries says more about fundamental economic inequities in those countries than the small role that energy costs actually play in the lives of most consumers.

of efficiency investments. For this and other reasons, consumers and small businesses tend to treat future energy savings with a high implicit discount rate. Coupled with the low share of energy costs in total costs for most end uses, this leads to low investment in energy efficiency. Producers of energy-using equipment for consumers (homes, appliances, cars) are aware of consumers' habits and are therefore averse to raising the first costs of their goods. Neither consumers nor producers are likely to take account of the long-range implications of their actions. The result is a vicious circle that leads to a kind of stalemate. Low energy prices make the situation worse. The consequences of this stalemate are not large for either individual consumers or for producers, but they are for society as a whole.

Government can help the situation through a variety of policy instruments, as described in Chapter 11. These include efficiency standards or agreements, financing programs, incentives, and information programs. Policies may favor specific solutions, such as thermal performance requirements for new buildings, or may provide broader stimuli, such as fees that vary with the fuel economy of new automobiles. Policies may aim directly at influencing the efficiency of new energy-using equipment and promoting retrofit of existing systems, or may use indirect approaches that affect how energy-using capital is designed, built, and run. Policies should be holistic in design. For example, dealing with the components of energy-efficient homes without considering the housing *process* – the housing market and financing, urban planning, the construction industry, occupant or buyer tastes, etc. – may lead to design of efficient homes for which there are neither builders nor buyers.

Synergisms between pricing reform, incentive programs, and regulations are important. It is hard to promote efficient automobiles if company car policies or low gasoline prices encourage purchase of large, powerful cars. And it is difficult to promote efficient buildings that incorporate technologies with long payback times when other policies encourage the short-term turnover of buildings for speculative purposes. Harmonizing energy-efficiency goals with other policy goals is needed. In the end, the best energy-saving policies are part of good housing, transport, and industrial policies.

12.4.3 Increase support for energy-efficiency research and development

Government-sponsored research and development (R&D) has played a major role in supporting energy efficiency improvement in the OECD countries, particularly in the buildings sector (Geller et al., 1987). It is generally agreed that government activity should be concentrated in areas where the incentives for and availability of private investment are limited, which typically means sponsoring research and development of concepts and technologies that are too risky or whose payoff may be too long-range for the private sector. Energy-efficiency R&D should be seen as a component of general R&D policy designed to enhance national productivity and competitiveness.

For commercialization of R&D results, it is important to involve prospective users in all stages of the R&D process, from formulation to demonstration. Collaborative, cost-shared research between government and industry has proven successful in many cases. Government can also play a role in stimulating private sector R&D through tax incentives and other means. Such support should not be viewed as a substitute for government-sponsored R&D, however, especially for work on fundamental, "technology base" activities. As Geller et al. state, "Industry depends on such government-sponsored research to provide a technology base on which it can build innovative, marketable new products."

12.4.4 Encourage utilities to promote energy efficiency

Since improvements in energy efficiency can often save energy at lower cost than new energy supplies, encouraging electric and even gas utilities to look upon energy efficiency as a resource is a key strategy. A variety of utility demand-side management (DSM) programs in the United States and Europe are 'harvesting' energy-efficiency resources by coaxing households and businesses toward higher energy efficiency. Such programs are also being implemented in a number of developing countries. Utilities are recognizing that DSM can meet energy and capacity needs. For DSM programs to take root and flourish, however, it is necessary to establish a planning framework in which they are compared to supply-side investments to cost-effectively meet customer energy-service needs. Such "integrated resource planning" is gradually becoming more common in the United States, and some utilities are relying heavily on DSM resources to meet future needs.[8]

The attractiveness of DSM to utilities depends on their particular situation, especially the relationship between their average and marginal cost of generating capacity. Many DSM programs can save energy at less than short-run operating costs. DSM programs can also be financially

---

[8] For a discussion of key issues related to integrated resource planning, see Hirst & Goldman (1991).

attractive in the medium- and long-run due to deferment of capital expenditure for new supply. Despite these advantages, a reduction in revenues in the short run can occur and discourage utilities from aggressively pursuing efficiency options. In addition, the institutional history of most utilities, and the experience of its management and personnel, are heavily weighted on the supply side, so there is often a reluctance to engage in demand-side activities.

Ways of overcoming this reluctance vary depending on the nature of the utility industry in each country. Where utilities are state owned, which is the case in most of the world, a clear political mandate from a high level may be effective. Where they are privately owned, as is the case in most of the United States, financial incentives are needed.[9] In both cases, establishing incentives for utility managers is also helpful. The goal should be to get utilities to see themselves as comprehensive "energy service companies" rather than simply as energy supply companies. DSM activities need to be integrated in overall resource planning and become a normal part of doing business. This change in organizational mission, which is well underway in various parts of the United States, can be encouraged by well-designed incentives.

12.4.5 Provide leadership

The proper role of government in economic affairs is undergoing re-evaluation in much of the world. Clearly though, some of the actions that are needed, such as internalization of externalities, will only occur to a significant extent through government regulatory and tax policy. There are also important roles for government at the sub-national level. State and local governments can implement innovative policies and programs for which a national political consensus may not exist, thus providing a "laboratory" for evaluating such policies and programs. Success at the state or local level may lead to replication in other jurisdictions, and eventually to national implementation, as occurred in the United States with appliance standards that were initially adopted by California and New York.

The current world-wide trend toward greater reliance on market-oriented policies presents the private sector with an opportunity to take initiative of its own in addressing major environmental problems. Large corporations have enormous influence over the kinds of products that are

[9] For an overview of regulatory reforms and incentives to encourage least-cost planning and DSM activity in the US, see Moskovitz (1990).

bought and used. Recently industry initiatives, especially in Europe, have begun to take environment, climate, and sustainability seriously and have coordinated a far-reaching discussion of these issues within the board-rooms of some of the world's largest corporations. Professional societies for building designers and engineers have also taken a leading role in educating their members and promulgating energy-efficient practices in design and operation of buildings and equipment.

While private initiatives are very important, it would be unwise to overestimate how far they will go in the highly competitive and cost-conscious world of business. While it seems likely that cultivating a "green image" will be increasingly popular for business, government must establish basic ground rules, and offer positive reinforcement to those in the private sector who show responsible leadership. Government must also set a tone for private action. The opposition of the US government to serious action to deal with the threat of global warming, for example, has made it difficult to push ahead in the international way that the problem calls for.

Continuity in government policy is an important but elusive goal in democratic societies with competing interests. Since many of the insti-tutional and behavioral changes that are needed will take years to come to fruition and require substantial investment, at least a strong sense of policy direction is needed. International agreements, while difficult to negotiate, will assist in providing such direction to national governments by making it more difficult to abandon agreed goals. The controversy over how much efficiency can be accelerated to restrain greenhouse gas emissions is not a reason to delay action. The goal is to understand which government actions work best at the least cost to society.

12.5 Considerations for developing and former East Bloc countries

There is no question that current energy use in most developing and former East Bloc countries is very inefficient, that this inefficiency causes large economic losses, and that efficiency improvement could save resources, enhance productivity, and benefit the local environment. From a global perspective, the benefit from improving energy efficiency is higher, at the margin, in the LDCs and former East Bloc than in most OECD countries (McKinsey, 1989). While most effort to date has focused on retrofit in industry, it is perhaps most important to encourage energy efficiency in new products, buildings, and factories, particularly in those countries growing rapidly or restructuring.

Initiatives to improve energy efficiency in these countries should be supported by institutions that finance energy sector development, such as the World Bank. An attractive path is to divert funds and expertise that would have been spent on supply projects into energy efficiency investments. In the power sector, the amount of capital required to save a kilowatt of demand is often much less than the amount needed to supply a new kilowatt to end users. Utility demand-side management programs can reduce the power sector's capital needs, and also help to alleviate power shortages and reduce energy imports. Encouraging integrated resource planning and gathering the information needed to reliably incorporate energy-efficiency resources into planning should be high priorities.

Raising the level of management skills is critical to accelerating improvements in energy efficiency. Training is a key ingredient in improving design and utilization in large buildings and factories. However, these skills are not likely to develop much faster than skills in other areas. Realistically, initiatives to save energy should not demand much more sophistication than can be mustered in other parts of the economy. Many of the barriers to more efficient energy use in developing and former East Bloc countries are manifestations of fundamental problems such as pervasive subsidies, capital scarcity, restricted competition, poor housing and transportation systems, maldistribution of wealth, political instability, and outdated industry. Energy use may not be any less efficient than use of other resources. Trying to fix the energy symptoms of more fundamental problems is difficult, so it is important to carefully identify those rigidities in the social system that inhibit people from using energy more efficiently. Pushing technical fixes onto developing or former East Bloc countries may not be very successful unless they fit into the local context. Since the structure of society in these countries is changing rapidly, "fit" itself is a moving target. Building energy efficiency considerations into the wider solutions to the fundamental problems of development will yield the greatest benefits in the long run.

Increasing energy efficiency beyond what may occur as a natural part of the development or reconstruction process can assist that process. Pricing reform is crucial to give energy users proper signals. Strategic energy-efficiency programs are important, as is development of local human resources and technology.[10] But the scale of effort focused on energy efficiency must be balanced with other needs. Program goals and structure

[10] For a recent discussion of assistance priorities for developing and Eastern European countries in the area of energy efficiency, see Levine et al. (1991).

must recognize that many of the basic problems causing energy inefficiencies will not easily or quickly disappear. It is crucial, therefore, to shape
energy efficiency strategies that are not "quick fixes" but rather provide a
lasting boost that supports the development process itself.

Internal reforms and better access to advanced technology are crucial,
and are also complementary. Energy-efficient technology and lessons in
efficient management from the OECD countries should be transferred
rapidly to other countries, but care must be taken that they are appropriate.
Transnational corporations have a key role to play, since they are in the
best position to accelerate transfer of know-how and technologies among
countries. Discouraging adoption of Western consumer habits (or at the
least not subsidizing them) may be difficult, but is nonetheless important.
Along with accelerating improvement of energy efficiency, this will save
resources that would otherwise be devoured by the energy sector.

While LDCs must shape their own energy-efficiency agenda, there is no
doubt that assistance from the wealthy countries could be beneficial. With
respect to the threat of global warming, we agree with the Statement of the
New Delhi conference cited earlier: "Having caused the major share of the
problem and possessing the resources to do something about it, the
industrial countries have a special responsibility to assist the developing
countries in finding and financing appropriate responses."[11] An important
indirect way in which the OECD countries can help the LDCs and former
East Bloc is by taking strong action to put their own house in order. Strong
action would set an example for the rest of the world. Moreover, policies
in the OECD countries that encourage development of energy-efficient
technologies would have profound effect world-wide. The more rapidly
energy-efficient technologies in the OECD countries are commercialized,
the faster they will be available world-wide. Other means of accelerating
the transfer of energy-efficient technologies from rich to poor must also be
pursued.

12.6 Conclusion: steps to restraining energy demand

The potential for improving energy efficiency is enormous, but exploitation
of this resource has slowed in recent years. This is regrettable for several
reasons. First, not incorporating higher efficiency now often means passing

---

[11] Smith (1991) presents the case that the industrial countries, being responsible for most of
the greenhouse gases added to the atmosphere since 1850, owe the rest of the world a
"natural debt".

up opportunities that will be more expensive or even impossible to implement in the future. This is especially true for long-lived capital, such as new buildings. Second, reduced research and development into new efficiency options will make it more difficult to accelerate the pace of efficiency improvement in the future. Finally, the flow of more efficient technologies to the non-OECD countries will be hindered by the slowdown in efficiency improvement in the OECD countries.

Well-designed policies can help recapture the momentum that has been lost. A key element of the context in which policies will be implemented is the global wave of privatization of productive activity. Accompanied by reduction of trade barriers, this will boost competition and encourage movement of energy-efficient technologies and practices. In many cases, efforts to improve overall economic productivity will lead to improved energy efficiency as well. However, an increasing share of energy demand is used for personal activities that are not "competitive," or in providing personal and business services for which energy costs are insignificant. The competitive pressures in the global market may not directly affect this growing share of energy use. The contrast between Japan's "productive" sectors, which responded to higher energy costs with many energy-saving measures, and the "consumer" sectors, which showed little savings, illustrates the importance of this distinction. Much of the equipment used in buildings and transportation is manufactured on a world market that is increasingly competitive among firms, but energy efficiency is generally not a primary consideration for producers. Both national and international actions will be needed to encourage producers and consumers to place higher priority on energy efficiency.

Successful intervention requires understanding the context of energy-use decisions. New policies should reflect the lessons of past failures and successes, and bear in mind that saving energy is not an end in itself. It is important to recognize that policies to promote energy efficiency interact with policies in other areas. Trying to tackle energy inefficiency in isolation will lead to less than desired results. Below are some key steps for stimulating more careful use of energy.

Rationalize energy pricing and gradually internalize environmental externalities. Experience shows that promoting energy efficiency without addressing reform of energy prices is difficult. Eliminating subsidies of energy prices is a first step, with care taken to cushion the impact on those who are truly harmed by doing so. Environmental costs should be gradually internalized into prices through both regulation and taxation.

Governments should begin with those costs that are most clear and agreed upon. Uncertainty about costs is inevitable, but is not a reason to put off some degree of internalization. For costs that are global in nature, such as global warming, those countries that are most able to afford internalization must lead the way.

Improve present energy-using capital. Sound policies and programs are needed to overcome market barriers and short time horizons. The key areas to address are building retrofit and industry. Utility programs can do much to encourage electricity-saving measures, but other efforts will be needed to bring about more efficient use of fuels.

Implement energy-efficiency standards or agreements for new products and buildings. Energy-efficiency standards or agreements for buildings and various products can give the market a societal perspective as well as provide clear signals to manufacturers and builders. For certain types of products (such as home appliances), regional standards may be called for. At an international level, seeking agreements with major world manufacturers is a promising avenue to explore.

Encourage higher energy efficiency in new products and buildings. Incentives are needed to accelerate development, commercialization, and market penetration of highly efficient technologies. It is important to involve all actors in the market: manufacturers, intermediate suppliers, retailers and, of course, consumers. Innovative strategies that help create a market for cutting-edge technologies should be tested.

Promote international cooperation for R&D and technology transfer. Actions are needed to accelerate both indigenous development and transfer of energy-efficient technologies for developing and former East Bloc countries. Care must be taken not to force technologies or solutions into these countries, however. Transfer of technologies and policy approaches among developing countries should be encouraged by strengthening communication and institutional linkages.

Adjust policies that encourage energy-intensive activities. Energy-efficiency policies, however successful, are not the only ones that matter. It is necessary to adjust policies that artificially stimulate or subsidize travel, automobile use, large, single-family dwellings, urban sprawl, and energy-intensive industries. Discouraging energy-intensive activities (such as use of single-occupant automobiles) or encouraging less energy-intensive ones (such as use of rail in freight transport) can have larger impacts on energy use than many energy-efficiency policies. Changes in land-use planning can play a major role in reducing travel distances and reliance on the automobile.

Promote population restraint worldwide. Even if global per-capita energy use were to level off, it is very likely that the related problems would increase more rapidly than population simply because of the scale of supply that would be entailed and the pressure to harvest less attractive energy resources (Holdren, 1991). In the near and medium term, population growth in the industrial countries, slow as it is, contributes the most to global energy/environment problems, since the per capita energy use in these societies is so much higher than in the developing countries. In the longer run, however, the high rate of growth in the developing countries, combined with increase in per capita energy use, will have an enormous impact globally as well as locally. While technological innovation will allow higher standards of living with less impact per person, restraining population growth will permit breathing room to develop and disseminate new techniques. Without such restraint, many of the benefits from the energy-efficiency efforts will be overwhelmed by the impact of greater numbers of people.

Considerations about population growth are an important reminder that restraining energy use is not simply a technological problem. While there are many technological advances that will restrain energy use, present trends in human activity are increasing it. People, not machines, make the decisions that affect energy use. Insight into the human dimension of energy use is key to better understanding future energy trends and how to act effectively to manage them. Energy-efficiency analyses show that there is much to do, but there is a great need to learn how to reap the potential most effectively.

A key lesson from our work is that problems related to energy efficiency are closely connected to broader issues of economic and social development. How the built environment, transportation and communications infrastructures, and agriculture and industrial sectors of economies evolve means as much, if not more, to energy demand and energy efficiency as the efficiencies of individual technologies. This is especially important in the developing and former East Bloc countries, where much of the future energy-using infrastructure is still to be built.

The problems associated with energy use – the impacts on the environment, human health, and economic development – are interlinked with challenges in other areas. Fortunately, many solutions are also interconnected. It is essential to move beyond a narrow focus on technologies, to discover the synergisms between solutions in different areas, and to strengthen them where possible. While each country must take responsibility to put its own house in order, greater international cooperation will be needed to ensure that the global home is a livable one for future generations.

References

BP (British Petroleum). 1990. Discussion paper: Economic instruments – a business perspective. London, UK.

Bye, B., Bye, T. & Lorentsen, L. 1989. Discussion paper: SIMEN, studies of industry, environment and energy towards 2000, No. 44. Oslo, Norway: Central Bureau of Statistics.

Geller, H., Harris, J. P., Levine, M. D. & Rosenfeld, A. H. 1987. The role of federal research and development in advancing energy efficiency: A $50 billion contribution to the US economy. *Annual Review of Energy*. **12**, 357–395.

Goldemberg, J., Johansson, T. B., Reddy, A. K.N. & Williams, R. H. 1987. *Energy for a Sustainable World*. Washington, DC: World Resources Institute.

Gupta, S. & Pachauri, R. K. (eds.). 1989. *Proceedings of the International Conference on Global Warming and Climate Change: Perspective from Developing Countries*. New Delhi, India, February 21–23, 1989. New Delhi: Tata Energy Research Institute.

Hirst, E. & Goldman, C. 1991. Creating the future: Integrated resource planning for electric utilities. *Annual Review of Energy*, **16**.

Holdren, J. 1991. Population and the energy problem. *Population and Environment*, **12**, 3.

Howarth, R. B. 1991. Intergenerational competitive equilibria under technological uncertainty and an exhaustible resource constraint. *Journal of Environmental Economics and Management*, **20**.

Howarth, R. B. & Norgaard, R. B. 1990. Intergenerational resource rights, efficiency, and social optimality, *Land Economics*, **66**, 1–11.

IPCC (International Panel on Climate Change). 1990. *Climate Change: The IPCC Scientific Assessment*. New York, NY: Cambridge University Press.

Levine, M. D., Gadgil, A., Meyers, S., Sathaye, J., Stafurik, J. & Wilbanks, T. 1991. *Energy Efficiency, Developing Nations, and Eastern Europe: A Report to the US Working Group on Global Energy Efficiency*. Washington, DC: US Working Group on Global Energy Efficiency.

Lovins, A. & Lovins, H. 1991. Least cost climate stabilization. *Annual Review of Energy*, **16**.

Manne, A. & Richels, R. 1990. CO_2 emission limits: An economic cost analysis for the USA. *Energy Journal*, **11** (2), 51–74.

1990a. The costs of reducing US CO_2 emissions: further sensitivity analyses. *Energy Journal*, **11** (4), 69–78.

McKinsey & Co. 1989. *Protecting the Global Environment: Funding Mechanisms*. Prepared on behalf of the Dutch Government for the Ministerial Conference on the Environment, Nordwijk, Holland.

Mitigation Panel, Committee on Science, Engineering, and Public Policy, National Academy of Sciences. 1991. *Policy Implications of Greenhouse Warming*. Washington, DC: National Academy Press.

Moskovitz, H. 1990. Profits and progress through least-cost planning. *Annual Review of Energy*, **15**, 399–421.

Nordhaus, W. D. 1991. The cost of slowing climate change: a survey. *Energy Journal*, **12** (1), 37–65.

Pearce, D., ed., 1991. *Blueprint 2: Greening the World Economy*. London: Earthscan Books (International Institute for Environment and Development).

Schipper, L. J., Barlett, S., Hawk, D. & Vine, E. 1989. Linking life-styles and energy use: A matter of time. *Annual Review of Energy*, **14**, 273–320.

Smith, K. S. 1991. Allocating responsibility for global warming: The natural debt index. *AMBIO*, **20** (2), 95–6.

US EIA (Energy Information Administration). 1991. *Annual Energy Outlook with Projections to* 2010, DOE/EIA-0383(91). Washington, DC: U.S. Department of Energy.

US OTA (Office of Technology Assessment). 1991. *Changing by Degrees: Steps to Reduce Greenhouse Gases*. Washington, DC: Congress of the United States.

WCED (World Commission on Environment and Development). 1987. *Our Common Future*. Oxford, UK: Oxford University Press.

WEC (World Energy Conference), Conservation and Study Committee. 1989. *Global Energy Perspectives 2000–2020,* Montreal, Canada.

Williams, R. 1990. *Will Constraining Fossil Fuel Carbon Dioxide Emissions Really Cost So Much?* Princeton, NJ: Princeton University.

Appendix A
Data sources for the OECD countries

Below we describe the principal data sources for the OECD countries covered in Chapters 3–6. Because of the disaggregated nature of our work, we make only limited use of international energy use data, and then only for presenting certain regional or world-wide aggregate figures (as in Chapter 2). The International Energy Agency (IEA) does present data on energy use by sector within a common framework. While useful for some purposes, there are problems with using these data for disaggregated analysis. It is difficult to find the original source of a particular datum, or check whether definitions match from country to country. One cannot disaggregate from "road use" of transport fuels the components that correspond to autos, buses, and trucks. Separating the residential sector from the service sector is problematic; the IEA balances give a separation, but our examination has shown that this separation almost never corresponds to that published by authorities in each country. In general, the disaggregated energy consumption data in national sources often do not match the data given in the IEA balances.

A.1 Manufacturing
Denmark
The energy and value-added data are based on the input–output series from Danmarks Statistik.

France
The energy data are from the Centre d'Etudes et de Recherches Economiques sur l'Energie. The value-added data are from the Institut National de la Statistique et des Etudes Economique. The data were converted from the French NCE system to the ISIC classification as follows:
Paper and pulp: *Papier carton* (NCE 35).
Chemicals: *Chemie and Textiles artificiels* (NCE 23–28).
Stone, clay, and glass: *Plâtre chaux cements* (NCE 20); *Matériaux de construction* (NCE 21); and *Verre* (NCE 22).
Iron and steel: *Siderurgie* (NCE 16) and *Première transformation de l'acier* (NCE 17).
Nonferrous metals: *Metaux non ferreux* (NCE 18).

Germany (West)

The energy data are from the Arbeitsgemeinschaft energy balances. The value-added data are from the Deutsches Institut für Wirtschaftsforschung. The data were converted from the German SYPRO classification system to the ISIC framework as follows:

Paper and pulp: *Zellstoff- und Papierzeugung* (SYPRO 55) and *Papierverarbeitung* (SYPRO 56).

Chemicals: *Chemische Industrie, Spalt-, Brutstoffe* (SYPRO 40).

Stone, clay, and glass: *Steine und Erden* (SYPRO 25); *Feinkeramik* (SYPRO 51); and *Glasgewerbe* (SYPRO 52).

Iron and steel: *Eisenschaffende Industrie* (SYPRO 27); *Eisen- und Stahlgießereien* (SYPRO 2910); and *Ziehereien und Kaltwalzwerke* (SYPRO 3011, 3015).

Nonferrous metals: *NE-Metallerzeugung* (SYPRO 28) and *NE-Metallgießereien* (SYPRO 2950).

Japan

The energy data are from the energy balances prepared by the Japanese Institute for Energy Economics. The value-added data are based on production indices and 1980 value-added statistics from the Ministry of International Trade and Industry. The use of petroleum feedstocks in the chemicals sector, which is included in the energy balances, was eliminated by subtracting the energy content of naptha used in this sector.

Norway

The energy and value-added data are from Statistisk Sentralbyrå. Energy statistics were not available for 1971–2, 1974–5, and 1977. Figures for these years were estimated by linear interpolation based on data for adjacent years.

Sweden

The energy and value-added data are from Statistiska Centralbyrån.

United Kingdom

The principal energy data are from the Digest of UK Energy Statistics (DUKES). Information from the British Paper and Board Industry Federation was used to separate energy use in paper and pulp from the more aggregate paper, printing, and publishing sector. Energy use data in the nonferrous metals sector, which accounts for only a small share of UK manufacturing energy use, were obtained from the Energy Efficiency Office for 1974 and 1980 and from DUKES for 1984–8; figures for other years were interpolated. The value-added data are based on the industrial production series in the Annual Abstract of Great Britain. Because of changes in the level of detail reported in this source, data for the paper and pulp, ferrous metals, and nonferrous metals sectors had to be estimated based on output trends in the paper, printing and publishing and primary metals industries for 1971–4 and 1987–8.

United States

The 1971–85 energy data are from the national energy accounts prepared by the Department of Commerce. The 1988 data are from the *Manufacturing Energy Consumption Survey* conducted by the Energy Information Administration (EIA).

Since no published data are available for the years 1986 and 1987, figures for these years were interpolated based on 1985 and 1988 energy intensities. The 1985 and 1988 statistics are not fully comparable since the 1988 survey includes small establishments not covered in the earlier series. EIA estimates that the divergence is roughly 4%. The value-added data are based on the Federal Reserve Board's production indices and value-added statistics. These were converted from 1977 to 1980 currency using the ratio of nominal to base year (1977) value-added for the manufacturing sector as a whole.

A.2 Transportation

France

Energy use data are derived from the following sources: Observatoire de l'Energie: *Tableaux des Consommations d'Energie en France* (Edition 1990), *L'Energie dans Les Secteurs Economiques*, and Didier Bosseboeuf of l'Agence Française de la Matrîsse de l'Energie (AFME).

Activity data are from various sources. Air passenger (p–km) and seat activity (seat–km) data refer to Air Inter, which handles approximately 95% of all domestic flights. Rail activity data for both intercity (p–km) travel and freight (t–km) refers to SNCF. Bus activity (p–km) assumes a load factor (LF) of 23 for years 1970–1980 (which is about the 1983–87 average). It is estimated by multiplying this LF with known v–km numbers.

Vehicle use data are based on the following assumptions: (a) automobile use (km/car/yr) for years 1970, 1971, and 1973 is estimated assuming a LF of 1.85 and using activity (p–km) and stock data; and (b) gasoline-powered automobile use was estimated, assuming that diesel cars in 1970 went 2.4 times as far as the average car, which narrowed to 2.0 times by 1988 (refer to Observatoire de l'Energie).

Automobile energy use includes liquid petroleum gas (LPG). The 1970–1972 data for both gasoline and diesel powered automobiles are estimated by multiplying toe/vehicle and stock of vehicles. Air energy use is fuel used for domestic flights by Air Inter. After 1985, a new means of accounting for diesel energy use for buses was adopted. Rail electricity use data of SNCF and RATP are converted from primary to delivered energy.

Assumptions for energy use include: (a) 1970–1972 data for gasoline-powered automobiles are based on the 1974 ratio of tons of oil equivalent (toe) and vehicle-kilometers (v–km); (b) For these same years, it is assumed that fuel economies (MJ/v–km) were about constant for both diesel and gasoline cars in years, 1970 and 1973. This assumption was made to approximate average fuel economy estimates supplied by Didier Bosseboeuf; (c) 95% of air energy use is for passenger use (which is derived from Air Inter's energy intensity figures (MJ/p–km) for domestic flights; and (d) passenger share of rail transport assumes one passenger-kilometer (p-km) uses as much energy as 1.25 ton-kilometers (t–km), which coincides with 1988 data.

Germany (West)

The primary source of data on transportation and energy use is: Deutsches Institut für Wirtschaftsforschung: *Verkehr in Zahlen* (various editions, mostly 1990). Additional supporting data for rail and air travel are from: Deutsches Institut für Wirtschaftsforschung: *Detaillierung des Energieverbrauchs in der BRD im HuK, Industrie und Verkehr nach Verwendungswecken*; and Deutsches Institut für Wirtschaftsforschung. *Der Endenergieverbrauch im Sektor Verkehr nach Subsektoren sowie nach Verwendungsarten und Verkehrsbereichen* (1984).

Italy

Major sources of data include: ANFIA, *L'automobile in cifre*, 1988; AGIP Petroli; Ministero dei Trasporti, *Conto Nationale Trasporti (Anno 1988 e prime anticiazioni per il 1989)*; Ministero dei Trasporti, *Piano Generale Trasporti*; ISTAT: *Sommario di Statistiche Storiche*; and International Road Federation (IRF), *World Road Statistics*.

Energy use data come from the following sources: AGIP Petroli; Unione Petrolifera; Ministero dei Trasporti, *Piano Generale Trasporti*; Ministero dell'Industria, Commercioled Artigianato, *Bilancio Energetico Nazionale*.

Automobile vehicle use data include average kilometers traveled by both gasoline and diesel cars. Truck vehicle use data include 3-wheeled trucks. Intercity activity data refer to freeways and trunk roads. Pipeline activity data include pipelines greater than 50 kilometers.

*Intra*city passenger and freight movement data exist only for rail. All other intracity movement (bus, car, truck) are estimates from AGIP Petroli.

Energy use from coal in rail transport applies the conversion factor of 7500 kcal/kg (except for 1970 and 1972, which applies 7410 and 6500 kcal/kg, respectively). Assumptions on energy use include: (a) diesel passenger share used in calculating total energy use in rail transport assumes transporting 1.25 persons is equivalent to 1 ton; (b) passenger share of jet fuel use is estimated at 97% which is similarly used for other countries; and (c) jet fuel domestic share energy use is estimated at 18% for 1973 and grows at 1% per year. This assumption allows consistency with AGIP Petroli's modal intensity figures.

There are some inconsistencies in the energy use data: (a) the public sector diesel consumption drops significantly from 1978 and 1979, suggesting that the 1970–1978 time series may include diesel fuel consumption for heating purposes; (b) truck energy use data, which come from Ministry of Transport, are missing for a number of years (1970–1971, 1973–1977, 1979–1986, and 1988) and therefore have been interpolated. If one tries to calculate energy use, weighted by activity (v–km), different numbers result. The question concerns how the Ministry of Transport arrived at their calculations; (c) data on energy consumption of jet fuel in air transport for years 1976–1978 were adjusted to correct for inconsistency; and (d) end-use energy data from the Ministry of Industry appear to be high. It is uncertain if the data include other uses, like heating or cooking.

Norway

Estimates of passenger- and tonne-km activity are published in *Samferdsel Statistikk* (Transportation Statistics) and in publications from Transport Økonomisk Institute (TØI) in Oslo. Estimates of automobile use stem from surveys taken in 1967, 1973, 1981, and 1985–88, *Eie og Bruk av Bil*. Numbers of vehicles are published in *Samferdsel statistikk* and in *Bil og Vei*, the publication of the Norwegian Road Authority (Veg Direktorat). "Cars" (biler) includes virtually all vehicles, but "person biler" represents automobiles for private and business use. Energy use is poorly documented. The Bureau of Statistics publishes "Road", "Rail", "Ship", and "Air" energy use by fuel in their yearly *Energistatistikk* and *Energiregnskap*. Data from 1976 to 1980 and 1980 to 1986 contain many detailed breakdowns of individual transportation mode's energy use (and activity). Esso (A. Kvamme, *priv. comm.*) has made their own research into the matter, breaking both the automobile and truck fuel markets into considerable detail. Because the Esso data cover the longest period (1970 to present) and make the most detailed attempt to balance all the various liquid fuels markets, we use the data they kindly provided to match energy use, activity, and energy use per vehicle-km. TØI has estimated the fuel economy of new cars by examining the most popular models sold and their test fuel consumption.

Sweden

The data on energy use come from two sources, the National Energy Administration (STEV); and the Transportation Council (TPR). In 1977 SIND (the predecessor to STEV) prepared a forecast of energy use in Sweden, that was based in part upon detailed breakdowns of energy use in the transportation sector provided by the predecessor of TPR. These were "updated" in subsequent energy studies published by STEV. TPR has continually published data on passenger- and tonne–km, as well as on v–km. The Central Bureau of Statistics publishes data on the characteristics of the vehicle stock. The Swedish Automobile Association publishes a yearbook with other details of the vehicle stock, such as the number of cars by weight.

In the 1980s J. Wajsmann of TPR began a systematic bottom-up analysis of energy use in the transportation sector. His unpublished analyses have been provided to STEV for their own yearly breakdowns of Swedish energy use. In these he examines the number of vehicles, km driven and consumption of fuel per km for four types of cars (gasoline private cars and taxis, and diesel private cars and taxis), buses, and trucks. He covers domestic air travel and inland shipping, as well as many smaller users of liquid fuels. Data on electricity use for the railways and local transit are published by the Central Bureau of Statistics' *El och Fjärrvärme Försörjning* (Electricity Supply Statistics). Wajsmann's analyses cover 1980, and 1983 to 1989. The match with the 1970–76 data is not perfect, but acceptable for our purposes. Using data on the stock of vehicles and modal activity, we have reconstructed 1978 and 1981–82 energy use patterns and interpolated remaining years between 1976 and 1983. We have also estimated automobile v-km and fuel economy for 1970–1976, since the SIND data and their TPR source contain very little information on these two parameters. However, *Energyprognoseutredning* (1974) provides a detailed breakdown of transportation energy use in 1970 and some information for 1973. Assembling these together we believe we have created a reasonable picture of the 1970–76 period that can be compared with the period from 1980 to the present.

United States

The transportation data come from three major sources: Oak Ridge National Laboratory (ORNL), the US Department of Transportation (DOT), and the Rand Corporation. Virtually all of the time-series data beginning from 1970 to the present are extracted from ORNL's *Transportation Energy Data Book: Edition* 11, January 1991. Data prior to 1970 primarily come from various editions of ORNL's *Transportation Energy Conservation Data Book*, DOT's *Summary of National Transportation Statistics Final Report* June 1975, *National Transportation Statistics Annual Report August* 1979, and FHWA *Statistical Summary to* 1985.

Energy use data are from ORNL's *Conservation Data Books*. Assumptions for vehicle use (v–km) and energy use include: (a) light trucks have the same mileage as automobiles; (b) all light freight vehicle use is assumed to be for intracity transport; (c) domestic air is estimated at 87% of total v–km; and (d) prior to 1963, all trucks are assumed to use gasoline.

Load factor (LF) estimates include the following: (a) automobile LF is estimated at 2.2 persons from 1960 to 1970. It then decreased to 1.87 by 1977 and 1.7 by 1983, at which it remains to the present; (b) motorcycle LF is estimated at 1.1 persons; (c) personal truck LF is estimated at 110% of the automobile LF; (d) intracity light truck LF is estimated at .25 tonnes/truck; (e) intracity mid-size trucks is estimated at 5 tonnes/truck; and (f) school buses is estimated at 20 persons.

Splicing the pre-1970 and post-1970 data was fairly trivial, since both sources gave common data for 1970. Energy use data that required splicing included buses (transit and intercity), air (certified total and general aviation domestic and passenger share), rail (total and freight), water and pipeline.

Two areas of concern are: (a) a discrepancy exists between between automobile stock cited in ORNL (Polk) and DOT FHWA. The former survey shows fewer cars than FHWA; and (b) there is a growing population of light trucks used solely for personal travel. TIUS survey data (reported in ORNL and used in the time-series data on stock and activity) show the share of trucks used for personal travel growing from approximately 25% in 1960 to 68% in 1988. This equates to a 77% share of light trucks used for personal travel in 1988.

A.3 Residential sector

LBL has published an extensive series of analyses of household energy uses in the main OECD countries. The most important recent reports are Ketoff & Schipper, 1990, Tyler & Schipper, (1990), Schipper, Ketoff & Kahane, (1985), Schipper & Ketoff, (1985) (See Chapter 5 list of references.)

The principal data sources come from national authorities. Additionally, the information on the characteristics of dwellings or households comes from censuses and public or private surveys, which also provide information on ownership of energy-using household equipment. Characteristics of electric appliances and data on other electricity uses are from national electricity utility authorities and in some cases a leading electricity utility in each country. Most of these sources also provide data on the ownership and characteristics of household equipment.

Denmark

Principal data are from Energistyrelsen (Danish Energy Agency). Household characteristics and equipment are from the Bygning og Bolig Register (BBR), a computerized data base covering all buildings in Denmark (Danmarks Statistik). Electric power data and equipment ownership come from DEFU and the yearly reports of Danmarks Elværkers Forening.

France

Principal data are from the Agence Francaise pour la Matrîse d'Energie, published in part in their yearbook, and from unpublished sources. Electricity consumption details and equipment characteristics were kindly provided by Electricite de France.

Germany (West)

Data are from energy balances for Germany (Arbeitsgemeinschaft Energiebilanzen), and a large number of private surveys of fuel and electricity-using equipment cited in Schipper, Ketoff, and Kahane (1985).

Italy

No data on residential energy use are institutionally collected and processed with regularity. The only information which exists has been generated by the national electric utility – ENEL, the natural gas distribution company – Italgas, the major heating-oil supplier – AGIP Petroli, and the major appliance manufacturer – Zanussi. We reconstructed historical series of structure and intensity of major residential energy uses between 1972 and 1988 on the basis of mostly unpublished data from these four sources, plus a number of specific studies or analyses produced in the last two decades. These include: Colombo et al. (1977), Il Rapport WAES Italia, Franco Angeli, Roma; ENI (1979), Gli Usi Finali dell'Energia in Italia, Rome; Ketoff, A. (1984), "Facts and Prospects of the Italian End-Use Structure", Proceedings of the II Global Workshop on End-Use Energy Strategies, Saõ Paulo, Brazil.

The most detailed documentation, particularly on ownership and diffusion of different energy-using devices, comes from the periodical ENEL surveys of their customers, which are over 90% of all residential customers.

Japan

The principal sources for energy demand data are the Japan Institute of Energy Economics (IEE), which publishes yearly energy balances for the energy industries and the Ministry of International Trade and Industry. These balances are then broken down into end-uses by fuel. IEE also provided us with details of equipment ownership and electricity use. The Japan Environmental Research Institute provided us with additional analyses of residential energy use, and information on dwelling characteristics.

Norway

Energy data are from Energistatistikk, Central Bureau of Statistics, modified to separate correctly the residential and services sectors. Equipment data are from various surveys (*Boforholdunderßokkelsen*, *Forbrukksunderßokkelsen*, private surveys from Esso, *Energiunderßokkelsen* from the CBS, and several other reports).

Sweden

The principal sources are *Energistatistik*, *Energistatistik foer Smohus*, *Energistatistik for Flerbostadshus*, as interpreted and updated continually by LBL. Other equipment data are from the Swedish State Power Board's household surveys (1971, 1973, 1975, 1979, 1982, 1985, 1987), from private surveys conducted by SIFO, and from Carlsson (1985).

United Kingdom

Principal sources are the *Digest of UK Energy Statistics* and the Building Research Establishment, as well as unpublished equipment surveys from the Electricity Council, British Gas, Audits of Great Britain, Powergen, and the Building Research Establishment.

United States

Energy consumption data are from DOE Energy Information Agency's *State Energy Data Report*, adjusted using the *Residential Energy Consumption Surveys* by Meyers (1987). Equipment data sources include RECS, *Annual Housing Survey*, and sources in Meyers, 1987.

A.3 Service sector

The data on service sector GDP and population are drawn from the OECD National Accounts, and the total figures are within one percent of the the IMF statistics used for general reference. For all countries except the UK, the service GDP is based on the subsector data from OECD. Floor area data is used where available, although the definitions of what is counted vary in different countries.

Denmark

The energy figures are from Energistyrelsen, and the Electricity Supply 10 Year Overview put out by the Danske Elværkers Forening. The electricity figures for the years after 1984 are based on the energy balance table in the Electricity Supply 10 Year Overview. These crude energy balance figures are comparable with the earlier time series.

Floor Area – The figures shown as 'used' area are from Peter Bach's Energistyrelsen sheet. They are the only time series we found. For 1986, 1987, and 1988 there is information on net and gross floor area from the Byging og Boliregister BBR runs from Danmarks Statistik. The BBR figures for net floor area are comparable to the time series from Energistyrelsen. BBR describes net as excluding basement and unused area, and we termed this category 'used' since it is not clear whether it is conditioned or not, although it probably corresponds to heated area in some other countries.

France

The energy data for France are based on the Observatoire d'Energie Report on the Residential and Service Sectors. The figures are climate and stock corrected. The 1988 energy data are estimated from the combined residential and services data.

Germany (West)

The energy statistics are from VDEW, which includes military uses. We excluded oil for power which covers vehicle fuel.

Floor Area – No data available.

Italy

The energy figures are from "The Final Uses of Energy in Italy: 1970–1988" by V. D'Ermo et al. The Italian information does not break down the nonelectric fuels for the service sector; the only information provided is for residential and services sector combined. The figures for oil, gas, and solids are estimated based on subtracting IES residential figures from the Italian totals with adjustments based on percentages so that the figures add up to the total nonelectric fuel figures presented by the Italians.

Floor Area – The D'Ermo information gives total services floor area. There is no discussion of exactly what is included in this floor area. We have treated it as total floor area.

Japan

The energy data are based on estimates from the Institute of Energy Economics (IEE) in Japan. IEE estimates floor area and energy consumption each year based on information from many sources. The energy estimates are based on consumption figures from utilities and suppliers. The most reliable figures are those for electricity.

Floor Area – The floor data are gathered from many relevant sources. For example, floor area for education would come from the Department of Education. The time series for the floor data is estimated from these many sources.

Norway

The energy figures up to 1986 are drawn from IES estimates. The 1987 and 1988 figures are drawn from data presented in the *Særtrykk fra Statistik Ukehefte* for energy use by fuel source and SIC categories.

Floor Area – Based on data in Sagen's 1986 and LBL estimates.

Sweden

The energy data are from the Swedish National Energy Administration time series from June 1990. These are estimated based on the Swedish Bureau of Statistics and information from Predeco (a consulting company).

Floor Area – Continuing our option of choosing the time survey over detailed survey data, the floor area data we used are from the *Energy Statistics for Residential and Services* for 1978–1987. The data for other years are estimated from other sources including Carlsson's data, the Statistic Meddelanden estimates, and the detailed survey data.

United Kingdom

The time series for the energy statistics is based on figures from the Digest of UK Energy Statistics. These data are based on gross calorific values for energy, not the net values that are the international convention.

The service sector GDP for Great Britain were only available in current year values. These data were used to determine the service sector share of total GDP. This percentage was applied to the total GDP numbers which were available in constant year values. Thus subsector GDP detail is not included in our database.

United States

The energy data are from EIA's State Energy Data Report, with an estimated 200PJ of oil deleted to account for residential apartments counted in commercial accounts.

Floor Area – We have chosen the more complete time series based on F. W. Dodge information instead of the more detailed national survey (CBECS) information. These figures are low compared to the CBECS information, but because the emphasis is on comparisons between current years and 1973, for which there is no CBECS data, we have used the consistent data source.

A.4 Energy prices

Energy prices for the OECD countries are taken from the International Energy Agency's quarterly publications, *Energy Prices and Taxes*. This publication explains data sources and limitations carefully. For the years before 1980, we used a compilation of prices from 1960 to 1980 carried out by Ms. Pat Baade, then of the U.S. DOE Energy Information Administration. While these data were never officially published, they are well explained and referenced, and map almost perfectly into the IEA series, which begin in 1978. Checking both sources with data from many of the countries we studied proved that they can be chained together.

Prices were compiled in physical units and local currency, then converted to real local (1980) currency using either the consumer price index or the GDP deflator. These were then converted to GJ or kWh and converted to US dollars using 1980 purchasing power parities, as given by the OECD.

Appendix B
Notes on OECD energy intensity scenarios

These notes explain assumptions and derivations used in the scenarios presented in Chapter 10. Data for the United States and Japan are taken directly from our database, while those for Europe (and Canada, Australia, and New Zealand) are extrapolated from the seven European countries we have studied extensively.[1]

B.1 Manufacturing and other industry

Between 1972 and 1985, final energy intensity in manufacturing declined by 3.4%, 3.4%, and 2.3%/yr in the United States, Japan, and Europe-6, respectively, with the effects of structural change eliminated. (Structural change caused a further decline in aggregated manufacturing energy intensity in all three regions.) The decline for fuel intensities was nearly 4%/yr in the United States and Japan and 3%/yr in Europe-6, while that for electricity was 1.5%/yr in Japan, 0.8%/yr in the United States, and -0.6%/yr (an increase) in Europe. In the aggregate, fuel intensity fell 3.2%/yr, electricity intensity 0.5%/yr.

It is possible to provide estimates of the potential for more efficient energy use by calculating the impact on fuel and electricity intensities by the year 2010 of (1) recent historical rates of decline in energy intensity (Trends); (2) accelerated decline in energy intensity (Moderate); and (3) rapid decline in energy intensity brought about by both R&D (Vigorous) and an increase in the rate of decline in aggregate manufacturing energy intensity through structural change.

In "Trends", the intensity of fuel use falls at 3%/yr, that of electricity by 0.3%/yr, reflecting the slight plateau of efficiency improvements that set in after 1985. The "Moderate" case represents an acceleration of "Trends" by approximately 1%/yr for fuel intensity (to 4%/yr) and 0.2%/yr for electricity intensity (to 0.5%/yr). This rate is slightly above the average for the 1972–1985 period but below that which occurred between 1979 and 1985, reflecting what we believe to be the reaction of manufacturing to real energy prices that are 25–50%

---

[1] In the course of our work, we have examined many aspects of the structure of energy use in other OECD countries: manufacturing in Canada, households and services in Canada, Holland, and Belgium, and transportation in Canada and Holland. From this preliminary work we believe we can generalize our findings to include these countries. Of the remaining countries, Austria and Switzerland have consumption patterns not unlike those of West Germany, while Finland resembles Sweden in many respects. The most important omissions are in fact Spain and Portugal, OECD countries growing rapidly, and Australia, a very energy-intensive country. We have also omitted Luxembourg and Ireland. The "omissions" however, account for less than 15% of OECD energy use.

higher during most of the 1990–2010 period. The "Vigorous" case represents the outcome when fuel intensity declines by 4.5%/year and electricity intensity by 1%/year, still below the rate of decline for fuel in the 1979–1985 period, but 0.1%/yr more rapidly than that which occurred for electricity during that period. Energy prices rise to between 50% and 100% of their 1990 values for most of this period and DSM-type implementation programs and R&D targeted to industrial energy uses would be intense.

B.2 Transportation

In 1985, fuel use for major freight and travel modes of travel was 38 EJ. Current trends in the intensity of each mode would reduce this by 1.3%/year to slightly over 32 EJ if 1985 patterns of activity were held constant. Automobile fuel economy could increase to about 8.5 l/100 km (27.9 mpg) from 10 l/100 km, principally because of improvements in the United States. The energy intensity of other modes would fall 15–25% by 2010 were present trends to continue.

Two kinds of policies are critical to achieving the savings set out in the "Vigorous" scenario, and probably important to the "Moderate" scenario as well. First, a variety of measures will have to "convince" both automobile companies and car drivers/owners to move away from power and speed as characteristics that drive the new-car market. Second, overall transport policy must examine the various regulations, hidden costs, and subsidies that shape the present transport infrastructure as well as the demand for travel and freight transportation services. This is particularly true for policies that affect land use. Many present policies favor short-distance travel by personal vehicles and stimulate the overall demand for travel as well.

Automobiles and light trucks

In 1985, automobiles and personal light trucks consumed approximately 12.5 EJ in the United States, 1.2 EJ in Japan, and 5.5 EJ in Europe, virtually all of it as petroleum products. The most attractive technical options would increase fuel economy in Europe/Japan mpg from 9 l/100 km (27.6 mpg)[2] in Europe and Japan and 14 l/100 km (17.2 mpg) in the United States in 1985 to 6.9 l/100 km (US approx. 35 mpg) in all regions. The approximate incremental cost to do this is less than $1000/car in 1985 dollars (Difiglio et al., 1990). Compared with OECD automobile and light truck energy use of 21 EJ in 1985, the savings in the "Moderate" scenario would be 6.3 EJ (United States) and 1.7 EJ in Europe and Japan, leaving consumption at 13 EJ.

In the "Vigorous" scenario, the fuel-saving options in existing prototypes would be transferred to the next generation of production automobiles. While the present prototypes are clearly not what consumers want at today's fuel and car prices and driving costs, the advanced diesel motors, lighter materials, and some reduction in power or acceleration could yield savings. We speculate that the cost of this move, with present technology, is probably an additional $1000-$2000/car if size remained constant, significantly less if some sacrifice in size and performance is permitted. This cost is probably "high" as perceived by car buyers and car makers, but could fall if a concerted effort were made. We do believe that this "car of the future" – one that gets well below 4 l/100 km (above 60 mpg) with only moderate sacrifices in comfort and performance relative to new cars sold today and moderate increase in cost – can be developed if the ideas behind the present prototypes are incorporated in vehicles that in other respects resemble the present day fleets of

[2] Values given refer to *actual, on-the-road* performance.

Northern Europe. In all, fuel economy increases in the "Vigorous" scenario from 6.9 l/100 km (35 mpg) to around 4 l/100 km (60 mpg). The remaining consumption of gasoline and other liquid fuels falls to 36% of its present value.

The "Vigorous" scenario could also involve *structural* changes in travel: improvement of the driving cycle through avoidance of congested areas and periods effectively boosts real fuel economy. An important advance could encourage improvements in driving, namely the development of both smart cars and smart roads. By this we mean systems such as Prometheus, which would reduce congestion and thereby improve fuel economy significantly. For such systems to work, however, the majority of cars, and the major guideways, must be outfitted with the proper equipment, no small feat. Clearly this can only be achieved if automobile producers, national authorities, and local authorities cooperate so that cars, road, transit, traffic, and the layout of cities are coordinated.

Is it realistic to foresee reduced driving for purposes of saving energy? The relatively constant distances most cars are driven each year appear to have risen slightly in many countries after the dramatic drop in oil prices in 1985. Few experts foresee a significant decrease in per capita driving distance. However, an increasing share of driving is in short trips, reducing the average length of each trip. This is particularly significant in the United States, where the average trip by car is *shorter* than in Europe. But fuel economy in trips of under 5 km is less than half of "steady-state fuel economy" and pollution control is also less effective. Indeed, US-style catalytic converters are almost useless until warm. Were the trips eliminated predominantly short ones, the savings in fuel would be more than proportional to the miles not traveled. Road and fuel pricing policies, as well as higher parking fees, might bring this savings about. Of course, the impact of a successful introduction of very fuel-efficient cars would be to blunt the impact of higher fuel costs and higher intensities under cold start. Therefore, these driving reductions are uncertain, and not counted in the scenarios.

We have not proposed that modal shifts make a significant contribution to energy savings. One reason is that the average energy intensities of automobiles have been converging on those of busses and rail. The main reason is that, outside of peak hours, most transit systems have low load factors. In the short run, obtaining energy savings from mode switching will be contingent upon filling up existing vehicles. Such change is not simple in the short run. One reason is that about 2/3 of driving is not regular, as in commuting, but irregular, for shopping and other family business and leisure. Regular bus or rail travel is not nearly as well suited for these purposes as it is for commuting. A reasonable goal, therefore, might be to focus on peak load commuting, where queues are getting worse, and on some off-peak shopping trips to central cities, where traffic is also congested and low fares could attract riders, as is the case now in London and Gothenburg. (The low fare program in Stockholm was discontinued after less than one year.) A longer-run shift to transit from individual modes, stimulated principally by changes in land use and introduction of road pricing, could reduce auto use. Were this to occur, an additional EJ might be saved (not counted in the scenarios), at a cost of approximately 0.3 EJ of primary energy for other, principally electrified modes.

Freight trucks

The intensity of freight, in MJ/tonne-km, increased in many OECD countries between 1972 and 1985. Energy use for trucking has increased steadily in most OECD countries. In 1985, trucks in the United States consumed approximately 5.0 EJ, those in Europe approximately 3.2 EJ, and in Japan, 1.1 EJ. Intensity has been increasing because of poor handling (i.e., declining load factors), congestion, higher speeds in intercity traffic, and an increase in short-haul freight. Improvement in the vehicles themselves might offer a 5% reduction in energy intensity, while some improvements in load factors a 5% reduction in vehicle–miles travelled. These improvements are incorporated into "Trends", and they save approximately 1 EJ, leaving consumption at 8.4 EJ. In the "Moderate" scenario, a 20% reduction in vehicle intensity (both through technology and reduced congestion) over 1985 levels, and improved handling and optimization (using computers) to improve load factors and reduce empty runs will reduce energy use to only 7 EJ. In the "Vigorous" scenario, we assume a 50% reduction in freight energy intensity compared with 1975, lower freight haulage, and a slight shift back to rail, reflected by reducing tonne–km by 10%. Trucks wind up using only 4.7 EJ, while other modes gain about 0.25 EJ.

The "other modes" include rail and inland water-born transport. We assume that the intensity of these modes falls 20% in "Trends", 25% in "Moderate", and 30% in "Vigorous". Electricity use in "Vigorous" is 50% higher than in the base case because of a significant increase in electric-transit and rail use. Rail includes electric inter- and intra-city lines.

Aircraft

Orderly replacement of old planes with newer ones (757, 767, 777, MD11, AB320) will likely reduce intensity in "Trends" by 25%. Accelerated purchases of newer aircraft (possibly required by noise-abatement laws), reduction in airport congestion, increased load factors, as well as further improvement of existing designs reduces intensity by 40% in "Moderate". The "Moderate" scenario also envisions some penetration of the ultra-high bypass fan jet and the propfan engine, the latter not currently marketable at 1990 oil prices. The reduction in average intensity is 40% by 2010. In the "Vigorous" scenario, there is high penetration of the propfan engine. The overall intensity reduction in "Vigorous" is 50%. In addition, replacement of some air travel with high-speed rail in those markets where justified (Europe, Japan, the Northeastern United States) could reduce air travel intensity by eliminating many short air traffic stages in congested areas. However, we have not used this substitution in our calculations.

B.3 Residential sector

In 1985, OECD households used approximately 22.1 EJ of final energy, of which approximately 2.7 EJ was oil or LPG, 10 EJ was gas, 0.5 EJ coal, 1.5 EJ wood, 6 EJ electricity, and 0.4 EJ district heat.

The "Moderate" and "Vigorous" scenarios require strong policy leadership, leading to efficiency standards on new equipment, requirements for property upgrading at time of sale, financing, incentive programs to raise interest in advanced lighting, and Golden Carrot programs to stimulate design, manufacture, and sale of advanced appliances. The time required to carry out measures on every home would be at least 20 years, except that the replacement of incandescent bulbs would only take a few years, the time for all of the existing incandescents to burn out. Private companies that supply building materials, household equipment, and construction companies as well must play a strong role in the implementation of such efforts.

We have not proposed any structural changes here except to "reduce" hot water consumption and house area per capita by 10%. These changes are incorporated into hot water and space heating energy use in the "Vigorous" scenario. Appliances, particularly refrigerators, remain at their 1985 sizes/capacities.

Space and water heating

Entrance of new, more efficient homes into the stock reduces average heating intensity in the "Trends" scenario (Fig. 10.4). This scenario also envisions a continued but slow pace of retrofitting of existing homes. These steps, along with modest improvement in the thermal integrity of new homes, result in a reduction in fossil fuel intensity of 15%. The energy intensity of appliances declines, driven mainly by the US efficiency standards and the efficiency programs in place in Europe and North America in 1990. The intensities of electric space and water heating (and cooking) also fall. This contributes to a drop in electricity intensity (averaged over all uses) of slightly over 20%.

The "Moderate" scenario foresees a far higher penetration of retrofit measures, as described in Chapter 9. Retrofit of 50% of homes built before 1975 results in replacement of all single glazing with double (and double with triple), conversion of boilers to the latest condensing types, and addition of full controls on heat distribution systems. Most changes would be done in conjunction with renovation. The result of this effort is a decline in the intensity of fuel-based space and water heating of 40%. Electric space heating declines by 35% compared with its 1985 level. This implies a rate of improvement slower than that which occurred between 1972 and 1985. Sustained over 25 years, however, space heating intensities in most OECD countries approach those of Scandinavia.

In the "Vigorous" scenario, the most advanced techniques are applied to essentially all homes. Gas-fired heat pumps would become popular and the intensity of fuel-based space and water heating declines by 75% (electric heating, 50%) relative to 1985, to below the averages for the present stock in Scandinavia.[3]

What would these extensive retrofits cost? If figures for these kinds of changes calculated for the Danish building stock by the Ministry of Energy serve as a guide, the reductions in "Trends" are economic at present energy prices. Reductions in the "Moderate" scenario are profitable at 1980/81 price levels, which were considerably higher than those of 1990. Those in the "Vigorous" scenario make economic sense only when taken at the time other work is being carried out on buildings, and only when environmental taxes of approximately 50% have been added to the energy prices of the "Moderate" scenario.

New homes and buildings in most cold countries already surpass the efficiency levels proposed for "Trends". Bringing these new homes to the level of "Moderate", the present level of thermal integrity in Scandinavia, should entail minimal cost if sufficient development time is allowed. Building to the level of "Vigorous" means reaching the level of "lavenergi hus" (low energy houses) in Denmark or Sweden. Improving construction methods (including training of workers) is important if new homes are to achieve the low energy requirements of Scandinavian homes, and adjusting financing to the Swedish system, whereby the extra measures to save energy are included in mortgages without forcing builders or buyers to omit other features of the home, is crucial.

---

[3] In these examples, electric space heating shows a smaller decline than that for fueled because a) electrically heated homes are generally newer and better insulated than those heated by fossil fuels, b) there is no combustion equipment that can be improved upon (heat pumps are widely used in the countries where they make the most sense, Japan and the United States). This leaves mainly measures to improve building shells in electrically heated homes as the principal energy-saving strategy.

Electric appliances

Presently electric appliances consume about 2.85 EJ of final energy.[4] In "Trends", the average electricity intensity of the stock in 2010 falls to that of the average *new* model in 1985 (cf. Fig. 10.5), about 75% of the 1985 stock intensity - with size remaining approximately constant. The savings would be 25% of present consumption, or 0.75 EJ of final energy. In the "Moderate" scenario, more advanced appliances appear: ambient temperature washers, advanced refrigeration and cooling systems, heat/moisture recovery from dryers. These use 50% less electricity than those in the present market, i.e., virtually all appliances sold between 1990 and 2010 use as little electricity as those with the *lowest* unit consumption on the market in 1990, yet provide similar services. (The approximate first cost increase is very small, at most \$50–\$100 per appliance.) The savings from this step are 1.4 EJ, compared with the 1985 pattern. In the "Vigorous" scenario, appliances are even more advanced, and those for refrigeration somewhat smaller. Consumption per appliance falls to 35% of its 1985 value, and savings will be 1.85 EJ compared with 1985 consumption. *Since electricity is saved, primary energy savings are more than three times this figure.*

Household lighting

Presently, 0.66 EJ (180 TWh) are consumed as final energy for household lighting.[5] In the "Trends" scenario, the four most-used bulbs (which likely account for 50% of kWh burned) are replaced with compact fluorescents in every home. Savings would be 75% of 50% of 180 TWh or 0.16 EJ.[6] First cost is high relative to what bulbs cost today, about 4 x \$15/home, but this investment pays back well since the bulbs last at least 5000 hours each. In the "Moderate" scenario, we replace 12 bulbs in the house that account for approximately 80% of the lighting electricity consumed. The savings are 0.39 EJ relative to "Trends". In the "Vigorous" scenario, every bulb is replaced. Savings are 0.50 EJ. We presume that because of the increased scale of use, manufacturers lower the cost of these bulbs significantly in the "Moderate" and "Vigorous" scenarios. *Since electricity is saved, primary energy savings are more than three times this figure.*

B.4 Service sector

Current consumption is about 9.6 EJ of fuel (mostly oil and gas) and 6.4 EJ of electricity. District heat is significant only in Scandinavia (over 20% of space heated). Electricity has a share of over 40% of final energy in the United States, Canada, Scandinavia, and France. Fuel intensity declined by around 40% between 1972 and 1985, while electricity intensity increased.

The overall share of space heating has fallen steadily because of improved efficiency and because of the growing importance of lighting, cooling/ventilation, and electronics, all of which are powered by electricity. (Waste heat from electricity-using processes provides for a not-insignificant share of space heating needs in many buildings.) In the "Trends" scenario, space and water heating intensities will fall by 15% by 2010, electricity intensity by 10%. Given the rapid

---

[4] 800 TWh of site electricity: 100 TWh Japan, 400 TWh in the United States, and 300 TWh in Europe.

[5] The average OECD home uses about 600kWh/year for lighting: 400kWh/yr on the European Continent, 500kWh/yr in Japan, 750kWh/yr in Scandinavia, and 1000kWh/yr in North America for a total of about 180 TWh.

[6] If CFL bulbs "produce" the light now generated by 480kWh/yr, the bulbs use only 120kWh/yr.

increase in energy services provided by electricity, the decline in electricity intensity represents a significant improvement in efficiency, spurred by many local programs currently in effect.

As was the case with the residential sector, we believe that the "Moderate" and "Vigorous" scenarios could only occur under strong policy leadership.

Space and water heating

Improvements in space heating controls, boilers, etc., in existing buildings, better heat recuperation and other built-in energy saving features in new buildings will save 35% of electricity and 40% of fuel in the "Moderate" scenario, and at least 55%/50% in the "Vigorous" scenario. The savings in fuel are slightly higher than those for electricity because improvements can be made to the boiler or furnace that are not present in electrically heated buildings. If economic large-scale heat pumps are developed, the savings in electricity or fuel may be even greater.

To achieve these savings, gas utilities, heating equipment manufacturers and maintenance companies, and other parties would have to participate to both finance and carry out the important retrofit activities. Some local authorities might consider more stringent building codes, including rules that apply to existing buildings when offered for sale.

Electricity uses

Improvements in lighting, ventilation, etc., lead to savings of 35% in the "Moderate" scenario, and 50% in the "Vigorous" scenario. Important synergies occur, since reduced lighting and other electric loads reduces cooling requirements in the United States, Japan, and Southern Europe. On the other hand, some of the non-heating load reduces the amount of heat that has to be supplied by the heating system. We have borne this in mind when estimating the overall savings potential. *Since electricity is saved, primary energy savings are more than three times the absolute value of the savings indicated here.*

While "Trends" represents the application of relatively inexpensive common-sense measures to most buildings, the "Moderate" incorporates new technology, and the "Vigorous" implies an all-out campaign to improve the efficiency of electricity use in virtually every building in the OECD. The resulting energy intensities from even the "Vigorous" scenario still lie above those of the most efficient buildings as reported in many surveys. In practice, it would take two decades or more to implement the "Moderate" or "Vigorous" scenarios, as much of the work would be carried out at the time of sale or renovation.

References

Difiglio, C., Duleep, K. G. & Green, D. 1990. Cost effectiveness of future fuel economy improvements. *The Energy Journal*, **11** (11), 65–86.

Index

369

Index compiled by Indexing Specialists, Hove, East Sussex.